T0269549

CAMBRIDGE LIBRARY COLLECTION

Books of enduring scholarly value

Darwin, Evolution and Genetics

More than 150 years after the publication of On the Origin of Species, Darwin's 'dangerous idea' continues to spark impassioned scientific, philosophical and theological debates. This series includes key texts by precursors of Darwin, his supporters and detractors, and the generations that followed him. They reveal how scholars and philosophers approached the evidence in the fossil record and the zoological and botanical data provided by scientific expeditions to distant lands, and how these intellectuals grappled with topics such as the origins of life, the mechanisms that produce variation among life forms, and heredity, as well as the enormous implications of evolutionary theory for the understanding of human identity.

Scientific Papers of Asa Gray

Born in the state of New York, Asa Gray (1810–88) abandoned a medical career to pursue his true interest in botany. He sought the mentorship of the influential American botanist John Torrey and their collaborative efforts in classifying North American flora according to biological similarities paved the way for Gray's professorship at Harvard University after years of research. Gray was also one of the few scientists to whom Charles Darwin revealed his early ideas of evolutionary theory. After Gray's death, his fellow botanist Charles Sprague Sargent (1841–1927) compiled the lesser-known writings of a prolific author whose user-friendly *Manual of the Botany of the Northern United States* and other works inspired generations of botany enthusiasts. The two-volume collection appeared in 1889. Volume 1 contains Gray's reviews of important scientific publications, illuminating the development of botanical literature between 1834 and 1887.

Cambridge University Press has long been a pioneer in the reissuing of out-of-print titles from its own backlist, producing digital reprints of books that are still sought after by scholars and students but could not be reprinted economically using traditional technology. The Cambridge Library Collection extends this activity to a wider range of books which are still of importance to researchers and professionals, either for the source material they contain, or as landmarks in the history of their academic discipline.

Drawing from the world-renowned collections in the Cambridge University Library and other partner libraries, and guided by the advice of experts in each subject area, Cambridge University Press is using state-of-the-art scanning machines in its own Printing House to capture the content of each book selected for inclusion. The files are processed to give a consistently clear, crisp image, and the books finished to the high quality standard for which the Press is recognised around the world. The latest print-on-demand technology ensures that the books will remain available indefinitely, and that orders for single or multiple copies can quickly be supplied.

The Cambridge Library Collection brings back to life books of enduring scholarly value (including out-of-copyright works originally issued by other publishers) across a wide range of disciplines in the humanities and social sciences and in science and technology.

Scientific Papers of Asa Gray

VOLUME 1

EDITED BY
CHARLES SPRAGUE SARGENT

CAMBRIDGE
UNIVERSITY PRESS

University Printing House, Cambridge, CB2 8BS, United Kingdom

Cambridge University Press is part of the University of Cambridge.
It furthers the University's mission by disseminating knowledge in the pursuit of education, learning and research at the highest international levels of excellence.

www.cambridge.org
Information on this title: www.cambridge.org/9781108083669

This edition first published 1889
This digitally printed version 2015

ISBN 978-1-108-08366-9 Paperback

SCIENTIFIC PAPERS

OF

ASA GRAY

SELECTED BY

CHARLES SPRAGUE SARGENT

VOL. I.

REVIEWS OF WORKS ON BOTANY AND RELATED SUBJECTS

1834–1887

London

MACMILLAN AND CO.

1889

The Riverside Press, Cambridge, Massachusetts, U. S. A.
Printed by H. O. Houghton and Company.

INTRODUCTION.

ASA GRAY'S first scientific paper was published in 1834; his last was written in 1887, a few weeks before the end of his life. The number of his contributions to science and their variety is remarkable, and astonishes his associates even, familiar as they were with his intellectual activity, his various attainments, and that surprising industry which neither assured position, the weariness of advancing years, nor the hopelessness of the task he had imposed upon himself, ever diminished.

Professor Gray's writings may be naturally grouped in four divisions. The first in importance contains his contributions to descriptive botany. These with few exceptions were devoted to the flora of North America, and although it did not fall to his lot, as it did to that of some of his contemporaries, to elaborate any one of the great families of plants, the extent and character of his contributions to systematic botany will place his name among those of the masters of the science.

His works of a purely educational character are only second in importance to his writings on the flora of North America; and their influence upon the development of botanical knowledge in this country, during the half century which elapsed between the publication of the first and the last of the series, has been great and must long be felt. No text-books of science surpass them in the philosophical treatment of the subjects they embrace, or in the beauty and clearness of their style.

A series of critical reviews of important scientific publications, and of historical accounts of the lives and labors of botanical worthies, may be conveniently grouped in the third division of Professor Gray's writings; while in the fourth fall a number of papers which owe their existence to the discussions which followed the publication of Mr. Darwin's "Origin of Species," — discussions in which Professor Gray took in this country the foremost position.

It is not proposed to republish the works of descriptive botany, although some of the early and most important of these memoirs are out of print and quite beyond the reach of the great mass of botanical students. The value of these papers, however, is historical only, as all that they contain of permanent usefulness has already been incorporated in standard works upon the science, or will be used in due time to lighten the burden of those upon whom has fallen the task of completing the "Flora of North America." There is even less reason for reprinting any of the earlier editions of the text-books. The last editions contain their author's latest views upon the science, and are still within the reach of students. Works of this character change necessarily as knowledge increases, and the value of every edition of a text-book, except the last, is merely historical.

The philosophical essays, or the most important of them, which grew out of the discussion of the Darwinian theory, have already been republished by their author, and another republication of these papers is therefore not proposed at this time, although it is impossible, without having read them, to understand rightly Professor Gray's influence upon the intellectual movement of his time.

There remain the reviews, the biographical notices, and a few essays upon subjects of general interest to botanists. They have long been out of print and have not been incorporated in any recent publication. It was believed therefore that a reissue of these papers, or a selection from them, would be a useful contribution to botanical literature, and a proper tribute to the memory of their author; and for these reasons these volumes have been prepared. Many of the reviews are filled with original and suggestive observations, and, taken together, furnish the best account of the development of botanical literature during the last fifty years that has yet been written.

There can hardly be a question with regard to Professor Gray's value as a critic. His reviews represented the opinion of a just and discriminating mind, thoroughly familiar with all sides of the question before it, critical rather than laudatory, loving the truth and its investigators, but the truth above everything else. No other naturalist of his reputation and attainments ever devoted so much time to literary work of this sort, or continued it so uninterruptedly for so many years; and in our time the criticism and advice of no other botanist has been so eagerly sought or so highly valued by his contemporaries.

The selection of the articles for republication has been an embarrassing and difficult task. The amount of material at my disposal has been overwhelming, and desirable as it might have been to republish it all, it has not been possible to do so within reasonable bounds. More than eleven hundred bibliographical notices and longer reviews were published by Professor Gray in different periodicals; and it was necessary in preparing these volumes to exclude a number of papers of nearly as great interest and value as those which are chosen.

I have endeavored in making this selection to present, as far as it is possible to do so in a series of papers written independently of each other during a period of more than fifty years, a history of the growth of botanical science during a period which must remain one of its great eras — a period marked by the gradual change of ideas among naturalists upon the origin and fixity of the species which has broadened the field of all biological investigation; by the establishment and systematic arrangement of vast herbaria gathered from all parts of the world; by the introduction of improved and more philosophical methods of investigation in the laboratory; and by the growth of popular appreciation for the value of scientific training. I have tried, in making a selection of these articles, to display as far as possible the mental grasp of their author and his varied attainments in all departments of botany; and to include the reviews of those works which Professor Gray himself believed had played in the two continents, during his time, the most important part in elevating the science to which his whole life was devoted.

The second and third volumes of this series will contain a few essays of general interest, and a selection of the biographical sketches of the principal botanists who have died in recent years.

C. S. S.

BROOKLINE, April, 1889.

CONTENTS.

REVIEWS.

LINDLEY'S NATURAL SYSTEM OF BOTANY.[1]

THE cultivators of botany in this country are generally acquainted with the former edition of this work through the American reprint, edited by Dr. Torrey, and published by Messrs. Carvill of New York, in the spring of 1831. Dr. Lindley's treatise was, at the time of its appearance, the only introduction to the Natural System in the English language, if we except a translation of Achille Richard's "Nouveaux Elemens de la Botanique," which was published about the same period. It is unnecessary to state that a treatise of this kind was greatly needed, or to allude either to the peculiar qualifications of the learned and industrious author for the accomplishment of the task, or the high estimation in which the work is held in Europe. But we may very properly offer our testimony respecting the great and highly favorable influence which it has exerted upon the progress of botanical science in the United States. Great as the merits of the work undoubtedly are, we must nevertheless be excused from adopting the terms of extravagant and sometimes equivocal eulogy employed by a popular author, who gravely informs his readers that no book, since printed Bibles were first sold in Paris by Dr. Faustus, ever excited so much surprise and

[1] *A Natural System of Botany;* or a systematic view of the Organization, Natural Affinities, and Geographical Distribution of the whole Vegetable Kingdom ; together with the uses of the most important species in Medicine, the Arts, and rural or domestic economy. By John Lindley. Second edition, with numerous additions and corrections, and a complete list of genera and their synonyms. London : 1836. (American Journal of Science and Arts, xxxii. 292.)

wonder as did Dr. Torrey's edition of Lindley's "Introduction to the Natural System of Botany." Now we can hardly believe that either the author or the American editor of the work referred to were ever in danger, as was honest Dr. Faustus, of being burned for witchcraft, neither do we find anything in its pages calculated to produce such astonishing effects, except, perhaps, upon the minds of those botanists, if such they may be called, who had never dreamed of any important changes in the science since the appearance of good Dr. Turton's translation of the "Species Plantarum," and who speak of Jussieu as a writer who "has greatly improved upon the natural orders of Linnæus." [1] We have no hesitation, however, in expressing our conviction that no single work has had such a general and favorable influence upon the advancement of botanical science in this country, as the American edition of Dr. Lindley's Introduction to the Natural System. This treatise, however useful,

[1] Dr. Lindley is quite right in his remark that the chief difficulties the student has to encounter in the study of botany, upon the principles of the Natural System, have been very much exaggerated by persons who have written upon the subject without understanding it. To refer to a single instance. In the fifth edition of the Manual of Botany, by Mr. Eaton, an account of the Natural Orders of Jussieu is given, in which the genera Ambrosia and Xanthium are referred to *Urticeæ;* and in a note it is added, "Some botanists place the last two genera in the order *Corymbiferæ* also in the Linnæan class *Syngenesia.* I see no good reason for these innovations." Now Linnæus, in his artificial arrangement, certainly did place these genera (and also Parthenium and Iva) in *Monœcia Pentandria;* but the *innovator* in this instance is Jussieu himself, who never referred these two genera to *Urticeæ*, but places them in his order *Corymbiferæ* (*Compositæ*), where they truly belong. The descriptions of Natural Orders in Eaton's Manual, purporting to be taken from Jussieu, bear a very remote resemblance indeed to the ordinal characters of the admirable Genera Plantarum of that author, while the occasional criticisms on its supposed errors afford the clearest proof that the work was not understood by the author alluded to. It should be recollected that, previously to the reprint of Dr. Lindley's Introduction, Mr. Eaton's Manual was the only work professing to give a view of the Natural System within the reach of the great majority of the botanical students of his country, excepting, perhaps, the American edition of Smith's Grammar of Botany.

was indeed not absolutely indispensable to the favored few, who, aided by the works of Jussieu, Brown, De Candolle, the elder and younger Richard, etc., were already successfully and honorably pursuing their studies and investigations; but to the numerous cultivators of botany throughout the country, who could seldom be expected to possess, or have access to, well-furnished libraries, and to whom the writings of these great luminaries of the science were mostly unknown except by name, this publication was a truly welcome acquisition, conferring advantages which those alone who have pursued their studies under such unfavorable circumstances can fully appreciate.

A second and greatly improved edition of this work having appeared within the past year, it occurred to the writer of these remarks that a cursory notice of it might not be unacceptable to the readers of the "American Journal of Science." We do not intend, in these observations, to engage in a defence of what is called the Natural System of Botany, but take it for granted that the science can by no other method be successfully and philosophically pursued; or, to employ the forcible language of Linnæus, "Methodus naturalis primus et ultimus finis botanices est et erit . . . Primum et ultimum in hoc botanicis desideratum est." The few persons who remain at this day unconvinced of its advantages are not likely to be affected by any arguments that we could adduce. A somewhat larger number may perhaps be found in this country who admit the importance and utility of the natural arrangement in the abstract, but decline to avail themselves of the advantages it affords in the study of plants, because, forsooth, it is too much trouble to acquire the enlarged views of vegetable structure which are necessary for the application of its principles. It would almost seem, from the views and practice of such botanists, that they considered it the chief object of a classification to afford the means of ascertaining the name of an unknown plant by the slightest examination of its structure, and with the least possible expenditure of thought.

In the first edition, Dr. Lindley entered into some detailed explanations to show the fallacy of the common opinion that the artificial system of Linnæus is easy, and the Natural System difficult of application. The sentiments of the public have undergone so great a change upon this subject within the last five or six years that he finds it no longer necessary to adduce these considerations, and accordingly commences at once with a development of the principles on which the Natural System is founded, namely, "That the affinities of plants may be determined by a consideration of all the points of resemblance between their various parts, properties, and qualities; that thence an arrangement may be deduced in which those species will be placed next each other which have the greatest degree of relationship; and that consequently the quality or structure of an imperfectly known plant may be determined by those of another which is well known. Hence arises its superiority over arbitrary or artificial systems, such as that of Linnæus, in which there is no combination of ideas, but which are mere collections of isolated facts, not having any distinct relation to each other." (*Preface*, p. vii.)

We have never met with a more clear and succinct account of the principles upon which the primary divisions of the vegetable kingdom rest than that comprised in the following extract. Those acquainted with the first edition will perceive that the author has changed his opinions respecting the number of these primary divisions or classes; the *Gymnospermæ*, or Flowering plants with naked ovules (comprising the *Coniferæ*, *Cycadeæ*, and, according to Brongniart and Lindley, the *Equisetaceæ*), and the *Rhizanthæ*, as originally established by Blume, being here admitted to the rank of independent classes. Their claim to this rank, however, can as yet be hardly considered as fully established.

"One of the first things that strikes an inquirer into the structure of plants is the singular fact that while all species are capable of propagating their race, the mode in which this important function is accomplished is essentially different in different cases. The great mass of plants produce flowers which are succeeded by fruits, containing seed, which is shed or scattered abroad, and grows into new individuals. But in

Ferns, Mosses, Mushrooms, and the like, neither flowers, nor seeds properly so called, can be detected; but propagation is effected by the dispersion of grains or spores which are usually generated in the substance of the plant, and seem to have little analogy with true seeds. Hence the vegetable world separates into two distinct groups, the Flowering and the Flowerless. Upon examining more closely into the respective peculiarities of these two groups, it is found that flowering plants have sexes, while flowerless plants have none; hence the former are called Sexual and the latter Asexual. Then again the former usually possess a highly developed system of spiral or other vessels, while the latter are either altogether destitute of them, or have them only in the highest orders, and then in a peculiar state: for this reason flowering plants are also called Vascular, and flowerless Cellular. More than this, all flowering plants when they form stems, increase by an extention of their ends and a distention or enlargement of their sides; but flowerless plants appear to form their stems simply by the addition of new matter to their points; for this reason, while the former are principally Exogens or Endogens, the latter are called Acrogens. Flowering plants are also for the most part furnished with respiratory organs or stomates, while flowerless plants are to a great extent destitute of them. No one then can doubt that in the vegetable kingdom, two most essentially distinct divisions exist, the Flowering and the Flowerless, and that these differ not in one circumstance only, but are most essentially unlike in many points both of organization and physiology.

"In like manner, Flowering plants are themselves divisible into equally well-marked groups. Some of them grow by the addition of the new woody matter to the outside of their stem beneath the bark; these are Exogens: others grow by the addition of new woody matter to the inside of their stem near the centre; these are Endogens. But Exogens have two or more cotyledons to their embryo, and hence are called Dicotyledons; while Endogens have only one cotyledon, and are, therefore, Monocotyledons. Exogens have the young external wood connected with the centre by medullary processes; Endogens, having no occasion for such a provision, are destitute of it. In Exogens the leaves have their veins disposed in a netted manner; in Endogens the veins run parallel with each other. The number of parts in the flower of an Exogen is usually five or its multiples; in an Endogen it is usually three or its multiples. In germination the young root of Exogens is a mere extension of the radicle; but of Endogens it is protruded from within the radicle; hence the former have been named *Exorhizæ*, and the latter *Endorhizæ*. In this case, then, as in the last, we have two groups differing entirely from each other in their germination, the structure of their stem and leaves, their mode of growth, the arrangements of the parts of the flower, and in the organization of their embryo. It is impossible, therefore, not to recognize such groups also as natural.

"To this separation of the vegetable kingdom into Exogens, Endogens, and Acrogens, or by whatever synonymous names these groups may be known, many botanists confine themselves. But there are two others, of subordinate importance, perhaps, but nevertheless characterized by circumstances of a similar nature, and therefore, I think, to be esteemed of equal dignity with them. In true Exogens and Endogens, the fertilizing principle is communicated to the young seeds through the medium of a stigma which terminates a case or pericarp in which they are inclosed. But in some plants otherwise Exogens, the fertilizing principle of the pollen is applied immediately to the seeds, without the intervention of any pericarpial apparatus, and they bear the same relation to other Exogens as frogs and similar reptiles to other animals. These plants, therefore, are separated as a distinct class, under the name of Gymnosperms. Like the other groups of the same grade, these are also found to possess peculiarities of a subordinate nature. For instance, they have in many cases more cotyledons than two, whence they have been called Polycotyledons; their radicle usually adheres to the albumen in which the embryo lies, and that circumstance has given rise to the name *Synorhizæ*. The veins of their leaves, when they have any veins, are either simple or forked; in which respect they approach Endogens on the one hand, and Acrogens on the other. And finally, their vascular system is very imperfect compared with that of other Exogens of an equal degree of development.

"The other group, called *Rhizanthæ*, is far less correctly known, but it seems to stand as it were between Endogens and Acrogens of the lowest grade; agreeing with the latter in the absence or very imperfect state of the vascular system, in a general resemblance to Fungi, and in the apparent seeds being mere masses of sporules; but apparently according with Endogens in the ternary number of their floral envelopes, and in the presence of fully developed sexes.

"Certainly there is no possibility of obtaining such important primary groups as these by any kind of artificial contrivance." (*Preface*, pp. x.–xii.)

The grand natural divisions of the natural kingdom are, therefore, perfectly obvious, and may be very clearly defined. With our present knowledge of vegetable structure no great difficulty is experienced in characterizing the orders of natural families, and all subordinate groups. The great desideratum has ever been to effect such an arrangement of the orders under the primary classes that each family should be placed next to those which it most nearly resembles. This might easily be accomplished, if the idea once so strongly insisted upon by poets and metaphysicians, of a chain of beings, a regular gradation, by a single series, from the most perfect and complicated to the most simple forms of existence, had any foundation in truth. On the contrary, nothing is more evident than that almost every order, or other group, is allied not merely to one or two, but often to several others, which are sometimes widely separate from each other; and, indeed, these several points of resemblance, or affinity, are occasionally of about equal importance. A truly natural lineal arrangement is therefore impracticable, since by it only one or two out of several points of agreement can be indicated. As this method is, however, the only one that can be followed in books, all that can be done is to arrange the orders in such a manner as to offer the least possible interruption to their natural affinities. The number of orders is so large that practical convenience seems to require their arrangement into groups subordinate to the primary classes; and when manifestly natural assemblages cannot be recognized, we are obliged to employ those which, being less strongly marked, and distinguished by a smaller number of characters, are apparently of a more artificial nature. The arrangement employed by the learned Jussieu, in his celebrated "Genera Plantarum," although to a considerable extent artificial, has been almost universally adopted, until within the last few years.

In this method Dicotyledonous plants are primarily divided into three groups: the first including those with a polypetalous corolla; the second, those with a monopetalous corolla; and the third, those destitute of a corolla. These sections are subdivided (as also the Monocotyledons) by means of characters taken from the insertion of the stamens (or corolla), whether hypogynous, perigynous, or epigynous. The arrangement here pursued, which is too well known to require further notice, is substantially adopted by De Candolle, the difference being more in appearance than in reality. Dr. Lindley discarded these subdivisions in the first edition of his work; but the new distribution of the orders therein proposed possesses few advantages, and indeed seems not to have satisfied the author himself. In the same year with the publication of the work just mentioned, the "Ordines Plantarum" of Bartling appeared, in which a more natural arrangement of the orders is attempted by the formation of aggregate or compound orders, as originally proposed, and in several instances successfully accomplished, by Robert Brown. An analogous plan was pursued by Agardh in his "Aphorismi Botanici" (1817), and again in his "Classes Plantarum" (1825); but these attempts, however ingenious, do not seem to have obviated, in any considerable degree, the inconveniences of lineal arrangement.

We now return to our author, whose views upon this subject have been materially modified since the original publication of his Introduction of the Natural System. The method now employed was first sketched in the "Nixus Plantarum" (1832), and afterwards in the "Key to Structural, Physiological, and Systematic Botany"[1] (1835), and is more fully developed and illustrated in the work before us. He now admits, as we have already seen, five primary classes, two of which, however, are much smaller than the others and of subordinate importance, and may be considered as transition classes, namely, *Gymnospermæ*, which connect Exogens with the higher Acrogens, and *Rhizanthæ*, which form the transition from Endogens to Acrogens of the lowest grade. The great class *Exogenæ* (*Dicotyledones* of Jussieu) is divided into three subclasses, namely: —

1. *Polypetalæ;* those with the floral envelopes consisting of both calyx and corolla, the latter composed of distinct petals.

2. *Monopetalæ;* those with the petals combined in a monopetalous corolla.

3. *Incompletæ;* those always destitute of a corolla, the calyx also often incomplete or absent.

Thus far this mode of subdivision is nearly the same with that of Jussieu; Dr. Lindley, however, neglecting altogether the character afforded by the insertion of the stamens, divides the polypetalous orders into seven, and the *Monopetalæ* and *Incompletæ* each into five sections or groups. As a specimen of this plan, we copy the names of the groups of the first subclass, with their synoptical characters.

1. *Albuminosæ.* Embryo very considerably shorter and smaller than the albumen.

2. *Epigyinosæ.* Ovary inferior, usually having an epigynous disk.

3. *Parietosæ.* Placentation parietal.

4. *Calycosæ.* Calyx incompletely whorled; two of the sepals being exterior.

5. *Syncarposæ.* None of the characters of the other groups, and with the carpels compactly united.

[1] This excellent little work consists of an augmented edition of the author's Outlines of the First Principles of Botany, with a revised translation of the Nixus Plantarum.

6. *Gynobaseosæ.* Carpels not exceeding five, diverging at the base, arranged in a single row around an elevated axis or gynobase. Stamens usually separate from the calyx.

7. *Apocarposæ.* None of the characters of the other groups, but with the carpels distinct, or separable by their faces, or solitary.

Next, every group is divided into smaller groups, each of which includes one, two, or several orders. These minor groups are called *Alliances*, and are distinguished by the termination *ales.* Thus, under the Albuminose group, we have: —

Alliance 1. *Ranales*, comprising the *Ranunculaceæ*, *Papaveraceæ* (with its suborder, as Lindley, following Bernhardi, considers it, *Fumarieæ*), *Nymphæaceæ*, (to which *Hydropeltideæ* is improperly joined), and *Nelumbiaceæ;*

Alliance 2. *Anonales*, which comprehends the Nutmeg tribe, the *Anonaceæ*, *Magnoliaceæ*, etc.;

Alliance 3. *Umbellales*, including the Umbelliferous tribe, with the nearly allied *Araliaceæ;*

Alliance 4. *Grossales*, consisting chiefly of the *Grossulaceæ* or currant tribe; and lastly,

Alliance 5. *Pittosporales*, which strikes us as a singularly heterogeneous assemblage, bringing together into one group the *Vitaceæ*, *Pittosporaceæ*, *Olacaceæ*, *Francoaceæ*, and *Sarraceniaceæ.*

All the subclasses and groups, both of Exogens and Endogens, are subdivided in a similar manner; but we cannot here proceed further with our enumeration. It will be borne in mind that the chief object of an arrangement of this kind is to facilitate the study of the natural orders, by dividing the extensive primary classes into sections of convenient size, and to dispose these groups, and the orders they comprise, as nearly in accordance with their respective affinities and relationships as a lineal arrangement will allow. It is impossible, in the present state of our knowledge, to say how far the views of our author will ultimately be approved. Every attempt of the kind must necessarily be very imperfect, so long as the structure of only a limited portion of the whole vegetable kingdom has been attentively and completely examined;

and the author is well aware "that this part of the work will require many great changes and improvements before it can be considered at all established." Notwithstanding the objections to which it is liable in many particulars, we agree with the author in the opinion, "that even in its present state it will be found to be attended with many advantages, and that every step which may be taken in determining the limits of the natural groups subordinate to the primary classes must be a decided gain to the science. So rapid is the advance of our knowledge of the vegetable kingdom, and so numerous are the new types of structure that present themselves to the systematic botanist, that it is to be feared lest another chaos should be brought on by the masses of imperfectly grouped species with which the science will soon abound."

The names of natural orders, as first established, do not appear to have been framed in accordance with any uniform rule, as to derivation or mode of termination. They were sometimes intended to express some characteristic feature (Ex. *Leguminosæ*, *Labiatæ*, *Cruciferæ*, *Umbelliferæ*, *Coniferæ*, etc.), but more commonly some genus was selected as the type of the family, which was designated either by the plural of the genus simply (as *Myrti*, *Lilia*, *Irides*, *Euphorbiæ*), or with a slight prolongation (as *Orchideæ*, *Jasmineæ*, etc.), or with the termination still further modified (as in *Cyperoideæ*, *Aroideæ*, *Boragineæ*, or *Ranunculaceæ*, *Rosaceæ*, *Cucurbitaceæ*, etc.). The derivation of the name of the order from some prominent genus is now the universal practice; and for the sake of uniformity as well as to distinguish such names from those of genera in the plural number, the termination *aceæ* is given to orders, and that of *eæ* to suborders, etc. The advantages of uniformity in this respect is manifest, and Dr. Lindley therefore insists upon the adoption of the rule in all cases. In the "Key to Botany," published the year previous to the appearance of the second edition of the present work, the termination in *aceæ* is employed, not only in the names of orders formed from those of genera, but also in the few still in use which relate to some peculiarity in the habit of the family. Thus instead of *Cruciferæ*, *Umbelliferæ*,

Coniferæ, etc., we have *Cruciaceæ*, *Umbellaceæ*, and *Conaceæ*. These are, however, very properly abandoned in the work before us, in which the author inclines to give up the old and familiar names of these orders, and to substitute those formed in the customary manner from well-known genera. *Brassicaceæ*, *Apiaceæ*, and *Pinaceæ* may certainly be as good names as any other when we once get accustomed to them, but it seems hardly necessary to make any change in names of this kind. Dr. Lindley, as we have already seen, gives to the names of Alliances the termination *ales*, and to groups that of *osæ*. The chief advantage of this system is, that the name of any group at once indicates its rank and importance.

The value of this work is greatly increased by the complete list of genera (so far as known at the time of publication), with the principal synonyms, appended to each order and properly arranged under their several suborders, sections, etc. This laborious and difficult task is upon the whole very faithfully executed. We observe, however, several errors, typographical and otherwise, which are not noticed in the appendix; and in a few instances the same genus is referred to two different orders. The whole catalogue will doubtless be rendered more perfectly accurate in a future edition.

The whole number of genera comprised in this enumeration, exclusive of synonyms, is 7840. Sprengel's "Systema Vegetabilium," which was finished in 1827, contains (exclusive of the appendix) only 3593 genera, or not quite half the number now known; while the twelfth edition of "Systema Naturæ" (the last of Linnæus himself) comprises 1228 genera, or only about a third more than are now known in a single family.

This great and rapid increase is perhaps chiefly owing to the discovery of new plants; but it is also attributable in a good degree to the more accurate knowledge of those already known. In either case, it is the natural result of the progress of discovery; and instead of embarrassing the student, as is often supposed, does in reality render the study of the science much more clear and satisfactory. Notwithstanding the great increase of genera within the last few years, it may be safely said that at no previous period could a really useful knowledge of the vegetable kingdom be acquired with so little labor. In hazarding this remark, it is of course taken for granted that the student will avail himself of all the advantages of modern physiological botany and of the natural system: for so rapid has been the discovery of new and strange forms of structure, for which the artificial arrangement of Linnæus makes no provision, that the student who takes that system as his guide has indeed a hopeless task before him.

The essential characters of the orders appear to have been very carefully revised in this edition, as also the remarks upon their affinities, geographical distribution and sensible properties. Did our limits allow, we might call the attention of our readers more particularly to this part of the work. We cannot bring our remarks to a close, however, without suggesting what we consider a very desirable improvement upon the manner in which the seed is described, not only in this, but in almost all modern systematic works. It is very necessary that an organ which affords such important characters, both as to its situation in the fruit, and particularly as to its internal structure, should be described with the greatest possible clearness and precision, and in a uniform manner. The prevalent fault of which we complain is thus noticed, as long ago as the year 1811, by that most acute botanist, the late L. C. Richard.

"Cæsalpinus, Adanson, Jussieu, and Gærtner, always take into view the direction of the embryo relative to the pericarp merely. This method appears to me improper: first, because it does not indicate with precision that direction which is most important to be understood; secondly, because the pericarpic direction of the embryo is often difficult to be ascertained, and is sometimes variable or even wholly different in the seeds of the same fruit. I have already shown by numerous examples in my 'Analyse du Fruit,' that the best method is to indicate the direction of the seed relative to the pericarp, and of the embryo relative to the seed."

In very many descriptions, the direction of the embryo relative to the seed can only be inferred from the pericarpic direction, or, which is still more objectionable, the same structure is described by very different language in different instances, thus rendering unnecessarily complicated an investigation which of itself is not usually difficult. We may adduce as an example the five orders comprised in the alliance *Ranales*, which stands at the commencement of Dr. Lindley's treatise. We have no means of ascertaining, from the essential character of any one of these orders, either the spermic direction and position of the embryo, or the situation of the chalaza and micropyle relative to the hilum, from which the former may be inferred. It is commonly stated that the embryo is situated at the base of the albumen; but it is not specified whether the radicle is next the hilum (as in *Papaveraceæ*, *Nymphæaceæ*, etc.), or points in the opposite direction (as in *Nelumbiaceæ* and *Cabombaceæ*); a matter of essential importance, since the seeds result in the one case from the ripening of anatropous, and in the other of orthotropous, ovules.

The students of botany in this country are greatly indebted to the learned editor and the enterprising publishers of the first American edition of this work. May we hope to have our obligations increased by the reprint of this greatly improved edition?

DE CANDOLLE'S PRODROMUS.

THE second part of the seventh volume of De Candolle's "Prodromus"[1] — with which our notices may appropriately commence — was published at the very close of last year, and comprises the following orders, namely, *Stylideæ*, *Lobeliaceæ*, *Campanulaceæ*, *Cyphiaceæ* (a very small order, founded on the Cape genus Cyphia, and here first proposed by Alphonse De Candolle), *Goodenovieæ*, *Gesneriaceæ*, *Napoleoneæ*, *Vaccineæ*, *Ericaceæ*, *Epacrideæ*, *Pyrolaceæ*, *Francoaceæ*, and *Monotropeæ*. Of these, the *Lobeliaceæ*, *Campanulaceæ*, and *Cyphiaceæ* were elaborated by Professor Alphonse De Candolle, the well-known son of the distinguished author; the *Vaccineæ* by Professor Dunal of Montpelier; and the tribe *Ericeæ* (the Heath-tribe) was prepared by Mr. Bentham. It will be observed that De Candolle has disposed the *Ericaceæ* nearly in the manner first proposed in the "Théorie Elémentaire," considering the *Vaccinieæ*, *Monotropeæ*, *Pyrolaceæ*, etc., as so many distinct families; a view, however, which will not probably be ultimately adopted. Among the uncertain or little known Ericaceous plants, De Candolle has introduced the genus Pickeringia of Nuttall (which was founded upon *Cyrilla paniculata* of the same author, published in the fifth volume of this Journal); this, however, has been long since ascertained to be a species of Ardisia, which belongs to a very different order; and Mr. Nuttall has accordingly recently dedi-

[1] De Candolle. *Prodromus Systematis Naturalis Regni Vegetabilis*, etc. Pars VII., sectio ii. Paris, 1839 (American Journal of Science and Arts, xxxix. 168). — Pars VIII. Paris, 1844 (Ibid., xlvii. 198). — Pars IX. Paris, 1845 (Ibid., 2 ser., i. 174). — Pars XI. Paris, 1847 (Ibid., 2 ser., v. 449). — Pars XII. Paris, 1848 (Ibid., 2 ser., vii. 309). — Pars XIII., sectio ii. Paris, 1849 (Ibid., 2 ser., viii. 300). — Pars XIV., sectio ii. Paris, 1857 (Ibid., 2 ser., xxv. 290). — Pars XVI., sectio i. Paris, 1870 (Ibid., 2 ser., xlvii. 125). — Pars XVII. etc. Paris, 1873 (Ibid., 3 ser., vii. 66).

cated to Dr. Pickering a curious Leguminous plant from California. The genus Galax, De Candolle has appended to *Pyrolaceæ* (tribe *Galaceæ*), a view which seems to be confirmed by an unpublished plant from the mountains of North Carolina, which, in compliment to an assiduous and well-known American botanist, will bear the name of *Shortia galacifolia.*

The prior portion of the seventh volume (published in 1838), as well as the whole of the fifth (1836) and sixth (1837), is exclusively devoted to the immense family of the *Compositæ* (the class *Syngenesia* of Linnæus), which fills more than 1700 closely printed pages, the immediate preparation of which occupied the indefatigable author for seven years! We may take this family as a fair example of the increase in the number of known species within the last eighty years. The whole number of Syngenesious plants described by Linnæus in the first edition of the "Species Plantarum" (published in 1753), including the few *Compositæ* referred to other classes, is 555, which is about 150 less than the now described species of the single genus Senecio. We have not time nor space to enumerate the species of the order in succeeding systematic works, so as to show the progressive increase. Suffice it to say that the whole number known to Linnæus and published during his lifetime cannot exceed 800 species, while the number described by De Candolle is in round numbers about 8700, which are disposed in 893 genera. If to these we were to add the species which have been since published, or are being published in works now in progress, and also the very numerous unpublished species which exist in all large collections, making at the same time reasonable allowance for nominal species, the number of *Compositæ* at present known would scarcely fall short of 10,000, which considerably exceeds the whole number of both flowering and flowerless plants described by Linnæus or his contemporaries. Of the 8700 species given by De Candolle, more than 3000 are described for the first time in this work. In the general disposition of the order, the clear and simple classification of Lessing is to a great degree adopted. It is first divided into three great series, namely:—

1. TUBULIFLORÆ; those with the perfect flowers tubular and regularly five (or rarely four) toothed.

2. LABIATIFLORÆ; those with bilabiate, or two-cleft, perfect flowers.

3. LIGULIFLORÆ; which have all the flowers strap-shaped.

The first series includes about four fifths of the whole family, which are arranged in five tribes, namely, *Vernoniaceæ*, *Eupatoriaceæ*, *Asteroideæ*, *Senecionideæ* and *Cynareæ*. The second series consists exclusively of the *Mutisiaceæ* and the *Nassauviaceæ*, chiefly South American plants; a single species of Chaptalia is, we believe, the only North American representative. The third series, comprising the *Chicoraceæ*, so readily known by their milky juice, and by having all their florets ligulate, contains many North American representatives.

So many orders or separate genera of Monopetalous plants have been the subject of recent monographs, and so much valuable assistance is also engaged for the ensuing portions of the "Prodromus," that several volumes may be expected at no very distant period. It may not be improper to state that Mr. Boissier of Geneva is engaged in the preparation of the *Plumbagineæ;* Mr. Duby of Geneva will prepare the *Primulaceæ;* Professor Dunal of Montpelier, the *Solaneæ;* Mr. Decaisne of Paris, the *Asclepiadeæ;* and Mr. Bentham, the *Scrophularineæ* and *Labiatæ*.

We can at length announce the publication of this important volume (viii.); the first of the series under the editorship of the son of the great Genevan botanist,[1] and which he has appropriately dedicated to the memory of his illustrious father. We are glad to state, that arrangements have been made to expedite the publication of the succeeding volumes. The printing of the ninth, it is said, has already commenced, and its appearance may be expected in the autumn of the present year. It will contain the *Loganiaceæ*, *Bignoniaceæ*, *Cyrtandraceæ*, *Sesameæ*, and *Borragineæ*, from the notes prepared by the late Professor De Candolle; the *Hydrophyllaceæ*, by Alphonse De Candolle; the *Gentianaceæ*, by Grise-

[1] The elder De Candolle died September 9, 1841.

bach; the *Polemoniaceæ*, by Bentham, and the *Convolvulaceæ*, by Choisy. The tenth volume will be occupied with the *Solanaceæ*, by Dunal, and the *Scrophularineæ*, by Bentham; in the elaboration of which orders, these two distinguished botanists are now actively engaged. The first order in the volume before us, the *Lentibularieæ*, is prepared by the editor. The North American species of Utricularia are distributed into three sections, namely: 1. Megacista, where the verticillate foliage is floated by inflated petioles; 2. Lentibularia, where the capillary segments of the submersed foliage are utriculiferous; and 3. Oligocista, where the leaves are few, undivided, and disappear after flowering; the roots strike into the soil or mud, and generally bear the utriculi, when these are present. *U. resupinata*, discovered by B. D. Greene, Esq., and first mentioned in the Massachusetts Catalogue of Plants and Animals, 1835, is wrongly placed by De Candolle among the yellow-flowered species of the second section. It has purple flowers and should stand next *U. purpurea*. The name *U. Greenei*, Oaks, in "Hovey's Magazine" must stand as a synonym, as there is no good reason for changing the prior name imposed by the discoverer.

The order of *Primulaceæ* is elaborated by M. Duby of Geneva, who follows Endlicher in the general distribution of the family. From some inadvertence, *Glaux maritima* is not cited as an American plant. *Naumburgia thyrsiflora*, Mœnch, = *Lysimachia thyrsiflora*, Linn., and *L. capitata*, Pursh. The *L. revoluta*, Nutt., is referred to *L. longifolia*, Pursh. The common Samolus, of the southwestern United States, which has smaller flowers than the true *S. Valerandi*, is referred to *S. floribundus*, HBK. *S. ebracteatus* is not noticed as a plant of the United States, although it is common along our southern borders, nor is it distinguished even as a subgenus, although, on account of its nearly free ovary and want of sterile filaments, a recent writer (M. Baudo, in Ann. Sci. Nat., Dec. 1843) has separated it, to form his genus Samodia. In the *Myrsinaceæ*, elaborated by the editor, we meet with two North American species, both natives of Florida, namely, *Myrsine Floridana*, A. DC., and *Ardisia Pickeringia*, Torr. and Gr. To the small order *Theophrastaceæ*, A. DC., our author has joined Jacquinia, a West Indian genus, one species of which extends into Florida. In the order *Sapotaceæ*, the editor has proposed one new North American species of Bumelia. In *Ebenaceæ* we have only our Persimmon. From this the order *Styraceæ* (embracing *Symplocineæ* and *Halesiaceæ* of Don) is distinguished chiefly by the position of the cells of the ovary opposite the lobes of the calyx. Hopea is kept as a mere section of Symplocos; including a dozen Asiatic species as well as our *S. tinctoria*.

The order *Oleaceæ* is published from the manuscripts of the late Professor De Candolle. The American species of Fraxinus still require the labors of a monographer.

The order *Jasmineæ* is made to comprehend Bolivaria (of which there is at least one Texan species) and Menodora; and the family *Bolivariaceæ* is shown to have been founded upon misconceived characters.

For the elaboration of the *Apocynaceæ*, we are indebted to the younger De Candolle. The only North American genera are Amsonia (is not *Echites Fraseri*, Roem. and Schultes, the *A. ciliata*, Walt.?), Apocynum and Forsteronia (*F. difformis*, DC., = *Echites difformis*, Walt.).

The order *Asclepiadeæ* has been very faithfully studied by Decaisne. All the North American representatives belong to the tribe of true *Asclepiadeæ*, with the exception of Gonolobus, of which we have several species (one of them, collected by Dr. Short, forms the new *G. tiliæfolius*), and one, or possibly two species of Chthamalia, Decaisne. *Metastelma Fraseri* is probably a native of the West Indies, not of Carolina. *Enslenia albida*, we notice, is about to be figured in the forthcoming volume of Delessert's Icones; as also is Podostigma. Acerates includes ten, chiefly North American species. Asclepias is reduced to forty-four species, all of which are American, and the greater part extra-tropical. We are happy to learn that the plates of the fifth volume of the "Icones Selectæ" of the liberal Delessert — chiefly devoted to the illustration of the eighth volume of the "Prodromus" — are already in the hands of the engraver.

A year ago we had the pleasure to notice the eighth volume of this indispensable work, the 1st of the series under the editorship of Professor Alphonse De Candolle. The ninth volume, now before us, was issued on the 1st of January last; and the forthcoming portions are in course of preparation under such favorable circumstances that we may now confidently look for the appearance of a volume a year, and for the full completion of this "Species Plantarum," according to the natural system, at no very distant period. We have already mentioned the arrangements that are made to secure this desirable consummation, and by which the work becomes as it were a series of separate monographs, prepared by the most skilful hands, under the superintendence of a common editor. Every botanist is aware of the improvement of the successive volumes as they appeared from the unrivalled hands of the elder De Candolle; and a further improvement is manifest in the later portions, elaborated or revised by his son, especially in the introduction of characters drawn from æstivation, placentation, the structure of the ovule, and other points which have only quite recently been turned to special account by systematic botanists. A particular account of a volume which is or soon will be in the hands of every working botanist, cannot be necessary, and we have not time at present for special enumeration. The ninth volume commences with the *Loganiaceæ*, by Alphonse De Candolle. The genus Cœlostylis, Torr. and Gr., is correctly reduced to Spigelia. Under this order we have a tribe created for the long-vexed Gelsemium, which we suspect is not yet finally at rest. Next follows the *Gentianaceæ*, elaborated by Grisebach, whose recent monograph of that family, which forms the basis of the present arrangement, was duly noticed in this Journal. The order *Bignoniaceæ* is edited from the manuscripts of the elder De Candolle; as are also the orders *Sesameæ* and *Cyrtandraceæ*, which last has been reduced by Mr. Brown to *Gesneriaceæ*. The *Hydrophyllaceæ* are elaborated by Alphonse De Candolle, in which, by attributing generic importance to the presence or absence of the appendages or nectariferous scales within the tube of the corolla, the number of genera is perhaps too greatly increased. The *Polemoniaceæ* are admirably worked out by Bentham, who has reduced to sections of Gilia his Hugelia, Fenzlia, Linanthus, Dactylophyllum, Leptosiphon, Leptodactylon, and the Ipomopsis, Michx. The elaboration of the *Convolvulaceæ* by Professor Choisy does not appear to give entire satisfaction to botanists. The term "infelicissime intricatus" is perhaps still applicable to their family; and the genera are probably unduly increased in number. Of the *Borragineæ*, printed from the elder De Candolle's manuscripts, with valuable notes and additions by the editor, we have the first three tribes, namely, *Cordieæ*, *Ehretieæ*, and *Heliotropeæ*. But for the true *Borrageæ* we must wait until the appearance of the tenth volume, which is already in press.

The long-expected eleventh volume of this work was published at the close of November last, or rather in December, and is now in the hands of botanists throughout Europe and America. It comprises, first, the *Orobanchaceæ*, by Mr. Reuter of Geneva; the true genera of which are Epiphegus of the United States; Phelipæa, containing thirty-eight species of the Old World, and two of North America west of the Mississippi; Conopholis (for the *Orobanche Americana*, L.); Orobanche of over ninety species belonging to the temperate regions of the Old World; Boschniakia of two high northwest American and Siberian species; Clandestina of one south European species; Lathræa of two Old World species; and Anoplanthus, Endl. (Anoplon, Wallr.), which is made to embrace not only the three genuine species of North America, but also a separate section corresponding with Anblatum, Endl., of two species of Central Asia. The earlier recent name of this last genus is Gymnocaulis, Nutt.; the next is Anoplon, by Wallroth. This was changed by Endlicher to Anoplanthus, because there is an Anoplon in zoölogy — an insufficient reason — with which, however, we need not trouble ourselves, since all three must give way to the early and excellent name of Aphyllon, founded by Mitchell, and published in 1748, on the typical species afterwards called *Orobanche uniflora* by Linnæus. This name, the revival of which is demanded not only by statute law, but also for the avoidance of the intricately conflicting names recently imposed, was adopted in the MSS. for this family prepared for the "Flora of North America," and also in the writer's "Manual of the Botany of the Northern United States." Besides *A. uniflorum*, we have in the West and North, *A. comosum*, and *A. fasciculatum*. To the Genera affinia vel dubia, Reuter appends Obolaria, which is correctly described, except that the insertion of the ovules over the whole inner surface of the ovary is overlooked; the present writer's illustration of this genus not having reached Geneva until after this family was printed off.

The great family of this volume, occupying almost four hundred pages, is the order *Acanthaceæ*, which is contributed by Nees von Esenbeck. This chiefly tropical or subtropical family, founded less than forty years ago upon a small number of genera and species, now ranks among the largest of the monopetalous series, and is arranged by Nees under two suborders, eleven tribes, and 146 genera (including those of the appendix). We have few *Acanthaceæ* in the United States, so that an analysis of the family would not interest our readers. We are pleased to find that the writer's reference of the *Ruellia justiciæflora*, Hook. (the Eberlea of Riddell), to the genus Hygrophila, R. Br. (vide Pl. Lindheim, p. 22, note), is confirmed by Nees von Esenbeck. It appears that the species is also Mexican, and had been already described by Schlechtendal. Our Ruelliæ belong to Dipteracanthus and Calophanes. Our *Dianthera Americana*, L., with its allies, is included in the large genus Rhytiglossa, established by Nees (in Lindl. Introd. Nat. Syst., ed. 2) a few years ago, on some Cape of Good Hope species, from which the author suggests that the American species may differ generically. However that may be, we urgently protest against this suppression of the old Gronovian and Linnæan genus, Dianthera, which was founded upon our species; and which name, although unaccountably overlooked by Endlicher, who is usually careful, as well as by Nees, who is careless, as to questions of priority, must surely be continued for the genus, however bounded, which includes the *Justicia pedunculosa* and *J. humilis* of Michaux.

The remainder of the volume, contributed by Schauer of Breslau, comprises, first, *Phrymaceæ*, another of these inconvenient little orders established on a single genus of a single species, which, however, differs remarkably from *Verbenaceæ* as well as *Labiatæ*, by the monomerous ovary, with a single orthotropous ovule erect from the base of the cell, an embryo with the radicle superior, and convolute cotyledons; and finally, the *Verbenaceæ*, composed of three tribes (the *Verbeneæ*, *Viticeæ*, *Avicennieæ*), ten subtribes, and forty-two genera. The genus Verbena, with which we are principally concerned in the United States, comprises seventy-one species, besides a dozen dubious ones appended at the close.

We had nearly forgotten the small family *Myoporaceæ*, which is elaborated by Professor Alphonse De Candolle himself. These twelve genera are all Australian or Oceanic, except Bontia, which is a Caribbean genus of a single species.

The twelfth volume, which will contain the *Labiatæ* by Bentham, is promised for June.

The twelfth volume of the "Prodromus," delayed somewhat by the convulsions of the continent, will be welcome to botanists. It concludes the series of *Monopetalæ*, with the exception of the small family *Plantaginaceæ*, and the large one of *Solanaceæ*, which last, Professor Dunal — from whom it has long been due — appears to find by no means easy to elaborate. Of the present volume, all but one hundred pages are occupied by the *Labiatæ*, from the hand of the most unwearied and best of monographers, Mr. Bentham. The plants of this large order, arranged in 101 genera, are thrown into eight tribes, instead of the eleven in the "Labiatarum Genera et Species;" the *Satureieæ* now being made to comprehend the *Menthoideæ* and the *Melissineæ* of the earlier work, and the *Scutellarineæ* being merged in the *Stachydeæ*. The *Ocimoideæ*, comprising nineteen genera, are represented in extra-tropical North America solely by one or two species of Hyptis, which inhabit our southeastern frontiers.

Of the *Satureieæ* we have Mentha, Lycopus, a single Cunila, Pycnanthemum (the whole seventeen species), an anomalous Satureia (*S. rigida*, Bartr.), three species of Micromeria; while to Calamintha is now referred the former *M. glabella* and *M. Nuttallii*, as well as the *Gardoquia Hookeri*, Benth., with the new *C. canescens*, Torr and Gr., MSS., and *C. Caroliniana*, Sweet (Thymus, Michx.); also two species of Dicerandra (of which *D. densiflora* is a new one from Florida); the California Pogogyne; Hedeoma, including *H. ciliata* (*Keithia ciliata*, Benth., Lab.); and Collinsonia, of which six species are recognized.

Of the tribe *Monardeæ* we possess a small portion of the vast genus Salvia, which is increased to 407 species; Audibertia of California, of six species; Monarda, of six species (*M. didyma* and *M. fistulosa* being retained nearly as in the earlier monograph); and Blephilia, of two species.

The tribe *Nepeteæ* affords us Lophanthus; a single Dracocephalum; and a Cedronella.

The tribe *Stachydeæ* furnishes Prunella, for which Alphonse De Candolle restores, with satisfactory reasons given, the orthography, Brunella; Scutellaria, of eighty-six species, sixteen of them North American; Physostegia, of which two species are admitted; Brazoria, Engelm. — Gray (in which the synonym "*Physostegia truncata*, Hook. Botanical Magazine, t. 3494," should be cited under *B. scutellarioides*, as shown in Chlor. Bor.-Amer., and not under *B. truncata*); Macbridea and Synandra, each of a single species; and Stachys (from which Betonica is now excluded), of 168 species.

The tribe *Prasiæ* belongs entirely to the Old World and to the Sandwich Islands; and the tribe *Prostanthereæ* is exclusively Australian.

Of the *Ajugeæ* we have Isanthus, Trichostemma (in which a corrected view is taken of the inflorescence of the typical section), and Teucrium.

The small order, *Selaginaceæ*, contributed by Professor Choisy of Geneva, consists of eight almost exclusively South African genera, to which the sub-arctic genus Gymnandra is doubtfully appended. Of three Siberian species of this genus two are found on our northwest coast, two are Himalayan, and one has recently been found at Aucher-Eloy in the mountains of Armenia.

The order *Stilbaceæ*, prepared by Professor Alphonse De Candolle himself, consists of three genera each of a single known species, and of one with five species; all of them natives of the Cape of Good Hope.

The *Globulariaceæ*, by the same author, comprises the typical genus, with eight species, and a new one of a single species; all of Europe and of Eastern Asia, except one in the Canary Islands.

The order *Brunoniaceæ*, also by De Candolle, contains a single genus of two Australian species; both made known by the prince of botanists whose name they bear.

Mr. Boissier, the most active and promising botanist of the Genevan school, has elaborated the *Plumbaginaceæ*. The tribe *Staticeæ* comprises six genera, namely, Ægialitis, R. Br., of the shores of eastern tropical Asia and Australia; Acantholimon, Boiss., of forty-two Central Asian species, and Goniolimon, Boiss., of seven North-Asian species, — both distinguished from the following by their capitate instead of filiform stigmas: Statice itself, reaching to one hundred and ten species; Armeria, with fifty-two species, and Limoniastrum, of two Mediterranean species. The Statice of our own coast, *S. Caroliniana*, Walt., Mr. Boissier distinguishes from *S. Limonium* by its fistulous scape, stricter branches, pyramidal instead of corymbose panicle, the distant one-flowered spikelets, and the very acute calyx-lobes; the Californian plant he introduces is a new species. The tribe *Plumbageæ* consists of the Siberian Plumbagella, the European and tropical Plumbago, the Abyssinian Valoradia, and the African and North Indian Vogelia.

In the Corrigenda to the volume we notice that Bentham has corrected the orthography of Trichostemma, so printed in the "Genera" of Linnæus, and by mistake in the "Labiatarum Gen. et Sp.," to Trichostema, as written by Gronvius, by Linnæus in the "Hortus Cliffortianus," and as the derivation requires.

This half volume (Pars XIII., sectio posterior) has appeared very nearly at the date announced for it, last autumn, when the twelfth was published. It is the second part, anticipating the first, which is to contain the *Solanaceæ* and the *Plantaginaceæ*, two families which will finish the Monopetalous series, as this begins the *Apetalæ* or *Monochlamydeæ*. It comprises the *Phytolaccaceæ*, *Salsolaceæ* (*Chenopodeæ*), *Basellaceæ*, and *Amarantaceæ*, elaborated by Moquin-Tandon of Toulouse, and the *Nyctaginaceæ*, by Professor Choisy of Geneva. Of *Phytolaccaceæ* we have in the United States, only *Petivera alliacea*, which grows in Florida (probably not in "Carolina"), *Rivina lævis* (to which we are surprised to see *R. portulacoides*, Nutt., joined), and *Phytolacca decandra*, which last is now so widely dispersed over the world that its native country is uncertain.

The large family of *Salsolaceæ* comprises 72 genera, disposed nearly as in Tandon's "Chenopodearum Enumeratio," in two suborders and seven tribes, most of which are further divided into subtribes. Our genera of the *Cyclolobeæ* (those with the embryo nearly annular) are Aphanisma, Nutt., a Californian plant discovered by Mr. Nuttall; *Teloxys aristata*, which is credited to us because Linnæus referred his *Chenopodium Virginicum* to *C. aristatum*, but it is doubtful if we possess the genus; Cycloloma (*Salsola platyphylla*, Michx.); Chenopodium, to which Tandon now reunites the greater part of his Ambrina (*C. ambrosiodes*, *C. anthelminticum*, etc.), leaving in Roubieva only the original species, recently illustrated in this Journal by Mr. Carey; Blitum, to which the author now refers, as a section, his former genus, Agathophyton (*Chenopodium Bonus Henricus*, L.); Monolepis, Schrad. (*Blitum chenopodioides*, Nutt.); Atriplex, of which too many of the older species are credited to the United States; Obione, Gærtn., of which nine species are North American, including (apparently with sufficient reason) the Pterochiton, Torr., Grayia, Hook. and Arn., of a single species; Eurotia; a doubtful Kochia; a Corispermum (which Tandon seems not to know as also a native of this country); Salicornia, in which we have *S. herbacea?* *S. Peruviana* (Caro. Fraser), and *S.*

Virginica, to which last he evidently would refer *S. mucronata*, Bigelow, a name unknown to him (and he has also dropped, apparently by accident, the homonym of Lagasca, so that the point in which we are interested is not elucidated); Arthrocnemum (*A.? ambiguum* = *Salicornia ambigua*, Michx.) being still kept distinct. Of the *Spirolobeæ* we have in North America, Chenopodina, a genus newly founded for the *Chenopodium maritimum*, L., which was formerly referred to Suæda, besides which species Tandon also gives us *C. linearis*, Ell., which, however, he thinks may be a variety of *C. prostrata*, which again he thinks may not prove distinct from *C. maritima*, and *C. depressa* (*Salsola depressa*, Pursh); of Shoberia, we have *S. calceoliformis* (*Chenopodium calceoliforme*, Hook.), which is stated also to be found "near New York"; of Salsola, we have *S. kali* only. The singular genus Sarcobatus of Nees (the Fremontia of Torrey in the reports of Fremont's first and second journeys) is enumerated among the Genera exclusa, and said to be "dubiæ sedis." Probably the author had not seen the figure of the fertile plant published by Dr. Torrey. Acnida, following the aspect and inflorescence, is here referred to the *Amarantaceæ*.

The order *Basellaceæ*, familiar to us only by the *Boussingaultia baselloides*, which is cultivated as an ornamental climbing plant, contains six genera, entirely of tropical plants.

The order *Amarantaceæ* includes forty-five genera, arranged under three tribes. There are credited to this country: Celosia, one Californian species; Amaranthus, about nine species; Mengea of Schauer, a Californian species which has much the aspect of *Amarantus Blitum;* one or more species of Euxolus, Raf. (*Amarantus lividus*, L. etc.); Acnida, in which *A. rusocarpa* appears to be mixed up, in a manner that requires much investigation to unravel, with *Amarantus tamariscinus*, Nutt., which again, though entirely distinct from Acnida itself, nearly accords in character with Moquin-Tandon's section Montelia; Banalia, a new genus, one section of which includes an Oregon species (*Halomocnemis occidentalis*, Nutt., ined.); an obscure Polycnemum:

Gossypianthus, Hook., two Texan species; Iresine, two species; Alternanthera, one species (*Achyranthus repens*, L.); besides the *A.* (Cladothrix, Nutt.) *lanuginosa*, which Lindheimer and Wright find abundantly in Texas, and which will certainly stand as a separate genus, if a striking peculiarity in respect to its fruit, observed by Dr. Torrey, proves to be a normal condition. *Telanthera ficoidea* and *T. polygonoides* appear to be only introduced plants along our southern coast. Frælichia (Oplotheca, Nutt.) has three North American species. Phyllepidium of Rafinesque is not identified and probably never will be.

The remaining family, *Nyctaginaceæ*, includes eighteen genera, in three tribes. Of Mirablis, though no species are credited to us, we have one or more in Texas, as well as the three species of Nyctaginia, Choisy. Of Oxybaphus, six North American species are indicated; and the Peruvian *Alliona incarnata* comes also from California. Four species of Abonia are described, besides *A.?* (Tripterocalyx) *micrantha*, Torr., which Dr. Torrey has since raised to the rank of a genus. *Pisonia aculeata* is found on Key West. Boerhaavia furnishes us three or four species; and there shall remain some undescribed Texan representatives of the family.

The second part of the fourteenth volume of the "Prodromus" contains the *Thymelæaceæ* by Meisner, the *Elæagnaceæ* by von Schlechtendal, the *Grubbiaceæ* by De Candolle, resting merely on one of those outlying or anomalous genera which there is too great tendency to raise to ordinal importance, merely because the author knows not what to do with them,—and *Santalaceæ* by De Candolle. Of the first order we have only Dirca, peculiar to this country, and with no congener known. There is nothing to add respecting our three species of *Elæagnaceæ*. As to our few *Santalaceæ*, it is interesting to remark that one of our characteristic genera, Pyrularia (the Oilnut), is found to have two representatives in the Himalayas (Sphærocarya, Wall.), and apparently two more in southern India (Scleropyrum, Arn.). Also that a European species is introduced into our Comandra (the *Thesium elegans* of Rochel), and the genus itself shown to be hardly distinct from Thesium. And Darbya, Gray, published in this Journal twelve years ago, is reduced to a subgenus of Comandra, to which we are not disposed to object. But we take the new species of true Comandra (*C. pallida*) to be a mere variety of *C. umbellata;* which, by the way, we did not state to be eight or ten feet, but only as many inches in height. De Candolle thinks that the hairs which connect the anthers of Comandra, and of most Thesia also, with the perianth, belong to the latter, not to the former, as the generic name implies. Our own observations, and especially some made by Mr. H. J. Clark upon very young flower-buds, confirm this view. The discovery, announced in this Journal in 1854, that the striking genus Buckleya, Torr., is truly dichlamydeous in the female flowers, proves a capital fact for M. De Candolle; who draws from it the confident inference that the floral envelope which in all other plants of the order occurs alone, and has the stamens opposite its lobes, is corolla and not calyx, and consequently so in the *Loranthaceæ* and *Proteaceæ* also. Our author's views are presented in detail in an article, "Sur la Famille des Santalaceæ," in the "Bibliothèque Universelle," published last autumn, and they appear wellnigh convincing. An analogous case is found in Zanthoxylum (only here the suppression is the rare case), *Z. Americanum* plainly wanting that which in *Z. Carolinianum* is the corolla ("Genera Illustr." 2, p. 148). Nyssa offers a good instance of the limb of a calyx so reduced as to have escaped notice until four years ago. For what to De Candolle seem to be petals (p. 622, note in char. of order *Santalaceæ*), were seen to be so, and the observations recorded in the fifth volume of the "Memoirs of the American Academy," p. 336, and afterwards extended in the "Manual Bot. U. S.," ed. 2, p. 162 (1856). It is singular that De Candolle should remain so uncertain of the place of Nyssa in the Natural System. If he will compare it and Mastixia with Cornus he will surely be convinced that Nyssa is a true Cornaceous genus. So of Cevallia, the true place of which our author seems not to know, although given in the "Flora of North America" many years ago, under the sanction (we may add) of the very highest authority. Indeed, so plain is its relationship to Gronovia that Fenzl soon saw and corrected his mistake in referring the genus to *Calycereæ*. And if at this date any should doubt that these are Loasaceous plants, let them turn to the characters of Petalonyx, in the "Memoirs of the American Academy," v., p. 319.

Leaving these details, let us consider our pleasing prospect in respect to the continuation (at least through the *Dicotyledoneæ*) of the great work upon which the De Candolles, father and son, and other excellent botanists, have bestowed so much labor and talent. The great order of *Lauraceæ* was to have been included in the present volume. It would have extended the volume unduly. But, unfortunately, or fortunately, as the case may be, Professor De Vriese has gone to Java on a government mission without finishing the work; and the indefatigable Meisner now takes it in hand. It is to form the leading part of volume xv., the *Begoniaceæ* by De Candolle himself, and the *Aristolochiaceæ* by Duchartre being appended, and perhaps the *Euphorbiaceæ*, also by De Candolle, except the genus Euphorbia which Boissier undertakes. The sixteenth volume is intended to commence with the *Urticaceæ* proper, by Weddell, or the *Monimiaceæ* by Tulasne. We are pleased to learn that Professor Anderson of Stockholm is to elaborate the *Salicineæ*.

The other section of the sixteenth volume of De Candolle's "Prodromus" has just been issued. The two parts form indeed independent volumes, and are paged and indexed as such, so that for all time botanists will have to quote *D C.* Prodr. xvi. (I), p., etc., which is to be regretted, but there is no help for it. The present (prior) part, of 450 pages besides 65 intercalated ones, contains the *Buxaceæ* and some other plants excluded from the *Euphorbiaceæ*, by Dr. Müller; the *Empetraceæ* by Alphonse De Candolle himself (Empetrum reduced to one species, Corema of two, and a Ceratiola); *Cannabineæ* by the same (the *Ulmaceæ* and *Artocarpeæ* postponed not being ready), the *Urticaceæ* (i. e. the *Urticeæ*) by Weddell; *Piperaceæ* by Casimir De Candolle (the *Saurureæ* made a mere tribe, and the *Pipereæ* mainly included under Piper of

635 species, Peperomia of 389!); *Chloranthaceæ* by Solms-Laubach of Halle, reduced to three genera; and finally, *Garryaceæ* by the editor, comprising nine species of Garrya. It appears that the latter end of the volume was printed first, which explains the omission of *G. buxifolia*, a species discovered in northern California by Bolander, and published a year and a half ago. If one or two collaborators will now bring up their arrears, the editor may very soon have the great satisfaction of announcing the completion of the great Dicotyledonous series.

The seventeenth volume of De Candolle's "Prodromus" contains, in the first place, certain small outlying orders, or some of them genera that have to do duty as orders, which on various accounts have been left out of the "Prodromus" as it went on, namely, *Sarraceniaceæ*, *Phytocreneæ*, *Cardiopterideæ*, *Salvadoraceæ*, *Cynocrambe*, *Batidaceæ* (one Batis), *Lennoaceæ* (by Solms-Laubach), *Podostemaceæ* (by Weddell). Then *Nepenthaceæ*, by Dr. Hooker; *Cytinaceæ*, by the same; *Balanophoraceæ*, by Eichler; *Ulmaceæ*, by Planchon; *Moraceæ*, by Bureau; and a synopsis of the genera of *Artocarpeæ*. For the complete elaboration of the last family the volume has been a good while kept back, and has at length been issued, and (sad to say) the work concluded without it. It is to be hoped that whenever M. Bureau finishes his undertaking, the publishers of the "Prodromus" may print the *Artocarpeæ* uniformly with the rest, so that it may be appended.

Then follow a few pages of Genera omissa, with brief references, indicating that they are, or may be, so far as has been made out or conjectured; and finally, the wearied editor appends his "Prodromi Historia, Numeri, Conclusio." It is a terse and highly interesting account of this work, which (including the two preceding volumes of "Systema") occupied his celebrated and indefatigable father from the year 1816, or earlier, down to the end of his life, in 1841, and himself to the close of the past year; an enumeration of the contributors who have worked up particular families or genera; an enumeration of the orders, specifying the volume which contains each, and the number of genera and species described, 5134 genera and 58,975 species, which the missing *Artocarpeæ*, it is estimated, would bring up to 5163 genera and about 60,000 species of Dicotyledones. Among a few statistical data which are given, the ten orders are enumerated which contain the greatest number of genera, beginning with *Compositæ* and ending with *Cruciferæ.* Then the ten which most abound in species, which begin with *Compositæ* (8561 species) and end with *Umbelliferæ* (1016). *Leguminoseæ* are the second in both lists, and next *Rubiaceæ* in the former and *Euphorbiaceæ* in the latter. But the long interval between the publication of many orders, say between *Cruciferæ* and *Euphorbiaceæ*, much diminishes the value of such comparisons. The reasons which have prevented a more rapid publication of the volumes of the "Prodromus," especially since the work has been largely distributed among collaborators, are hinted at; and finally the regrettable announcement is made that the publication is now relinquished, at the close of the Dicotyledones. A full Index, down to genera and their sections, filling 170 pages, closes this great work.

We sincerely congratulate the editor upon the successful completion of this great undertaking at the limits he felt obliged to prescribe, and thank him heartily for his long and faithful service and many sacrifices. As it may be hoped that he has still years of good work in him, all will regret that he could not bear this burden through a few of them, while a half dozen collaborators, who might be named, elaborate the Monocotyledonous orders. But, as he declares that he should doubtless perish under it, we prefer the living botanist to the completed "Prodromus." We may expect from him original work instead of editorial drudgery, perhaps a new edition of his Geographical Botany, or new researches upon the same subject, investigated with his impartial judgment, under the new light which was just dawning when that comprehensive treatise was published.

Since these remarks were written we have received an interesting pamphlet, separately issued from the "Archives des Sciences" of the "Bibliothèque Universelle" for November, entitled, "Reflexions sur les Ouvrages Généraux de Botanique Déscriptive." In this M. De Candolle gives the history of the "Prodromus" and its forerunner with considerable fullness, explains more particularly his editorial trials and burdens, and the reasons why the work could not be made to get on faster, and gives his views as to the most practicable method of combining the labors of the botanist of another generation in the production of the new "Systema Vegetabilium" which will be demanded. An estimate is made of the time it must needs require, even with all the available monographers of the day enlisted in the service. The increased difficulties, or at least the augmented labor, of systematic botanical work, under the present demands of the science, are indicated. It appears that while in his father's time one could elaborate at the rate of about ten species a day, a faithful monographer now, under the modern requirements, can seldom exceed three or four hundred species per annum, that is, about a species a day! We suppose that the case on the whole is not overstated.

ENDLICHER'S GENERA PLANTARUM.

THIS is one of the most important works[1] of the age; and we are anxious to make it more generally known to the botanists of this country. It is not too much to say, that without this, and Lindley's introduction to the Natural System (or some equivalent work), no person who does not possess the advantage of a large library and an extensive general collection of plants, can obtain any correct idea of the present state of systematic botany. The work is published in parts, of eighty pages each, in an imperial octavo or a kind of oblong quarto form, closely printed in double columns. The eleventh fasciculus, which is the last we have received, reaches to page 880; but probably two or more additional numbers have by

[1] Endlicher's *Genera Plantarum secundum Ordines Naturales disposita.* Vienna, 1836–40. (American Journal of Science and Arts, xxxix. 176.)

this time appeared. It is stated in the original announcement that the work will not exceed ten or twelve numbers; we imagine, however, that four or five additional numbers will be required for its completion. It commences, like the "Genera Plantarum" of Jussieu, with the plants of the simplest or lowest organization (Thallophyta, Endl.); a plan which is now the most common and perhaps the most philosophical, but which is attended with many practical inconveniences to the tyro.

The first edition of the "Genera Plantarum" by Linnæus was published at Leyden in the year 1737; the second and third were published at the same place, the one in 1742, the other in 1752; the fourth and fifth were published at Stockholm; the latter (termed the sixth in our copy) in the year 1764, which is the last by Linnæus himself, is the edition generally cited, and was reprinted at Vienna in 1767. This last Stockholm edition forms the excellent model of all the succeeding editions, as they are termed, edited by various authors. It comprises 1239 genera, which in an appendix are reduced as far as possible to their proper natural orders. The first edition after the death of Linnæus is, we believe, that of Reichard, published at Frankfort in 1778, about the same time with the edition of the "Systema Plantarum" by the same author. To this succeeded the edition by Schreber (published also at Frankfort, 1789–1791, in two volumes), who is chiefly famous for having in this work changed all the unclassical names of Aublet and others for new ones made according to the Linnæan canons. Succeeding authors in plucking these borrowed plumes have despoiled him of some rightful feathers; as in the case of the genus Brasenia, for which most botanists have retained Michaux's name, Hydropeltis, which was published a dozen years later. The number of genera is here increased to 1769. About the same time (1791) an edition was published by Haenke at Vienna, which is apparently carefully digested. The latest edition of the "Genera Plantarum" which bears the name of Linnæus, and is arranged according to the artificial system, is that of Sprengel, and published at Göttingen in 1830 and 1831 (2 vols. 8vo), which is the latest complete work in which the known genera are characterized. He gives the date of the publication of each genus, and references to the principal figures. The whole number of the genera described is 4159.

The "Genera Plantarum secundum Ordines Naturales disposita" of the immortal Jussieu, with which a new era in botany commenced, appeared in the year 1789. This work has never been reprinted in France, and but once out of it, and is now very scarce. Until the commencement of Dr. Endlicher's work, a period of about half a century, it has remained the only Genera Plantarum according to the natural system. There is but one living botanist upon whom the task of preparing a new Genera of Plants would seem most appropriately to devolve; but since it cannot be expected from that quarter, we are glad it has been undertaken, and we may almost say completed, by so learned and careful a botanist as Dr. Endlicher. The only fault we have to notice is, that there is no mode of distinguishing directly the generic characters, which are compiled altogether from preceding authors, from those drawn from the plants themselves. An author can only be considered responsible for the latter; yet unless there be some means of distinguishing those which have been verified from the remainder, he becomes somewhat implicated in the mistakes of his predecessors. Dr. Endlicher being scarcely less distinguished as a classical scholar than as a botanist, this work is a perfect model of the classical style.

Simultaneously with this work, which it is in part intended to illustrate, the author is publishing an "Iconographia Generum Plantarum." It appears in quarto parts, with about twelve uncolored plates in each, executed in a very superior manner, with full analyses, which leave nothing to be desired in this respect. Seven or eight parts are already published. It is the cheapest illustrated work of the kind with which we are acquainted, and at the same time one of the very best.

HARVEY'S SOUTH AFRICAN PLANTS.

THIS volume[1] was written, printed in very handsome style, and published at the Cape of Good Hope. It was prepared, not, as we might suppose, for the purpose of making Cape plants better known to European botanists, but for the use of the students and lovers of flowers at the Cape. It is arranged, moreover, according to the Natural System, and is throughout a work of genuine science. Truly, if popular botanical works, based on the Natural System, are deemed most advantageous for students at the Cape of Good Hope, we may indulge the expectation that this method will in due time be universally adopted in Europe and the United States. Mr. Harvey, who, while occupied with his duties as colonial secretary, has been enabled to do so much for the botany of that rich and interesting region, both by his own researches and by encouraging the labors of others, was requested to recommend some introductory work on botany.

Had a mere introduction to the elements of the science alone been needed, the desideratum might easily have been supplied. "But I soon found," says Mr. Harvey, "on cross-questioning, that something very different was required. One lady told me that she knew already what 'calyx, corolla, stamens, and pistils, and all that' meant; and another had penetrated the mystery of Monandria, Diandria, etc., and did not want to be told that over again; what they desired was a book in which they could discover the names of every plant that struck their fancy in rambling through the fields, — in short, a Flora Capensis. Here I found myself completely at fault, for there seemed little use in recommending the Flora of Thunberg, or the more ancient writings of Burmann; for even could they be procured, which would not be without much difficulty, they would have proved perfectly useless to my lady friends, who, not being blue-stockings, could have derived little instruction

[1] *The Genera of South African Plants,* arranged according to the natural system, by William Henry Harvey, Esq. Cape Town, 1838. (American Journal of Science and Arts, xxxix. 173.)

from the crabbed Latin in which they were written." Mr. Harvey then conceived the idea of writing a Flora Capensis; but it at once occurring that such a work must consume a long series of years in preparation, he decided upon rendering that more prompt, though less complete assistance, which a work like the present is calculated to afford. "The Genera of South African Plants" is the result of this determination; for which the author deserves the thanks, not only of the lady friends whose benefit he had chiefly in view, but of all the cultivators of botanical science. Although much more time would be required for its preparation, the work would have been more valuable had Mr. Harvey placed still less dependence on preceding authors, and drawn his characters, in every practicable instance, from the plants themselves; but only those who are accustomed to prepare their works in this manner are aware of the vast amount of labor it involves. The general plan of the work, as the author informs us, is taken from Beck's "Botany of the Northern and Middle States of North America," and Nuttall's "Genera of North American Plants"; in the arrangement and characters of the orders, Dr. Arnott has chiefly been followed. The number of genera described is 1086, distributed under 135 orders. Many South African genera have been published in still more recent general works or particular memoirs, or in those which have not reached the Cape in time to be employed by Mr. Harvey, so that the number of Cape genera may be safely estimated at 1200.

SIEBOLD'S FLORA OF JAPAN.

THIS work[1] is, we believe, wholly arranged and prepared by Professor Zuccarini of Munich, from notes and specimens furnished by Dr. Siebold of Leyden, accumulated during his

[1] *Flora Japonica.* Sectio prima, Plantæ ornatui vel usui inservientes, digessit Dr. J. G. Zuccarini. Fasc. 1–10. Leyden, 1835–39. (American Journal of Science and Arts, xxxix. 175.)

long official residence in Japan. The admirable plates are executed in Munich: they are engraved upon stone after a peculiar method, which is now frequently employed, and are certainly not excelled in beauty or accuracy by any copper-plate engraving in the same style. The portion already published comprises only the ornamental or otherwise interesting plants, the general account of the Japanese flora being reserved for a future part of the work. The flora of Japan presents such striking analogies to that of the temperate part of North America as to render this work of more than ordinary interest to American botanists. To show this, we select from the forty-six species described and figured by Zuccarini, the following list, placing opposite the Japanese plant the related North American forms.

FLORA OF JAPAN.	FLORA OF NORTH AMERICA.
Illicium religiosum,	*Illicium Floridanum* and *parviflorum*.
Kadsura Japonica,	*Schizandra coccinea*.
Benthamia Japonica,	*Cornus florida*.
Corylopsis, two species,	*Hamamelis* and *Frothergilla*.
Aralia edulis,	*Aralia racemosa*.
Symplocos lucida,	*Hopea tinctoria*.
Styrax Japonicum, etc.	*Styrax*, several species.
Deutzia, three species,	*Philadelphus*.
Schizophargma Hydrangeoides, *Platycrater arguta*,	*Decumaria* and *Hydrangea*.
Diervilla, several species,	*Diervilla Tournefortii*.
Viburnum tomentosum,	*Viburnum lantanoides*.
Wisteria (or, as it should be, *Wistaria*) *Japonica*, and two other species,	*Wistaria frutescens*.
Paulownia imperialis,	*Catalpa cordifolia*.

While about half the species thus far published are nearly related to (chiefly characteristic) North American plants, only eight, besides those given above, belong to genera which have no representatives in this country. The list might be greatly extended by comparisons from other sources. Thus *Hoteia Japonica* of Morren and Decaisne (which belongs to the earlier established Astilbe, Don.), which was by Thunberg mistaken for *Spiræa Aruncus*, closely resembles our own *Astilbe decandra*, which has been more than once confounded with

Spiræa Aruncus. On some future occasion we hope to make a somewhat extended comparison between the flora of temperate North America, and that of Japan and Middle Asia.[1]

Professor Zuccarini, the author, in conjunction with Dr. Siebold, of the excellent "Flora Japonica" now in progress, has recently published the first part of a brief memoir, entitled, "Floræ Japonicæ familiæ Naturales, adjectis generum et specierum exemplis selectis: Sect. i. Plantæ dicotyledoneæ polypetalæ."[2] It is interesting to remark how many of our characteristic genera are reproduced in Japan, not to speak of striking analogous forms. Thus the flora of Japan has not only Wistaria, Lespedeza, Sieversia, Chimonanthus (in place of our Calycanthus), Philadelphus, several species of Rhus closely resembling our own, and two peculiar genera of *Juglandeæ*, but also a Pachysandra, some Berchemias, a Staphylea, and a peculiar genus of the tribe (Euscaphis) besides; not only a dozen Maples, but also a Negundo, a Stuartia, two Tilias, a Phytolacca, an Opuntia (surely not indigenous), a Sicyos referred to our own *S. angulata*, two Droseras, a Nelumbium, a Nuphar, and two species of Nymphæa, Gynandropsis, a real Dicentra (Dielytra) and an allied new genus, with several species of Corydalis, a Trollius, our own Coptis and two new ones like the western *C. asplenifolia*, an Isopyrum, two species of Aquilegia, one of them near *A. Canadensis*, a Cimicifuga, a Trautvetteria, an Illicium, some Magnolias, Kadsura and Sphærostemma in place of Schizandra, a Mitellopsis, two species of Astilbe (Hoteia), many Hydrangeas as well as peculiar Hydrangeaceous forms, a Hamamelis with two other characteristic genera of the family, some true Dogwoods, as well as Benthamia the analogue of our *Cornus florida*, some true Vines, and two species of Am-

[1] This short paper is of peculiar interest. It contains the earliest record of Professor Gray's investigations into the flora of Japan, and its relations to that of eastern North America — investigations which many years later enabled him to explain the distribution of plants through the northern hemisphere by tracing their direct descent from ancestors which flourished in the Arctic region during the latest tertiary periods, and established his reputation as a philosophical naturalist. — C. S. S.

[2] American Journal of Science and Arts, 2 ser., ii. 135.

pelopsis, three species of Panax, and four of Aralia, one of which is near our *A. nudicaulis;* and among Umbelliferæ are Hydrocotyle, Sanicula, Sium, Angelica, but what is most remarkable, Cryptotænia, Archemora, and Osmorhiza. Further cases of generic conformity abound in the remaining divisions of the vegetable kingdom; thus, for example, Diervilla, Mitchella, Maclura, Liquidambar, Torreya, and Sassafras, are represented in the flora of Japan.

VEGETABLE MONSTROSITIES.

THIS interesting treatise[1] on Vegetable Monstrosities is very properly prefaced by a statement of what is meant by the normal structure of plants, by vegetable individuality, and vegetable symmetry. The author proceeds to consider, first, those slighter deviations which are called varieties; and secondly, those more grave and mostly congenital anomalies which bear the name of monsters. As to the latter the author remarks, that nearly every monstrous or abnormal condition that has been observed is to be met with as the normal state of other vegetables; and that between a monstrous and a normal flower, the only difference often is, that the former is the occasional, and the latter the habitual state. "La monstruosité est donc, en général, l'application insolite, à un individu ou à un appareil, de la structure normale d'un autre appareil ou d'un autre individu. C'est un organisation transposée, c'est une loi changée de place. On l'a dit avec raison, la monstruosité ne se trouve pas en dehors de la nature, mais seulement en dehors de la coutume." It is clear, therefore, that while abnormal states may always be explained by the laws which regulate the normal structure, monsters themselves, as the etymology of the name indicates, often show us the true structure when it could not be certainly inferred from the

[1] *Elémens de Tératologie Végétale, ou Histoire abrégée des anomalies de l'Organisation dans les Végétaux.* Par A. Moquin-Tandon. Paris, 1841. (American Journal of Science and Arts, xli. 374.)

habitual condition. The author arranges monstrosities under four primary classes: those of volume, of form, of disposition, and of number. These are divided, the first class into monsters by diminution of volume (*Atrophy*), and by augmentation (*Hypertrophy*); the second class into monsters by alteration of form, whether irregular (*Difformation*) or regular (*Pelorias*), and monsters by the transformation of one organ into another (*Metamorphosis*); the third class into monsters by the abnormal connection of parts, or by the disunion of parts habitually united, and into those caused by change of situation, or displacement; the fourth class into monsters by diminution of number, or abortion, and those by augmentation of number. Under these heads the monstrosities of the different organs of plants are considered in detail, and in a philosophical and very interesting manner. This brief notice of the plan of Moquin-Tandon's work, we are confident, will suffice to commend it to the attention of the botanists of this country.

AGASSIZ'S ZOÖLOGICAL NOMENCLATOR.

THIS great work,[1] which must have cost an extraordinary amount of labor, is now almost completed. Trusting that some one of our able zoölogists will duly give an account of a work which is indispensable to every votary of their science, we propose at this time merely to call attention to the preface, published last year with the ninth and tenth fasciculi, and to express unqualified admiration of the manner in which a subject of interest to all naturalists, that of nomenclature, is there treated. While botanists are enjoying the benefits of a sedulous adherence to the wholesome rules imposed by the father of natural-history nomenclature, and of nearly unanimous agreement in the few changes which the progress of science and the multiplication of its objects have rendered needful, the zoölo-

[1] *Nomenclator Zoölogicus, continens Nomina Systematica Generum Animalium, tam viventium quam fossilium*, etc. Auctore, L. Agassiz. Soleure, 1842–46. (American Journal of Science and Arts, 2 ser., iii. 302.)

gists on the other hand, who have too generally allowed every one to do that which was right in his own eyes, are reaping in consequence a plentiful harvest of confusion. The difficulty of a reform increases with its necessity. It is much easier to state the evils than to relieve them; and the well-meant endeavors that have recently been made to this end are some of them likely, if adopted, to make "confusion worse confounded." Probably no living zoölogist is so conversant as Professor Agassiz with the actual state of the nomenclature of the animal kingdom, and so well qualified to judge of the practical working of proposed rules, which often involve consequences that the propounders never dreamed of. Our author's views are therefore entitled to great weight. We are glad to perceive that they entirely concur with those quite unanimously adopted in the other great department of natural history for which the Linnæan canons were originally framed. As these canons were the foundations of our nomenclature, Professor Agassiz has very properly reproduced them, *totidem verbis*, from the "Philosophia Botanica," adding now and then a short but pithy commentary. He then proceeds to examine the rules proposed by the Committee of the British Association, and shows that while some of them are mere iterations of the Linnæan canons, which should never have lost their authority, others are contrary to them, or threaten greater evils than they are intended to remedy. In most respects his criticisms concur with those already made by Dr. Gould in a former volume of this Journal (XIV. p. 1). We agree with Professor Agassiz in thinking these English canons worthy of adoption only when they agree with the letter or spirit of the Linnæan rules, which indeed they generally do. Those which conflict with them have not received, and probably will not receive, the general assent even of British naturalists. Hence, in our opinion, the American Geological Association has too hastily reaffirmed them, while they have, indeed, improved their form in several respects. It may be well to notice the comments of Professor Agassiz upon the more objectionable propositions.

Their first rule, "that the name given by the founder of a

group or the (first) describer of a species should be permanently retained," cannot be too firmly insisted on; for upon it rests the stability which is the most essential requisite of nomenclature. Their second rule, that since "the binomial nomenclature originated with Linnæus, the law of priority is not to extend to the writings of antecedent authors," restricts the former too arbitrarily, and conflicts, as Professor Agassiz states, both with the canons and the example of Linnæus, not less than with the conscientious practice of good naturalists ever since. Linnæus was not the founder of genera or of generic nomenclature, and, "far from making new names in every instance, he retained all names given by his predecessors, provided they could be received into his system." It is generally thought that Linnæus erred by adopting, not too many, but too few of the unobjectionable and well-established generic names of his predecessors, such as Tournefort, etc. Now when, in the natural progress of the science, a Linnæan genus is resolved into two or more Tournefortian ones, for instance, are the names of Tournefort to be excluded from use? In the breaking up of the Linnæan genus Lonicera, had not the Diervilla and Xylosteum (and if the division were to go farther, the Periclymenum and Caprifolium) of Tournefort, as well as the Symphoricarpos of Dillenius, an indisputable right to restoration? Indeed Linnæus was here plainly wrong in not adopting one of these prior names for the whole genus, instead of creating the new one. This, however, was to be submitted to; for, as Professor Agassiz remarks, "the names sanctioned by Linnæus are to be held as established above all others. Linnæus, for instance, received very few genera of Echinodermata. Nowadays this class numbers many, among which some of those founded by Klein, Link and Breynius, long anterior to Linnæus, hold their place with the modern ones of Lamarck, Müller, etc. But no one now prefers that new names should be made for such genera, rather than that such approved anterior ones should be brought into use again. I certainly see no cause why we may not call to life the names of former authors when we divide the genera of Linnæus." We think those naturalists blameworthy who do not. The third, fourth, and fifth of the British canons are accordant with Linnæan rules, and are regularly followed in botany. The next four relate to matters which follow as a consequence of the law of priority; but as to what relates to the use of synonymous names Professor Agassiz intimates that their rule is perhaps too absolute, and even contradictory to the Linnæan canon, § 244: "Nomina generica, quamdiu synonyma digna in promptu sunt, nova non effingenda."

The tenth rule, namely, "A name should be changed which has before been proposed for some other genus in zoölogy or botany, or for some other species in the same genus, when still retained for such genus or species," is not as well worded as the equivalent Linnæan canon, § 217, "Nomen genericum unum idemque ad diversa designanda genera assumptum, altero loco, excludendum erit." Mr. Agassiz remarks, greatly to our surprise, that the enforcement of this rule would demand the sacrifice of almost half the generic names made in recent times. In our opinion, while the same names ought not to be given both in zoölogy and botany, the time is passed when received names are to be changed on this account. While writers in the different departments of zoölogy alone have doubly employed the same name "in ten thousand instances," we must see that cases of this sort between zoölogists and botanists, occupying such widely separated fields, are inevitable, at least until as perfect lists of zoölogical names shall be compiled and kept up as is done in botany. Besides, it is now utterly impossible for any single naturalist, or any joint committee of botanists and zoölogists, to determine, in half the cases that arise, whether a particular genus is to be suppressed or retained in one department, so as to require or forbid a change of the posterior homonymous name in the other; hence the practical application of the Linnæan rule would now create tenfold more confusion than it can relieve. Each well-founded change of the sort does no more than obviate a possible inconvenience, while every needless one, in a genus of numerous species, draws after it a load of useless synonyms, which do not serve, like genuine synonyms, to tell the history of the genus and mark the progress of our knowledge. The whole subject is forcibly presented by Professor Agassiz, in another section of his preface (p. xxviii *et seq.*), where he states that he now knows three thousand generic names common to botany and zoölogy, which the Linnæan rule would require to be changed in one or the other department. But surely this number must comprise a host of synonyms long since laid on the shelf, as well as names of somewhat different formation or termination, although of the same derivation. In this case a small matter should give them impunity. If these changes must be made, no one could do the work for zoölogy better than Mr. Agassiz; but he affirms it to be a task quite beyond his power, and justly concludes that, "in the present state of the science, generic names ought not to be changed solely on account of their being employed in both kingdoms of nature." To this conclusion the American Association evidently accede.

As to generic names doubly or triply employed in the several classes of the animal kingdom (which, we are astonished to learn, already number nearly ten thousand), the necessity of applying the Linnæan canon is obvious, and would be imperative had not the evil reached such a height as to baffle remedy. The *summum jus* which demands the immediate change of nearly a moiety of the received zoölogical names would surely become *summa injuria* to the science, even if any naturalist were equal to the task of applying it. Justice must here be delayed in order that it may be rightly administered, and, as our author recommends, the business of gradually bringing this part of nomenclature under rule must be left to monographers and future systematists. But let those upon whom the *cacoëthes nominandi* is strong, obey our author's advice, desist from proposing new names in mere catalogues, and never attempt, while revising the genus which rightfully claims a particular name, to impose new names upon the homonymous genera in other classes, but leave that for their own respective monographers. It will be soon enough to give them new names, if such are needed, when the validity of these several genera is well made out.

Upon the 11th rule of the British Committee, namely, that "a name may be changed when it implies a false proposition which is likely to propagate important errors," Professor Agassiz remarks that the less this liberty is used the better, lest it should lead to licentiousness.

The 12th rule ordains that "a name which has never been clearly defined in some published work should be changed for the earliest by which the object shall have been so defined." This law, our author remarks, "has become very necessary, since dealers in natural objects have begun to arrogate the authorship of books collected from catalogues, and demand that authors shall receive their names for dividing species. It is the same with names which remain unpublished in public or private collections, and to which the proprietors or curators sometimes lay claim. But priority is to be conceded only to publication in a work which is accessible to the learned throughout the world. Yet while we strictly press the observance of this law in respect to the publication by learned men of the results of their observations, so much the more must we brand with infamy those impudent parasites who prowl about museums to pick materials for their *opuscula*, without mentioning the sources whence they have derived their spoil, and sometimes even furtively describing the species, the names of which they claim." The people alluded to well deserve this censure. On the other hand, not less blameworthy are those who purposely pass by, instead of courteously adopting, appropriate names under which naturalists often distribute their species in advance of publication. This felony is the more atrocious because remediless, and to be prevented by no rule except that of courtesy; for the public good requires that priority should be conceded to actual publication alone.

The two remaining laws (13th and 14th) are agreeable to, or identical with, Linnæan canons, and are approved by all good naturalists.

The rules recommended by the British Committee for the future improvement of nomenclature are next considered; and as they are far the most commendable and in general use among good naturalists, we shall only notice those that Mr. Agassiz criticises, or we have occasion to comment upon. The writer of the British Report has chosen to enforce the direction, to avoid harsh and inelegant or sesquipedalian names, by citing, as an example of the kind, the "*Enaliolimnosaurus crocodilocephaloides* of a German naturalist;" for which he is strongly censured by our author, who declares that no naturalist has ever proposed this name. Surely, if one is inclined "to cast stones into his neighbor's garden," as our author says, there is no lack of legitimate opportunity, nor necessity for fabricating hard names.

The British Committee condemns the future employment of generic names which have been superseded by the rule of priority. But this is contrary to the canon, § 245 — "Nomen genericum unius generis, *nisi supervacaneum*, in aliud transferri non debet" (and to *obs.* under § 244), no less than to the practice of Linnæus and of subsequent naturalists. For instance, Saururus of Plumier became a synonym of Piper, but this did not debar Linnæus from the subsequent application of the name to a new genus. Sisyrinchium of Tournefort being included in Iris, Linnæus gave the name to a different genus; nor did he hesitate to adopt the genus which Ellis had dedicated to Hales, on account of an earlier Halesia of Browne, which had already sunk to a synonym. Why should a good name be forever tabooed in such cases, and why not, if occasion offers, allow it to be remarried to a new genus? We should be careful, however, not to reproduce names which are likely ever to be resuscitated in their former relation.

The British Committee objects to the practice of giving to a genus the name which it bore as a species of a former genus. But, as Professor Agassiz justly remarks, when a species, which proves to be the type of a new genus, has a good proper name already, it seems quite as admissible to take that name for the genus and make a new one for the species, as to coin a new generic name, since either way a new name must be introduced: indeed it is preferable, because such Linnæan species frequently are found to comprise several, hitherto confounded, no one of which has a paramount claim to the specific name: *e. g. Cyprinus Gobio*, *C. Leuciscus*, *C. Barbus*, L. We go further, and maintain that the proper specific names are, *cæteribus paribus*, always to be preferred for genera in these cases, not only because that they are already familiar, but because they are most frequently old generic names which may claim under the law of priority. For example, *Lonicera Diervilla*, L. = *Diervilla*, Tourn.; L. *Symphoricarpos*, L. = *Symphoricarpos*, Dill.; *Rhamnus Paliurus*, L. = *Paliurus*, Dod.; R. *Zizyphus*, L. = *Zizyphus*, Dod.; *Rubus Dalibarda*, L. = *Dalibarda*, L.; and so of hundreds of proper specific names which have rightly resumed their generic rank.

The next proposition of the British Committee, namely, that specific names, even when substantive or borrowed from persons or places, should uniformly be written with a small (instead of capital) initial, is so contrary to long usage and offensive to good taste, that we are surprised that it should anywhere find favor. Mr. Agassiz pointedly condemns it. The only reason assigned for the change is, that some people might not be able to distinguish the specific from the generic name without the aid of typography. But, as Dr. Gould has already remarked in this Journal, such persons would be misled by almost anything; and the propounders of the rule should follow it consistently by writing their own cognomen with a small initial letter. We do not wonder that the Committee of the American Association refused to reaffirm this rule, as applied to proper names from persons; and we are quite sure that naturalists generally will not hesitate wholly to reject it; surely the committee would not approve the practice of a late botanical author of this country, who reduced the proper scientific names of Linnæus into adjective conformity, by writing "*Ranunculus flammulus*," instead of *R. Flammula*, "*Thymus serpyllus*" in the place of *Thymus Serpyllum*, and so on.

Professor Agassiz severely condemns the proposition to restrict the names of families to a uniform termination in *idæ*, and their subdivisions to *inæ*, without considering whether the words in question will receive that particular suffix kindly.

This is quite too straight-laced, and gives rise to many awkward forms, or

> "Sesquipedalia verba
> Vel nocitura sono, guttur læsura loquentis,"

which is not worth while to encounter needlessly, for the sake of mere technical uniformity, at least when they may be avoided by some liberty of choice in the mode of prolongation.

The proposition, D. of the British Committee, which directs that the name of the original propounder of a species should adhere to it when transferred to a different genus, is warmly defended by some naturalists in England and, in a modified form, in our own country also. Few naturalists are now so well qualified to judge of the practical operation of this scheme as Professor Agassiz. He declares his opinion that, if received, "it will introduce horrible and remediless confusion," and that no possible multiplication of synonymy is likely to lead to so many difficulties as this new practice. He therefore strenuously opposes it by arguments drawn from the precepts and practice of Linnæus, who meant the specific name to be subordinate to the generic, and never intended it to be inferred that he who applied to a plant or animal a certain name was therefore its discoverer, or even its first systematic describer. He affirms that Linnæus would have expressly rejected "*Tyrannus crinitus*, Linn. (*sp.*)," were the innovation proposed in his day, and have written *T. crinitus*, Swains., had he thought best to approve the division of his genus Muscicapa. On the other hand, if he disapproved the division, we may add, he would not have thanked a contemporary for making him seem to adopt it. The hardship is still greater when the question is not of the division of an old genus, but of the proper place of a species among admitted genera, when it is surely improper to cite an author as referring to one genus, while he expressly maintains that it belongs to another. In fact, the remedy is much worse than the disease which the English doctors would cure. Linnæus maintains that he is the true naturalist who understands genera; but from the new practice it will inevitably follow, as Professor Agassiz asserts, that the proper establishment and definition of genera which demand the highest powers of the naturalist, will be less esteemed than the mere distinguishing of species; a result which, far from promoting science, will especially retard the progress of that part of zoölogy in which there is most to be done, and in which the science of the animal is still far behind that of the vegetable kingdom. It is of the greatest importance that we should be able to thread our way back through entangled synonymy and mistaken references to the original sources. Here our difficulties would be greatly multiplied, unless two sorts of synonyms are used. For who, as Mr. Agassiz says, can find out what Linnæus has said of *Muscicapa crinita*, without a direct reference to the genus in which Linnæus himself placed it? And when, as often happens, the Linnæan species is mistaken, so that the *Tyrannus crinitus*, Linn. (*sp.*) according to Swainson, is not the *T. crinitus*, Linn. (*sp.*) according to some other author, the confusion becomes inextricable, unless we encumber ourselves with two modes of annotation, the old for expressing synonymy, and the new for the names really adopted. "Then the two modes will not agree with each other, nor can one know whither to turn himself. Surely the authors of this new rule cannot have considered these inconveniences, else they would have themselves discarded it. Therefore I entreat and pray them, by all the interests of the science they wish to promote, to abandon their proposition, and not to introduce a new schism into natural history, but to return again to the system of Linnæus, the most simple of all, and the least likely to errors and Babylonish confusion in nomenclature."

The Committee of the American Association more wisely adopted the mode, afterwards employed by Mr. Dana in his great work on Zoöphytes, namely, that of appending to the specific name the original authority for the species in brackets, and adding without brackets the name of the author who first described the species under the later received genus. To this plan there can be no objection, except that it is rather cumbrous, if it is to be used in every brief mention of the species, and, in our opinion, quite superfluous in a systematic treatise, where the synonymy is given in proper historical order. The recommendation to make subgeneric names agree in gender with that of the genus, Professor Agassiz thinks is of no consequence, unless the new annotation, just animadverted upon, should come into use. Besides, it would often interfere with the rule of priority, which requires synonyms, when they exist, to be adopted for sectional names. But he strongly commends the rule, that the etymology of names should always be stated by the proposer.

Justly does Mr. Agassiz condemn the practice of those who change the authority of a genus when they extend or narrow its bounds, or correct a faulty orthography. Thus he would write Lepidosteus, Lacep., although Lacepede wrote Lepisosteus; and especially would he write Perca, Linn. (Cuv.), not Perca, Cuv.

Hearty and just, also, is his censure of the custom of those French zoölogists who use vernacular appellations in scientific works, either to the exclusion of the systematic name, or in precedence of it.

Our author closes this part of his preface with some excellent reflections on the study of genera in the animal kingdom, and the need of a thorough reinvestigation of the grounds upon which natural families are constituted; remarks which we would gladly copy, if our limits allowed.

VON MOHL'S VEGETABLE CELL.

We desire in a special manner to commend this condensed treatise[1] not only to botanists, but to animal physiologists, to

[1] *Grundzüge der Anatomie und Physiologie der vegetabilischen Zelle.* Hugo Von Mohl: Braunschweig, 1851. English translation (*Principles of the Anatomy and Physiology of the Vegetable Cell*) by Arthur Henfrey, London, 1852. (American Journal of Science and Arts, 2 ser., xv. 451.)

medical students, and to all who would obtain a clear view of the present state of vegetable anatomy and physiology, — a knowledge of which, most interesting in itself, is almost indispensable to the correct understanding of the minute anatomy and physiology of animals. Professor Mohl is, without question, the first of vegetable anatomists, and his statements carry with them the highest authority on this class of subjects. We copy the short preface which he has contributed to the English translation, as it gives a clear view of the nature and scope of the work.

"Mr. Arthur Henfrey having informed me that he intends publishing an English translation of the present treatise, I take this opportunity of making known to the English reader the purpose I had in view in the preparation of the book. The following pages were not originally intended to appear as an independent work, or to give a summary of the wide subject of the Anatomy and Physiology of Plants, but appeared as an article in the 'Cyclopedia of Physiology,' published by Dr. Rudolph Wagner of Göttingen, drawn up to furnish students of Animal Physiology, and more particularly the medical profession, with a review of the anatomical and physiological conditions of vegetables (of the cell), in order to enable them to form a definite judgment upon the analogies which might be drawn between the structure and vital functions of animals and plants. This intention, together with the circumstance that I was compelled to crowd the whole exposition into the space of a few sheets, rendered it necessary to direct especial attention to the individual cell, as the fundamental organ of the vegetable organism. Since, however, the cell only presents itself in anatomical and physiological independence in the lowest plants, and since, in the more highly organized plants, both the structure and the physiological functions of the individual cells become subject to greater dependence upon the other parts of the plant, in proportion as the collective organization of the vegetable is more complex; moreover, since functions then present themselves, of which no trace can be found in the lower plants, it became requisite to take account of the plants of higher rank, and of the various organs which these possess. The treatise, therefore, contains, if an imperfect, still in many respects a more extensive résumé of Vegetable Physiology than might have been conjectured from the title.

"Unhappily, the Physiology of Plants is a science which yet lies in its earliest infancy. Few of its dogmas can be regarded as settled beyond doubt; at every step we meet with imperfect observations, and consequently with the most contradictory views; thus, for example, opinions are still quite divided regarding the doctrines of the development of the cell, of the origin of the embryo, and of the existence of an impregnation in the higher Cryptogams. Both in these and in other cases, the small compass of the present treatise forbids a more extensive detail of the researches upon which the opposing views are founded; I hope, however, that I have succeeded in making clearly prominent the chief points upon which these contests turn, and thus in facilitating the formation of a judgment by the reader; and I have never neglected to indicate the literature from which further instruction is to be derived."

It may be well to notice the views of so excellent an observer upon sundry points which have been more or less matters of controversy. As to the milk-vessels, or vessels of the latex, Mohl inclines to adopt the view that considers them as intercellular passages which have acquired membranous linings (p. 2). He denies that the membrane of nascent cells is soluble in water, as Schleiden states (p. 9). He briefly states the ground on which, in his controversy with Harting and Mulder, he successfully maintains that the primary cell-membrane is thickened by successive concentric layers of cellulose deposited on its inner face. The combination of spiral markings and pits on the wood-cells of Taxus and Torreya, as also in the Linden, is explained by considering the former to belong to a second layer or deposition within that to which the pits belong. This tertiary membrane or deposit forms the spiral fibre or band in the cells of the seed-coat of Collomia, the hairs of the achenium of Senecio, etc. (p. 18). The whole subject of spiral and other markings, rings, dots, slits, reticulations, etc., on the walls of cells is expounded in a masterly and convincing way. Mohl maintains (p. 28), as he has done in the "Botanische Zeitung," that cellulose forms the basis of all vegetable membranes in the higher plants, and that what Mulder regarded as peculiar compounds are combinations of cellulose with foreign infiltrated deposits, which interfere with the chemical reactions of cellulose, but which may be removed by previous maceration in caustic potash and nitric acid. He now maintains, in opposition to his early views (still defended by Schleiden), that the intercellular substance is a product or secretion of the cell, and not a universally distributed mass in which the cells are imbedded (p. 33). He shows that the thickened "cuticle" of Unger, Mulder, Harting, etc., consists of secondary layers of cell-membrane, deposited from the inside, and infiltrated with some substance that is colored brown by iodine; with the exception of an extremely thin external pellicle, the real cuticle of Brongniart, which is probably a secretion from the surface of the cell, like that which forms the outer coat of pollen-grains (p. 35). He insists that the layer of protoplasm, his primordial utricle, lining the cell is a soft and delicate membrane, and not a mere layer of mucilage (p. 37); the mode in which this is constructed and a partition formed by an inward growth at the fold, in the multiplication of cells by division, is very clearly explained (p. 50–53). Free cell-formation is said to occur, in Phænogamous plants, only in the embryo-sac, in which both the embryonal vesicle, or rudiment of the embryo, and the cells of the albumen, originate in this manner; in the Cryptogamia, only in the formation of spores in certain cases, as in the Lichens (p. 58); contrary to the view of Schleiden, who long maintained this to be the universal, and lately a general, mode of cell-production. Schleiden's original account of the process of the formation of a free cell from a nucleus is directly controverted in all essential points; the nucleus, according to Mohl, being always central, and at no time connected with the cell-membrane, but always enclosed in the primordial utricle. A nucleus, or mass of nitrogenous substance, first has the primordial utricle, or nitrogenous membrane, developed over its surface; and then the cell-membrane (of cellulose) is deposited upon this. In Cryptogamous plants, such masses, of larger or smaller size, may become coated with a cell-membrane without any proper nucleus appearing (pp. 57–60). Thus much for what relates to the anatomical condition of the cell.

Under the second head, the Physiological Condition of the Cell, our author treats, first, of the cell as an organ of nutrition, next, as an organ of propagation, and finally, as an organ of motion. He pronounces against the Knightian doctrine, that plants ultimately degenerate and perish when propagated for generation after generation by division (from the bud) (p. 64). That the crude sap, though absorbed by the parenchymatous tissue of the root, ascends through the woody tissue, and that the assimilated sap returns through the bark, and thence more or less into the wood by means of the medullary rays, is very neatly shown. "A few simple experiments leave no doubt about this. . . . If the bark of a plant, best of a tree, is cut through in a ring down to the wood, there is no interruption in the flow of sap to the parts situated above the wound; but if the wood is cut through, the greatest care being taken to avoid injuring the bark, that portion of the plant above the wound dries up at once. From the wood of the stem and branches the sap flows onward into the leaves, as is proved by the powerful expiration of watery vapor from them. Before the sap has reached the leaves it is incapable of being applied to nutrition; consequently the vegetation of a plant comes to a stand-still when it is deprived of its leaves. The sap ascending from the root to the leaves is thence termed the crude sap. It undergoes a chemical change in the leaves, rendering it fit to be applied to the nutrition of the plant. To this end the sap flows backwards from the leaves through the bark to the lower parts, as the following circumstances testify. If the bark is cut off the stem in a ring, the growth of the portion below the wound stands as it were still; the stem becomes no thicker; in the potato plant no tubers are produced, etc.; but, on the other hand, the growth above the wound is increased beyond the usual measure, thicker layers of wood are deposited, more fruit is perfected, this ripens sooner, etc. The deposition of starch which occurs in the cells of the medullary rays in autumn, goes to prove that the portion of assimilated sap which is not used for nutrition on the way to the root, runs back to the wood through these horizontal medullary rays; and thus the sap describes a kind of circle, not, indeed, in determinate vessels, but in a definite path leading through the different parts of the plant. It is difficult to see how in recent times the results of these experiments could have been so questioned, and the existence of the descending current in the bark denied. Certainly it is no improvement on the theory cast aside when the increased growth above the annular wound is explained by the artificial interruption of the upward current of crude sap, in consequence of which the sap contained in the upper part of the plant must soon become greatly concentrated and potential for development (Schleiden, 'Grundzüge,' 2d ed., ii. 513). When we can succeed in fattening an animal by depriving it of a portion of its accustomed food, this explanation may be received as satisfactory" (p. 70, 71). After some excellent points of criticism, Mohl concludes that the discovery of endosmose has not fully solved the problem of the movement of the sap in plants, although in all probability it does play an important and perhaps the principal part in its absorption and conveyance (p. 77). To the question, whether plants live on inorganic food alone or take in also organic matters, Mohl gives a sensible answer, rejecting the extreme view of Liebig, while still fully recognizing the great office and result of vegetation (p. 78). According to Mohl, however, it is proved that plants do not absorb the carbonic acid dissolved in water with the latter by means of their roots (p. 81); but this seems hardly reconcilable with several facts stated on the next page, from which it is justly concluded that carbonic acid is carried up with the ascending sap into the leaves. From the fact that plants perish so soon in air deprived of all oxygen gas, that sensitive leaves lose their irritability under such circumstances, etc., Mohl concludes, apparently with good reason, that the absorption of oxygen and the exhalation of carbonic acid in plants is a true physiological function, intimately connected with the life of the plant; and that this (rather than the opposite and predominant nutrient process, through which carbonic acid is decomposed and oxygen evolved) should be considered as the respiration of plants, if we use the term at all (p. 86). "The roots of plants, and not the leaves, take up the substances which furnish plants with nitrogen, while, on the contrary, the leaves play the essentially active part in the absorption of carbonic acid" (p. 88). The analogy of the milky juice of plants to the blood of animals, as propounded by Schultz, is thoroughly refuted by our author, and the flowing movement in the milk-vessels described by Schultz is positively denied to take place in an uninjured plant, except as produced by mechanical causes. That the milky juice is not a nutrient material, still less the nutrient juice, is also manifest (p. 96).

The cell as an organ of propagation is treated, first, as respects the multiplication of plants by division; second, by spores; and third, by seeds. The conjugation of certain Confervoid *Algæ*, such as Zygnema, is said to bear no analogy to sexual reproduction (p. 113); a conclusion which may be questioned. A good summary is given of the facts known respecting the free and spontaneous movements of the spores of the lower *Algæ* (p. 115); and also of the recent discoveries respecting the bisexual reproduction of the higher *Cryptogamia.*

The reprint of Henfrey's Report, in the January and March numbers of this Journal, has placed our readers au courant with the present state of knowledge on this interesting subject. It should be noticed that Mohl denies the existence of antheridia in lower *Cryptogamia*, or *Thallophytes;* but maintains that the small bodies, moving by two cilia, discovered by Decaisne and Thuret in the *Fucaceæ*, are more properly a second kind of spores, analogous to the small spores of the *Florideæ*, than of the nature of the seminal filaments of Ferns, Mosses, etc. (p. 117). The latter researches of Itzigsohn, Thuret, Tulasne, etc., however, lead rather to the conclusion that the lower *Cryptogamia* (except the very lowest) are likewise bisexual.[1] Under the head of Reproduction by Seeds, Mohl gives an interesting and critical account of the development and structure of the pollen and the ovule, and of the origin of the embryo. The latter arises by cell-multiplication of the germinal vesicle, a cell produced by free cell-formation in the embryo-sac usually before the pollen-tube has reached the latter. The germinal vesicle and the extremity of the pollen-tube are separated by the thickness of the parieties of the embryo-sac. The penetration of the pollen-tube into the latter or into an introverted portion of it, and the formation of the embryo from the apex of the pollen-tube itself, as taught by Schleiden, are wholly repudiated; and indeed the Schleidenian doctrine may now be considered as thoroughly demolished, by the direct observations of Amici, Mohl, Muller, Hofmeister, Unger, Henfrey, and Tulasne. The Cell as an Organ of Motion is considered as respects movements of individual cells through the agency of vibratile cilia, as respects the directions and curvature assumed by organs, and as respects movements by irritation of stimuli, etc., giving an excellent summary of our knowledge on these points, with much admirable criticism; which want of space prevents us from noticing in detail.

[1] Itzigsohn in *Botanische Zeitung*, May, 1850.—Here it is announced that the black dots on the surface of the frond of *Borrea ciliaris* contain antheridia, that is, cells from which escape animalcular-like corpuscles that move freely in water, and are similar to those of Mosses and Liverworts. Later, after stating that others had failed to detect those movements, he announces that they had been observed by Rabenhorst, after many ineffectual trials. He also (December, 1850, February, 1851) states that these "spermatozoids" do not manifest viral movements until after the maceration of the Lichen in water for several days.

Tulasne (*L. R.*) *Memoire pour servir a l'Histoire Organographique et Physiologique des Lichenes;* in *Ann. Sciences Naturelles*, 3 ser., xviii., No. 1, 2, 3, 4 (1852), with 16 plates.—A most admirable and complete memoir, elucidating in an unequalled manner the whole structure and morphology of the Lichens. It is to be hoped that the author will publish it in a separate form, as it introduces a new era in Lichenography. On the subject of the so-called antheridia (which alone we can here notice), M. Tulasne has recognized the universal occurrence of these bodies in Lichens, has ascertained their structure and development; but he has never detected any free movement of the corpuscles, except the general

ON THE GROWTH OF PLANTS IN GLAZED CASES.

THE first edition of this little treatise,[1] published in 1842, is doubtless well known to many of our readers; and some may remember Mr. Ward's original account of his interesting discovery of a method of growing every sort of plant in the dun atmosphere of the smokiest part of London, pub-

molecular or Brownian motion common to all minute particles. He therefore gives to the so-called antheridia the name of *spermogonia*, and to the contained corpuscles the name of *spermatia*. He unhesitatingly recognizes in them an apparatus of reproduction, doubtless analogous, at least in function, to those of the Florideous *Algæ*, in which the corpuscles are equally motionless, and of certain *Fungi*, and therefore probably representing male organs. Tulasne likewise calls attention to the fact that these dark tubercles or dots were particularly noticed by Dillenius, more than a century ago, in *Borrera ciliaris;* and that Hedwig, in 1784, expressed the opinion that they constituted the male apparatus of Lichens.

Decaisne & Thuret, *Recherches sur les Antherides et les Spores de quelques Fucus;* in *Ann. Sci. Nat.*, 3 ser., iii. p. 5.—Here the corpuscles known to the earlier Algologists, and considered by Agardh and Montagne as a second kind of spores (a view which Mohl adopts), are announced to be the spermatozoids of the antheridia of *Fucaceæ;* their active movements are described, and the discovery of the two cilia is announced by whose vibration the movement is effected.

Thuret, *Recherches sur les zoöspores des Algées et les Antherides des Cryptogames.*—These researches were communicated to the Academy of Sciences, Paris, and were rewarded by the great prize for natural sciences, in 1847. A copious abstract has been published in the *Ann. Sci. Nat.*, 3 ser., xiv. and xvi. (1850, 1851), with 30 plates. As to antheridia, bodies like the free moving corpuscles of the *Fucaceæ* are shown likewise to occur in all the *Florideæ*, except that they do not exhibit spontaneous movements; nevertheless, M. Thuret does not hesitate to attribute to both the same functions as those which the seminal filaments of the higher *Cryptogamia* fulfil. The Antheridia of Chara (in which Thuret first discovered the cilia by whose vibration the coiled filaments are moved), of the Liverworts, Mosses, and Ferns, are also admirably illustrated; but nothing of consequence is added to the facts mentioned in Henfrey's Report.

Léveille, in *Ann. Sci. Nat.*, 3 ser., xv. p. 119, has indicated the probable existence of the analogues of antheridia in *Fungi.*

[1] *On the Growth of Plants in closely glazed Cases.* By N. B. Ward. London, 1852. (American Journal of Science and Arts, 2 ser., xvi. 132.)

lished in the "Companion to the Botanical Magazine" in 1836. This new edition if reduced in size is increased in interest, and is embellished with tasteful illustrations on wood, several of them exhibiting approved forms of those glazed cases with which the name of our author is inseparably connected. The first chapter, on the natural conditions of plants, their relations to heat, light, and moisture, and the necessity of attending to the particular conditions, or combinations of circumstances, under which each species flourishes, is illustrated by ingenious and often novel observations. The second chapter treats of the causes which interfere with the natural conditions of plants in large towns, and gives some idea of the obstacles which prevent the cultivation of even ordinary plants in the open air of London, and to some extent in other large British towns. The third, on the imitation of the natural conditions of plants in closely glazed cases, tells us how a simple incident (the accidental growth of a seedling fern and a grass in a glass bottle, in which the chrysalis of a Sphinx had been buried in some moist mould), carefully and wisely reflected on, taught Mr. Ward how to overcome these obstacles, and thus to surround himself with his favorite plants, in beautiful vegetation, while living in one of the murkiest parts of London, and even to grow with complete success such ferns as the *Trichomanes radicans*, which is utterly uncultivable in any other way. A fourth chapter treats of the conveyance of living plants on shipboard; which brings to view one of the most important practical applications of Mr. Ward's discovery.

Sir William Hooker states that "the Wardian Cases have been the means, in the last fifteen years, of introducing more new and valuable plants to our gardens than were imported during the preceding century; and in the character of Domestic Greenhouses, *i. e.*, as a means of cultivating plants with success in our parlors, our halls, and our drawing-rooms, they have constituted a new era in horticulture." Formerly only one plant in a thousand survived the voyage from China to England. Recently, availing himself of our author's discovery, Mr. Fortune planted 250 species of plants in these cases in

China, and landed 215 of them in England alive and healthful. The same person lately conveyed in this way 20,000 growing tea-plants, in safety and high health, from Shanghai to the Himalayas. In fact, this mode of conveyance is now universally adopted, and has proved so successful, whenever properly managed, that it is no exaggeration to say that, probably, "there is not a single portion of the civilized world which has not been more or less benefited by the invention." An indispensable requisite to success in the transmission of living plants by this method is, that the glazed cases should be freely exposed to the light. Where this cannot be done, we must be content with the former method, of conveying plants in a passive condition, closely packed in peat-moss,—a plan, however, which is only partially successful in protracted voyages. Two additional and highly interesting chapters treat of the application of the "closed" plan in improving the condition of the poor; and on its probable future applications in comparative researches in vegetable physiology, and even in the treatment of diseases. To these, as to the other topics of the work, no justice can be rendered to our author's suggestions except by lengthened quotations, which the nature of this notice does not admit of. It must suffice to direct attention to this fascinating little volume. Those who read it and who have a true fondness for growing plants will scarcely be contented without a Ward case, of more or less pretension; which they will find an unfailing source of interest, especially during the long and total suspension of vegetation in our protracted winters. With proper management, and with the requisite amount of light, any plant may thus be cultivated. But we particularly recommend Ferns and Lycopodia, of the most delicate kinds, as requiring least care, and as making the prettiest appearance at all seasons. Most of these require little light; although our clear skies afford us this in abundance. So little bituminous coal is consumed, even in our largest cities, that the "fuliginous matter" with which all British towns are begrimed and rendered noxious to vegetation, here interposes no obstacle to rearing plants. Quite unlike England, the principal obstacle to the growth of delicate plants in our houses in winter, and in our grounds in summer, comes from the dryness of the air. For this, the Ward case affords a perfect remedy; as nothing is easier than to furnish a saturated atmosphere for those plants that require it, or to supply and retain the degree of moisture which suits any particular species.

HOOKER AND THOMSON'S INDIAN FLORA.

ONE half of this volume[1] is occupied by the Introductory Essay, in which a series of important general topics, akin to those discussed in the introduction to Dr. Hooker's New Zealand Flora, are treated with equal boldness and judgment, and with the same freshness and originality of illustration. These are arranged under six general heads, namely: 1. The object, scope, and design of the "Flora Indica." 2. General considerations connected with the study of systematic botany. 3. The variation and origin of species, the effects of hybridization, and the geographical distribution of species. 4. Summary of the labors of Indian botanists. 5. Sketch of the meteorology of India. 6. Sketch of the physical features and vegetation of the provinces of India. To which two maps are added: one of monthly isotherms, from Dove; the other a large and original map illustrating the physical geography of India and its botanical provinces. A complete alphabetical index to this part of the work is appended, as well as a detailed table of contents.

To enumerate even the principal points which are discussed would require a space which we are unable now to devote to this subject. Some of them we may hope to consider hereafter in other connections. Among the conclusions or suggestions that strike us as most true and timely are: The great

[1] *Flora Indica:* being a systematic account of the Plants of British India. Vol. I. By J. D. Hooker and Thomas Thomson. (*Ranunculaceæ* to *Fumariaceæ.*) With an Introductory Essay. London, 1855. (American Journal of Science and Arts, 2 ser., xxi. 134.)

want on the part of many naturalists of clear and logical views in respect to classification and system; — "the prevailing tendency on the part of students of all branches of natural history to exaggerate the number of species, and to separate accidental forms by trifling characters;" — the unphilosophical and detrimental character of "the modern system of elevating every minor group, however trifling the peculiarities by which it is distinguished, to the rank of a genus;" in other words, of considering every group of species to form a genus, — evincing a want of appreciation of the true value and nature of classification; — the fact that in the vegetable kingdom we do not discover that close and obvious connection between structure and function which is almost universally apparent in the animal kingdom, giving to physiology a greater influence over classification in zoölogy than in botany, and offering a guide to determining the relative value of structural characters in the one kingdom which is comparatively little available in the other, but yet may not safely be neglected.

Our authors assume, as most accordant with known facts on the whole, that species are distinct creations, and not arbitrary assumptions of the systematists; and they adopt that idea of species which alone appears to give them a perfectly clear and intelligible, distinct, objective existence in nature, namely, that they consist of individuals which have originated each from a common stock. They assume not only their original, but their continued definiteness in nature; but their variations, surprising as they often are, are restricted within certain limits, to which we may add that these limits are not *a priori* determinable. Among the causes inducing variation, or tending to produce a blended series of individual forms, if such did not exist from the beginning, they first consider the effects of hybridization; and remark that recent experiments have led to the following results:

"1. It is a much more difficult operation to produce hybrids, even under every advantage, than is usually supposed. The number of species capable of being impregnated, even by skillful management, is very few; and in nature the stigma exerts a specific action, which not only favors and quickens the operation of the pollen of its own species, but resists and retards the action of that of another; so that the artist has not only to forestall the natural operation, but to experience opposition to his conducting the artificial one.

"2. Even when the impregnation is once effected, very few seeds are produced; still fewer of these ripen; and fewest of all become healthy plants, capable of maintaining an independent existence.

"3. The offspring of a hybrid has never yet been known to possess a character foreign to those of its parents; but it blends those of each; — whence hybridization must be regarded as a means of obliterating, not creating, species.

"4. The offspring of hybrids are almost invariably absolutely barren, nor do we know an authenticated instance of the second generation maturing its seeds.

"5. In the animal kingdom hybrids are still rarer in an artificial state, are all but unknown in a natural one, and are almost invariably barren."

Perhaps some of these dicta are too unqualifiedly stated; indeed they are manifestly intended to affirm the results to which the whole evidence points, rather than those which can be said to be thoroughly verified.

The third proposition, however, is absolutely true; and, in connection with it, well do our authors say, that all we could legitimately conclude is, that were hybrids of the general occurrence which some botanists imagine, they would long ago have obliterated all traces of species as definite creations; whereas, exceptional in art, and not proven if not almost impossible in nature, they cannot be assumed to have produced any appreciable result. There is one point, however, which our authors do not take into consideration, but which should not be overlooked, namely, what is generally admitted as a fact, that a hybrid may readily be fertilized by the pollen of either of its parents; and that if hybrid plants are occasionally produced in nature, they would ordinarily stand a very good chance of being fertilized in this way. In such cases they are said to revert to the type of the species of the impregnating parent; but would they return exactly to that type, inheriting as they do a portion of the blood of a cognate species? And where — as not unfrequently occurs — two or more generally well-marked forms in nature are connected by certain occasional individuals of intermediate character, is it not very supposable that two species may have partially blended in this way? At any rate there is a *vera causa*, or what passes as such, which requires to be taken into account, as has not yet been done, so far as we know. This doubtless has operated in the case of cultivated plants, and contributed, along with other causes, to the inextricable blending of certain species. But we are not disposed to exaggerate its influence in nature; since we suppose, with Dr. Hooker, that wild plants rarely hybridize. Yet the possibility and even the probability of the occurrence must not be overlooked in a thorough discussion of the general question of the limitation and permanence of species.

However it may be as a blending influence, hybridization is far from being a considerable, or the most potent cause of the variation of species, since "the offspring of a hybrid has never yet been known to possess a character foreign to those of its parents." And we equally agree with our authors that the known facts of the case "especially warn us not to consider the influence of climate as paramount in determining the distribution of species or the prevalence of forms," or even as the most efficient cause of variation. What the cause is that the legitimate offspring does occasionally possess a character foreign to those of its parents we are wholly unable to say; but the fact is undoubted, and perhaps of more frequent occurrence than is generally supposed. It is usual to say that the abnormal forms originate only in cultivated or domesticated individuals: it were perhaps better to say that they were perpetuated, or are favorably situated for continuation and full development, only under these circumstances, on account of the greater segregation; for of the very various species of plants which are cultivated none are free from the tendency to "sport" into races, whether of ancient or of recent introduction. Why their existence is so transitory in nature, and so capable of being continued and further developed in domestication, it is not difficult to imagine. Our authors perhaps, in common with naturalists generally, do not sufficiently recognize the natural tendency to perpetuation of individual characteristics.

As regards ordinary variation between different individuals of the same species, the want of due consideration of what every good observer knows to be true, has indeed "mainly contributed to such an undue multiplication of species in the vegetable kingdom as botanists unfamiliar with large herbaria and exotic plants are slow to believe, and to the exaggerated estimates of the supposed known extent of the vegetable creation that gain common credence." Our authors believe that the number is swelled one third beyond its due extent by the introduction of bad species founded on habit, and on accidental variations produced by soil, exposure, etc.; and, we would add, on the imperfection of the materials from which the greater part of the species that crowd our books were originally described, most of them without due elaboration of already published species, and drawing after them an ever lengthening train of nominal species, founded on mere guesses at supposed differences from vague and incomplete descriptions, without any collection of specimens.

We have already exceeded our limits, while yet at the beginning of Drs. Hooker and Thomson's interesting and suggestive volume. We regret that we must omit all notice of their remarks upon habit as indicating specific difference, which, contrary to the general view, they regard as "most deceptive," and must pass over their important section upon geographical distribution in general, and its dependence upon specific centres. We only add, that whoever would attain a clear comprehension of the configuration, the diverse climates, and the general botanical geography of those extensive and widely varied regions which are comprised, and in most minds confused, under the general name of India, has only to study the admirable sections on the meteorology of India, and on the physical features and vegetation of its provinces, which occupy a large portion of the Introductory

Essay. The present commencement of the Flora itself, although comprising only fifteen natural orders, is also an inviting subject for extended comment and almost unqualified commendation.

DE CANDOLLE'S GÉOGRAPHIE BOTANIQUE.

THE "Géographie Botanique" [1] of De Candolle is not only one of the most important works of our day, but one which addresses, and will greatly interest, a much broader circle of scientific readers than any other modern production of a botanical author. It is, and probably long will be, the standard treatise upon a wide class of questions, highly and almost equally interesting to the botanist, the zoölogist, the geologist, the ethnologist, and the student of general terrestrial physics. To its production the author has devoted no small portion of the best years of his life; and it bears throughout the marks of untiring labor, directed by a remarkably sound, conscientious, and thoroughly systematic mind. Along with the admirable methodical spirit which is his by rightful inheritance, the younger De Candolle brings to these investigations a particular aptitude for numerical and exact forms, an intimate acquaintance with general physical science, and considerable ethnological and philological learning; which last is turned to good account in his chapters on the history of cultivated and naturalized plants. The result in the work before us — even if there were no other claims to the distinction — may fairly be said to go far toward inscribing the name of De Candolle anew in that select list of philosophical naturalists in which his father holds so eminent a position.

To give some idea of the topics considered in these volumes, and of the order of investigation (which proceeds in an admirable course, from the more simple, general, and better

[1] *Géographie Botanique raisonnée, ou Exposition des Faites principaux et des Lois concernant la Distribution Géographique des Plantes de l'Europe Actuelle.* Alphonse De Candolle. Paris and Geneva, 1855. (American Journal of Science and Arts, 2 ser., xxii. 429.)

known facts and principles towards the more complex, hypothetical and obscure), we will copy the titles of the chapters, twenty-seven in number; which are arranged in four books, and subdivided into articles, and these again into sections, to such an extent as to fill eight closely printed pages with the bare enumeration. Indeed, this repeated subdivision gives a rigid and rather tedious aspect to some parts of the work, and involves occasional repetitions; but it would not be easy to collocate well and clearly so vast an amount of material in any better way.

The First Book is occupied with some preliminary considerations upon the way in which temperature, light, and moisture act upon plants. Its three chapters treat of the relations of plants to surrounding physical conditions, and especially to heat and light; and contain the author's happy distinction between the temperatures actually operative in vegetation, and those which (being below the freezing point, etc.) are altogether null for vegetation, and ought to be eliminated from the tables of mean temperature, when these are viewed in relation to the northern and southern geographical range of species.

The Second Book is devoted to Geographical Botany, or the study of species, genera, and families, from a geographical point of view. Chapter iv. relates to the limitation of species upon plains and upon mountains, and the probable causes of their actual limits, applied both to spontaneous and cultivated plants; and there is good endeavor to show that the northern limit of species is fixed rather by the sum of heat available for vegetation during the growing season, than by the mean temperature of the year. Chapter v. treats of the shape of the area occupied by a species, a very curious point; and it seems that the area of species inclines to be circular or elliptical. Chapter vi. treats of the associations or disjunction of the individuals of a species in its area. Chapter vii. treats of the area of species as to extent of surface, considered as to the families they belong to, as to stations, as to size and duration of the plant, and as to the character of the fruit and seed, whether affording facilities to dispersion or not. Chapter viii.

considers the changes which may have taken place in the habitation of species, and discusses with great fullness the whole subject of naturalization, the obstacles in the way, the causes and means of transport, and the interchanges which have been effected between the New and the Old Worlds. Chapter ix. is a very long and interesting one, on the geographical origin of the principal cultivated plants, not only those intentionally, but also those unintentionally cultivated by man, — a chapter full of valuable matter, carefully collected and well discussed.[1] Chapter x. treats of disjoined species, — those occupying two or more widely separated areas, and not in intermediate stations. Chapter xi. discourses of the early condition and probable origin of the existing species; and brings out the various facts which go far to prove the geological antiquity of the greater part of existing species; and that their creation was probably successive. Chapter xii. treats of genera and their geographical distribution, and maintains the view (in which we by no means coincide) that genera are truly naturally-limited groups, even more so than species. Chapter xiii. is devoted to the distribution of the species of a genus within its area. Chapter xiv. treats of the extent of surface occupied by genera. Chapter xv. discourses of the origin and duration of genera. Chapters xvi.–xix. treat of families, as to their area, geographical limits, the distribution of species within the area of the family, etc.

The Third Book is devoted to Geographical Botany, or the characters of different countries considered as to their vegetation. Chapter xx., of the characters of the vegetation of a country; considered, in Chapter xxi., as to the relative numbers in the great classes respectively. Chapter xxii., comparison of different countries in respect to those natural orders

[1] It is singular that M. De Candolle should be so slow to abandon the idea that the aborigines of Carolina, or any other part of North America, cultivated or knew anything of the Potato, which, if Raleigh obtained them in Carolina, were certainly imported thither. But, though our aborigines had no Potatoes, they had Pumpkins or Squashes and Beans, which all writers upon the history of cultivated plants have overlooked except the late Dr. Harris.

which abound most in species; and Chapter xxiii., as regards their most characteristic natural families. Chapter xxiv., on the variety of vegetable forms in different countries and in the world at large, *i. e.*, the probable number of species, the proportion of genera to species, and of orders of genera and species. Chapter xxv., the division of the earth's surface into natural botanical regions. Chapter xxvi., sketch of the vegetation of the different countries in respect to the probable origin of their existing species, etc.

The Fourth Book, of a single brief chapter, consists merely of a summary of the author's general conclusions. We give these entire, for convenience availing ourselves of a translation in Hooker's "Journal of Botany."

"The plants now inhabiting the globe have survived many changes, geological, geographical, and, latterly, historical. The history of their distribution is hence intimately connected with that of the whole vegetable kingdom.

"To explain existing facts, it is fortunately unnecessary to adopt any conclusion upon the most obscure hypotheses of cosmogony and palæontology, or on the mode of creation of species, the number originally created, and their primitive distribution. Botanical geography can indicate certain probabilities, certain theories, but the principal facts in distribution depend upon more recent and less obscure causes. It suffices to understand and to allow certain facts and theories, which appear probable, namely, that groups of organized beings under different hereditary forms (classes, orders, genera, species, and races) have appeared in different places and at different times; the more simple perhaps at first, the more complicated afterwards; that each of these groups has had a primitive centre of creation of greater or less extent; that they have, during the period of their existence, been able to become more rare or common, to spread more or less widely, according to the nature of the plants composing them, the means of propagation and diffusion they are possessed of, the absence or presence of animals noxious to them, the form and extent of the area they inhabit, the nature of the successive climates of each country, and the means of transport that the relative positions of land and sea may afford; that many of these groups have become extinct, whilst others have increased, at least so far as can be judged by comparing existing epochs with preceding ones; and lastly, that the latest geological epoch, the Quaternary (that which preceded the existence of man in Europe, and which followed the last elevation of the Alps), has lasted many thousand years, during which important geographical and physical changes have affected Europe and some neighboring countries, whilst other regions of the globe have suffered no change, or have been exposed to a different series of changes.

"Thus the principal facts of geology and palæontology, reduced to the most general and incontestable, suffice to explain the facts of Botanical geography, or at least to indicate the nature of the explanation, which it requires the progress of many sciences to complete.

"The most numerous, the most important, and often the most anomalous facts in the existing distribution of plants, are explained by the operation of causes anterior to those now in operation, or by the joint operation of these and of still more ancient causes, sometimes of such as are primitive (connected with the earliest condition of the planet). The geographical and physical operations of our own epoch play but a secondary part. I have shown that in starting from an original fact, which it is impossible to understand, of the creation of a certain form, in a certain country, and at a certain time, we ought to be able, and sometimes are able to explain the following facts, chiefly by causes that operated previous to our own epoch: 1, the very unequal areas occupied by natural orders, genera, and species; 2, the disconnection of the areas that some of the species inhabit; 3, the distribution of the species of a genus or family in the area occupied by the genus or family; 4, the differences between the vegetations of countries that have analogous climates and that are not far apart, and the resemblance between the vegetation of the countries that are apart, but between which an interchange of plants is now impossible.

"The only phenomena explainable by existing circumstances, are: 1, the limitation of species, and consequently of genera and families, in every country where they now appear; 2, the distribution of the individuals of a species in the country it inhabits; 3, the geographical origin and extension of cultivated species; 4, the naturalization of species and the opposite phenomenon of their increasing rarity; 5, the disappearance of species contemporaneous with man.

"In all this we observe proofs of the greater influence of primitive causes, and of those anterior to our epoch; but the growing activity of man is daily effacing these, and it is no small advantage of our progressing civilization that it enables us to collect a multitude of facts of which our successors will have no visible and tangible proof."

An Appendix, indicating the researches now needed for the advancement of Geographical-botanical science, under several heads, addressed respectively to physicists and meteorologists, to geographers, to geologists, to vegetable physiologists, descriptive and traveling botanists, and to philologists, brings these most interesting volumes to a conclusion.

Our present object is to call the attention of American naturalists and natural philosophers to this work, not to criticise it. That would require much consideration and a wider range of knowledge than we can pretend to. There are, however, several topics upon which we are inclined to venture a few remarks, as fitting opportunities occur.

HENFREY'S BOTANY.[1]

THIS is a well-planned, compact, and comprehensive work, in which we may say, that the author has fairly accomplished his purpose, namely:—"to produce a good working text-book for the student, from which may be obtained a groundwork of knowledge in all branches of the science,

[1] *An Elementary Course of Botany; Structural, Physiological, and Systematic; with a brief Outline of the Geographical and Geological Distribution of Plants.* By Arthur Henfrey. London, 1857. (American Journal of Science and Arts, 2 ser., xxiv. 434.)

without the attention being diverted from the more striking features of the subject by details comparatively unimportant."

The work is divided into four parts. I. Morphology or Comparative Anatomy; treating, in successive chapters, 1st, of General Morphology; 2d, of the Phanerogamia, or the parts of Flowering plants and their modifications, and the laws which regulate them; 3d, Morphology of the Cryptogamia. Part II. Systematic Botany; treating, 1st, of the principles of Classification; 2d, of systems of Classification, and 3d, Systematic Descriptions of the natural orders, followed by an artificial analysis. Part III. Physiology; comprising, 1st, the physiological Anatomy of plants; 2d, general considerations on the Physiology of plants; 3d, Physiology of Vegetation; 4th, the reproduction of plants; 5th, Miscellaneous phenomena, under which are ranked the evolution of heat in plants, luminosity, and movements of plants. Part IV. Geographical and Geological Botany, very summarily disposed of in about forty pages.

It seems strange at first to interpose systematic botany between the morphological and the physiological; but if the anatomy and physiology of plants are to be completely disjoined from the study of the organs of the plant as a whole, the present arrangement is perhaps as good as any. It is adopted, as the preface shows, for the convenience of instructing medical students, who compose the principal part of classes in Great Britain as well as on the Continent; — for whom "one short course of lectures is devoted to this science, and three months is commonly all the time allotted to the teacher for laying the foundations and building the superstructure of a knowledge of botany in the minds of his pupils, very few of whom come prepared even with the most rudimentary acquaintance with the science." But the author remarks that "if the previous education of medical students prepared them, as it should, with an elementary knowledge of the natural sciences, we should make physiology the most conspicuous feature of a course of botany in a medical school."

While in England botany is scarcely an academical study,

here it pertains to collegiate and academical instruction where it is taught at all. In Europe not even an apothecary can be licensed without passing an examination in botany; in the whole United States, we believe, it forms no part, at least no regular part, of the medical curriculum; no medical school has a botanical chair; and no knowledge whatever of the science of the vegetable kingdom, which supplies the greater part of the *materia medica*, is required for the degree of Doctor in Medicine!

Professor Henfrey is chiefly known, and most highly esteemed, as a vegetable anatomist. Upon this subject he may speak with an authority which as a systematist, or even as a morphologist, he would not pretend to. We shall offer no apology, therefore, for making an occasional criticism, and for pointing out several errors in matters of detail. These are not intended to disparage the work, for if we had not formed a high opinion of it on the whole, we should not take this trouble.

As respects the first point noticed, our author, if wrong, is not alone. Still, we hardly expected him to teach that the radicle of the embryo is the true root; and we cannot let pass unchallenged his reiterated statement that in Monocotyledons, the radicle, or its inferior extremity, is never developed into a root in germination, but is abortive (pp. 14, 16, 18, 391, 537). Any one who will examine the germination of the seed of an Iris, an Onion, or even of a grain of Indian Corn, cannot fail to perceive that a primary root is developed, and that this is a direct prolongation of the extremity of the radicle. This, indeed, does not continue as a tap-root; neither does it in a great many Dicotyledons. In Squashes, Pumpkins, etc., there is no one primary root, but a cluster of rootlets from the first, all springing from the base of the stout radicle. In fact, this distinction between Monocotyledons and Dicotyledons is null. A character of certain monocotyledonous embryos, neither strictly peculiar to the class, nor by any means universal in it, should not be assumed as distinctive. As to the morphology of the radicle itself, we suppose that the germination of any of the larger *Cucurbitaceæ*, or of a bean, would suffice to convince any observer that the radicle is simply the first internode of the stem, giving birth to the primary root from its inferior extremity, usually, — and indeed, from the exceptional cases where it does not we should draw additional proof of its cauline nature. In fact, we know of no character in which a root differs from an internode of a stem in which it does not also differ from the radicle, excepting its tendency to direct its inferior extremity downwards. Again, should the statement, that "the radicle of a monocotyledonous embryo is never developed" be held to mean that the radicle never lengthens, we remark, no more does it in the Pea and some other hypogæous Dicotyledons; and we are not quite sure that the statement is absolutely true of all Monocotyledons.

Root-hairs or *fibrillæ* are mentioned (p. 19) as "often" occurring on young roots. Do they not always occur? Surely it cannot be true that: "the branches of the axial root are originally growths from the apex of the root thrown off to the side," (p. 538). By some slip of the pen, *Myrica Gale* is adduced as an instance of whorled leaves (p. 45).

On p. 49 the expression "over the petiole," instead of above or within it, would lead to a misconception.

Something more might be said about the tendrils of *Cucurbitaceæ* (which, besides, are not always single); but are the students of King's College really taught that, "tendrils of the vine are metamorphosed flowering branches arising in the axils of the leaves"? (p. 62.)

"In all seeds except in those of the few orders which present an incomplete or acotyledonous embryo, we do not find the young plant possessed . . . of a plumule" (p. 66). Even some much developed embryos, such as those of Maple and Morning Glory, do not show the plumule until after the full development of the cotyledons. It may be said, indeed, that the plumule is *in posse* when not *in esse*, but so it is no less in the cases excepted from the statement.

Very singular is the statement (on p. 68) that in England "the terminal bud of the Lilac is generally killed by the frost in the winter;" since in our much colder winter it is as completely hardy as the other buds whenever it happens to be formed, and, like them, is well developed before summer is over. As a general rule here, and we presume in England also, no terminal winter bud appears during the growing season, and so there is none to be killed by the frost of the following winter.

The deeply alveolate receptacle of the Cotton-Thistle is figured (on p. 78) as an illustration of a *paleaceous* receptacle.

Truly terminal flowers are said to be rare (p. 86): we do not quite understand this.

The interesting questions relating to the phyllotaxy and symmetry of the flower are clearly stated, but no new light is brought to bear upon them, — nor all of the old. The opposition of the stamens of *Rhamnaceæ* to the petals is, as usual, attributed to the probable suppression of an outer stamineal circle, although there is nothing in the blossom (as there is in *Geraniaceæ*, etc.) to base the supposition upon. And our author has overlooked the most natural of explanations for this and strictly like cases, the one moreover which tells directly against the doctrine of transverse chorisis, — namely, that in these cases of ante-position there is a return to normal phyllotaxy, *i. e.*, to the superposition of the corresponding elements of successive whorls, — a view first suggested, we believe, by Lestibudois.

"Real cases of collateral multiplication may probably be explained by comparison of a primary staminal leaf with an ordinary compound stem-leaf, and supposing the filament to subdivide like the petiole does [sic] in such cases." This is certainly the way we regard it; and as respects the application of this hypothesis to the stamens of *Cruciferæ*, we do not see what argument *Megacarpæa polyandra* brings against it; as the increase in the number of stamens is quite as explicable upon this as upon the ordinary theory. Indeed, our author's view that the glands represent suppressed stamens would seem to be negatived by this very case, since the glands have not disappeared with the increase of the stamens, but the contrary.

The abnormal fertile flowers of Viola and Impatiens are

not "achlamydeous," as our author states them to be (p. 90); generally they are not even apetalous.

In the botanical sense of the word, and as it is employed in the same sentence (p. 93), the petals of the vine cannot be said "to cohere above." The valvate petals are merely caducous for the most part before expanding, just as is more decidedly the case in many *Araliaceæ*. In passing, we remark that a valvate æstivation of the corolla in the latter is much less distinctive than our author supposes (p. 311). Aralia itself has the petals imbricated in the bud.

It is becoming common to regard the tube of a so-called superior calyx as a cup-like receptacle; and there appears to be reason for it in *Cactaceæ* and some other cases. Professor Henfrey would seem to apply this view universally; "for example, in *Rosaceæ*, *Umbelliferæ*, *Cucurbitaceæ*, *Compositæ*," [!] etc. But if applied to Rosa, why not to the *Sanguisorbeæ* and to other *Rosaceæ* with a calyx-tube lined with a disk bearing the stamens, etc.? And is the cup a receptacle in those *Melastomaceæ* which have an adnate ovary, but a calyx when the ovary is free? And how is it when the ovary and cup cohere only by the nerves of the latter?

For *paleæ* Professor Henfrey coins an English word, "pales" (p. 110), of which the singular would probably be "pale." We would propose to call them "palets."

There are convincing reasons why the perigynium of Carex cannot be regarded as a perianth, as our author takes it to be (p. 111).

It is not correct to say that the false dissepiments of Datura are formed "while the seeds are ripening" (p. 124); they equally exist in the ovary. And we doubt if the transverse false septa in Cathartocarpus and other *Leguminosæ* are "placental developments."

We are pleased to find that our author prefers to consider placentæ as belonging to the carpels rather than to the axis, although the close of paragraph 226 appears to imply the contrary.

We cannot agree that, "externally the campylotropous ovule resembles the anatropous, except that there is no rhaphe" (p. 130). No attentive student could fail to recognize the difference, especially in the families cited (*Cruciferæ* and *Caryophyllaceæ*).

Ripening must be regarded in a remarkably broad sense when it is stated with emphasis, "that the distinction between endocarps and epicarps, in the common stone-fruits, arises entirely during the ripening of fruit." Also: "it is well known that the easy separation of the pulp from the stone is a sign of ripeness." When are cling-stone peaches ripe? Again: "In Taxus . . . during the ripening of the seed a succulent cup-like envelope grows up around it" (p. 136). Is ripening synonymous with the formation and growth, as well as the maturing of the fruit?

Lindley's system of the classification and nomenclature of fruits is adopted, with some modifications. It is well to have such a system, as an analysis of the diversities of structure; but of the thirty-six kinds so carefully defined and named only fifteen or sixteen are ever used in descriptive botany, or ever will be, it is devoutly hoped. There is much inconvenience in practice, and little advantage in designating every possible modification of the same organ or set of organs by a distinct substantive name, or in distinguishing by separate technical names fruits formed of a simple ovary from those of a compound ovary, or fruits with an adherent from those with a free ovary. Why not call the gooseberry and the grape equally a berry, instead of restricting this name to the former and naming the latter a *nuculanium;* and why name the pod of an Iris a *diplotegia*, while that of a Lily is called a *capsule?* And while we term the pod of *Saxifraga stellaris* a *capsule*, and that of *S. tridactylites* a *diplotegia*, what name are we to apply to that of *S. aizoides*, which is only half-superior?

Probably a wrong example is adduced on p. 148, for we cannot believe that any species of Ranunculus has the rhaphe averse from the placenta in the ripe fruit. By an oversight, on the same page, the fruits of *Labiatæ* are spoken of as seeds.

As respects the systematic part, the chapters on the principles of classification, nomenclature, etc., strike us as sound and good throughout; and in the account of the natural orders a great amount of information, such as the medical student needs, is given in a comparatively small space. Errors or misconceptions will necessarily occur in the compilation of such an amount of materials, treating of structure, affinities, distribution, sensible properties, and medicinal or economical uses. They are not more numerous than was to be expected, and we are not disposed to make them the subject of criticism.

We may remark, in passing, that, as respects the morphology of the andrœcium in *Fumariaceæ*, the name of the writer of the present notice is referred to, by some misconception, as adopting Lindley's well-known hypothesis of the splitting of two stamens into halves; whereas he has maintained a very different view. And then this is mentioned as "offering a phenomenon of chorisis," which in that view is quite incomprehensible to us.

We were surprised at the statement that the bark and leaves of *Hamamelis Virginica* "are astringent and contain an acrid volatile oil" (p. 207). We trace it back to Lindley's "Vegetable Kingdom" (p. 784), and find, "The kernels of *Hamamelis Virginica* are oily and eatable. The leaves and bark are very astringent, and also contain a peculiar acrid essential oil;" and this, we find, comes from Endlicher's "Enchiridion." How did this bland and inert plant acquire such a reputation? Dr. Barton, who has figured it, says nothing of its possessing any sensible properties or useful qualities at all, except its use for divining-rods; nor do Pursh, Bigelow, Elliott, Darlington, etc., allude to any popular reputation of such qualities. No sign of any essential oil is to be detected in the foliage, and prolonged mastication of the leaves and bark while we write yields not the slightest trace of acridity and hardly any of astringency; no more, certainly, than a Beech leaf. We never heard of the seeds being eaten; and as they are "about the size of a grain of barley," or not much larger, and have a thick bony coat, they are not likely to become an important article of diet. After some search, we find the source of these extraordinary statements in the "Medical Flora" of the eccentric Rafinesque. He says the seeds are called Pistachio nuts in the Southern States, are rather oily and palatable, etc., but he neglects to mention their size. He adds, "The bark and leaves are somewhat bitter, very astringent, leaving a sweetish pungent taste. The smell is not unpleasant. It has not been analyzed as yet, but probably contains tannin, amarine, extractive, and an essential oil." To all this, Endlicher, on the strength of "the sweetish pungent taste," has added the acridity; and so one of the blandest and most useless of shrubs gets a world-wide and wholly factitious reputation for active medical qualities and esculent seeds; and even Dr. Griffith, who must have known the shrub, has been induced to give it a place in his "Medical Botany."

Our remaining remark relates to the random way in which mere analogies are mixed up with affinities in estimating or expressing the relationship of orders, etc., in this as in some other more notable works. It is, or at least ought to be, well understood, that mere analogy, *i. e.*, likeness in some one respect only, however striking the imitation, is no indication of relationship, but that relationship rests upon affinity, *i. e.*, upon agreement or similarity in the whole plan of structure, and especially of floral structure, whether general or particular, as the case may be. To speak, therefore, of "evident" and "most distinct" affinities between *Coniferæ* and *Lycopodiaceæ* is an example of this prevalent misconception of what affinity is. This is more intelligible, however, than the "approach" suggested of *Aquifoliaceæ* to *Loganiaceæ* and *Apocynaceæ*, while their resemblance to *Celastraceæ* is thought to be of small account; or that of *Umbelliferæ* to *Rubiaceæ*, *Saxifragaceæ*, and even to *Geraniaceæ*, to which the resemblances do indeed "seem rather superficial." Again, *Xanthoxylaceæ* (*i. e.*, *Rutaceæ*) are said to have considerable affinity to *Oleaceæ*, because Ptelea, in the former, has a samaroid fruit, as has Fraxinus in the latter. May we add, as quite as much to the purpose, that the common Xanthoxylums have pinnate leaves, and are popularly called Prickly Ash?

The study of affinities is neither guess-work nor divination, but a matter of logical deduction from structure, based upon scientific principles, — principles recognized and acted upon by sound botanists with considerable unanimity, although they have never been reduced to a system, nor expounded in detail, so as to make them matters of elementary instruction. Until this desideratum is supplied, the young botanist can do no better than to take as models the writings of Brown, and of those botanists who, according to their ability, have most closely followed the footsteps of this master in science.

Having continued this review far beyond our intention at the outset, we have small space left for noticing the best part of Professor Henfrey's treatise, namely, the third or Physiological part. Suffice it to say, that, in the important chapter on the physiological anatomy of plants, our author writes from the fullness of his acquaintance with the writings and doings of all the continental phytotomists, and also with the authority of an experienced original investigator. And, so far as we know, it comprises much the best résumé of vegetable anatomy and development now extant in the English language, at once succinct, clear, trustworthy, and well brought up to the present state of the science. Perhaps the succeeding chapters, on the Physiology of Plants generally, the Physiology of Vegetation, and on Reproduction, are equally commendable in their way; but we have as yet barely glanced over the pages. We like the following definition, and the ensuing paragraph upon the rôle of vitality in plants.

"The physiology of plants is that department of botany in which we investigate the phenomena of the *life* of plants, manifested in a series of changes taking place in the diverse parts of which each plant is composed" — (p. 475).

"The physiological phenomena which indicate vitality are always of more or less complex nature, and admit of being analyzed into a number of factors, of which a large proportion are found to be purely physical or chemical. A very considerable part of the changes which accompany the process of organization are the results of the action of physical and chemical forces, [and] capable of being explained up to a certain point, by the known laws of those forces. But in every case, after referring all the chemical and physical phenomena to their respective places, there remains a residual phenomenon to be accounted for, which is precisely the most important of all, — namely, that in living organic structures . . . the laws of inorganic matter are subdued under a higher influence, and caused to undergo modifications never occurring except in the presence of living matter, while — most important of all — the peculiar compounds of matter thus produced are not only made to assume forms, according to definite laws, totally unlike any forms of mineral matter, but [to] constitute bodies manifesting a continued interchange of material with the surrounding media, which, instead of resulting in decomposition, as in mineral bodies, effects a reproduction and increase of the already existing [organized] matter" — (p. 542).

In the paragraph on the longevity of trees (p. 549), we find renewed occasion to notice the longevity of unfounded statements, copied from one book into another long after the error has been pointed out. Here again the Adansonia of Senegal and the Wellingtonia or Sequoia of California figure as trees "whose age, deduced from the rings of growth of the stems, would amount to upwards of 3000 years." There is really no evidence to prove that the famous Baobabs described by Adanson are of such an age; and as to the Wellingtonia in question, an actual counting of the rings has shown that the tree was not half so old as it was vaguely computed to be.

The chapter on Reproduction appears to be excellent, as indeed we should expect. The geographical and geological part is necessarily very briefly treated.

NAUDIN ON THE GENUS CUCURBITA.

NAUDIN'S "Researches into the Specific Characters and the Varieties of the Genus Cucurbita"[1] are published in the 6th volume (4th series) of the "Annales des Sciences Naturelles," and are of no small interest, being founded upon a very conscientious investigation of nearly all the known forms, collected for the purpose, and cultivated under the author's eye at the *Jardin des Plantes.* These forms our author reduces to six species, and the alimentary sorts in cultivation to three, namely, *Cucurbita maxima, C. Pepo,* and *C. moschata.* The remaining three species are *C. melanosperma* of Braun, newly introduced from eastern Asia, and the two perennial and tuberous-rooted species, *C. perennis* and *C. digitata,* Gray, natives of our southwestern borders, the fruits of which are not esculent. Indeed, the Pumpkins and Squashes cultivated in Northern Europe, and with us, as now understood, belong to only two species, since the third, *C. moschata,* hardly comes to perfection north of the Mediterranean region. Of these, *C. maxima* is made to include *C. Melopepo;* and *C. Pepo,* comprising our Pumpkins and a large part of our Squashes, is made to include *C. ovifera, aurantia, verrucosa,* etc., and the species are defined by botanical characters, which apparently may be relied upon. The varieties of *C. maxima* fall into two main groups, characterized by their fruits, namely, the "Turbans," having crowned fruits, that is, the summit projecting beyond the adnate calyx-tube, a peculiarity found in no other species, and the crownless sorts, in which this peculiarity is not manifest. The innumerable varieties of *C. Pepo* are arranged in seven groups, according to the configuration of their fruits.

M. Naudin has not undertaken to discuss the questions respecting the birthplace of these plants. He remarks that *C. maxima* and *C. moschata* have been known in European gardens scarcely above two centuries; but that *C. Pepo* was perhaps known to the Greeks and the Romans in the time of Pliny.

[1] American Journal of Science and Arts, 2 ser., xxiv. 440.

The younger De Candolle, in his discussion of the history and origin of the principal cultivated plants, which forms a most interesting chapter of his "Géographie Botanique," although he is unable to assign them to any country as their home, confidently (perhaps too confidently) refers all the squashes and pumpkins to the Old World ; but not to India, because they have no Sanscrit name. He will not believe that any of them came from America, and appears to think little of the current statements that squashes or pumpkins were in cultivation by our aborigines before the European settlement of the country. On the other hand, our lamented Dr. Harris — who, during the later years of his life, assiduously studied this question, and who was very cautious in drawing conclusions — had become satisfied that the North American Indians, as far north even as to Canada, cultivated squashes and pumpkins, one or both, along with their maize, before the whites were established here. We are unable at this moment to refer to his manuscripts, or to what he had too imperfectly published upon this subject. But we well remember his laying much stress upon the narrative of Champlain ; and with good reason, as it appears to us on turning casually to the pages of "Les Voyages du Sieur de Champlain . . . ou Journal tres-fidèle des Observations faites et Découvertes de la Nouvelle France," etc., etc., edition of Jean Berjon, Paris, 1613, 4to ; also "Voyages et Découvertes faites en la Nouvelle France depuis l'année 1615, jusques à la fin de l'année 1618," — second edition, published by Collet in 1627, small 12mo, — to which volumes we desire to direct M. De Candolle's attention. In Champlain's narrative of his own voyage along the coast of what is now the State of Maine, in the year 1604, and the two voyages of Le Sieur de Mons along the coast of New England in 1605 and 1606, *Citrouilles* and *Courges* are repeatedly mentioned, along with maize (*Bled d'Inde*) and beans ; *e. g.*:

"Nous y vismes force *citrouilles*, *courges* & petum, qu'ils cultiuet aussi. . . . Pour les febues elles cõmẽçoiẽt à entrer en fleur, cõme faysoyẽt les courges et citrouilles" (p. 68).

"Ceux que nous auions enuoyes deuers eux, nous apporterent des petites citrouilles de la grosseur du poing, que nous mangeasmes en sallade comme coucombres, qui sont tres-bonnes" (p. 77).

See also pp. 83, 115, 116. Of course it does not follow that these esculents were natives of New England, any more than maize; but both may probably have been carried northward together. Whatever their origin, our Indians were found cultivating them together at this early date as well as in later times. According to Nuttall, the Indians along the whole Upper Missouri half a century ago were cultivating *Cucurbita verrucosa.* This common squash is, according to Naudin, a variety of *C. Pepo*, as also is *C. aurantia* (the *C. Texana* vel *ovifera*, Gray, "Pl. Lindheimerianæ"), which has every appearance of being indigenous in the western part of Texas, on the Rio Colorado and its upper tributaries. At least, this is the opinion of Mr. Lindheimer and of Mr. Charles Wright, two good judges. The latter personally informs us that, from the stations and localities in which alone it is met with, he could not suspect it to be other than an indigenous plant.

That the later Greeks and Romans possessed the bottle gourd or Lagenaria, and also some kind of summer squash, seems pretty clear; but we see no decisive reason for the opinion that they had any form of *Cucurbita Pepo*, as that species is now understood. According to De Candolle, the earliest figures referable to this species are, one of *C. ovifera* by Lobel in 1576, and one of *C. verrucosa* by Dalechamp in 1587, namely, about a century after the discovery of America, and long after maize had become well known in the south of Europe; and we have seen that some forms probably of this very species (undoubtedly originating in a warmer region) had by this time found their way in this country nearly as far north as the climate will permit of their cultivation. So that there appears to be about the same evidence for the American origin of some squashes and of pumpkins that there is for the American origin of maize.

A remaining argument brought by De Candolle against this view may also be turned the other way, namely, that no certain species of the genus is known as indigenous to America. He has equally allowed that none is known to be indigenous to the Old World. Now of the six species recognized by Naudin, two only are known in their natural wild state, and these are our southwestern species with perennial roots, namely, *C. perennis* and *C. digitata*, to which we add that *C. Pepo* itself (*i. e.*, *C. ovifera* or *aurantia*) grows wild in the same district with *C. perennis*, and has the same appearance of being indigenous there. We leave the subject with these incidental remarks, as we did not intend here to investigate this question, and will briefly allude to another subject, upon which Naudin's investigations have thrown new light.

It is generally thought that the cultivated *Cucurbitaceæ*, and especially that the species of Cucurbita, cross-breed with extreme facility. According to Naudin this is true of the races only *inter se.* A good illustration of the immediate and great variation from this cause in the fruit of *C. Pepo* is given in Naudin's third plate, where fifteen different forms of the fruit are figured, taken from as many individual plants raised from seeds of one fruit, which had grown in the vicinity of other varieties. It is by no means certain, however, that all these forms originated from direct crossing. But the species themselves strangely refuse to hybridize. Naudin carefully experimented with the five species in cultivation at the *Jardin des Plantes* (namely, all known, except *C. digitata*); and out of seventy distinct trials all but five were utterly ineffectual. In five instances the fruit set, indeed, but in none of these was a single seed containing the vestige of an embryo produced! What are we to think, then, of the universal belief that squashes are spoiled by pumpkins grown in their vicinity, or pumpkins by squashes; and even melons (which are of a different genus) by squashes? The fact of some such influence seems to be well authenticated. Dr. Darlington, one of the most trustworthy of observers, speaks of it from his own knowledge, thus: "When growing in the vicinity of squashes the fruit [of the pumpkin] is liable to be converted into a kind of hybrid, of little or no value. I have had a crop of pumpkins totally spoiled by that cause, the fruit becoming very hard and warty, unfit for the table and unsafe to give to cattle." — ("Fl. Cestrica," ed. 2, p. 555.)

Now that this is not the effect of hybridation is clear from the fact that the result appears in the fruit of the season, not in that of the next year, namely, in a generation originated by the crossing. A clue is perhaps furnished by Naudin's observations, that the ovary is apt to set and even develop into a fruit in consequence of the application of the pollen of another species, although, as the result proves, none of the ovules are fertilized. And he hazards the conjecture that the pollen may exert a specific influence first upon the ovary, inciting its farther development, and then upon the ovules. To test this conjecture he was to examine the action, if any there be, of the pollen of Cucurbita upon the ovary of melons. The past summer — which has been as unusually warm in western Europe as it has been cool in this country — must have favored such researches in Paris; and we may expect soon to hear of the result. Improbable as such an influence seems to be, it is hardly more so than the now authenticated fact that the graft of a variegated variety of a shrub or tree will slowly infect the stock, so that the variegation will at length break out in the foliage of the natural branches; — an old observation, which, according to the Gardner's Chronicle, has recently been verified in several instances.

WEDDELL'S MONOGRAPH OF URTICACEÆ.

DR. WEDDELL'S preliminary studies upon the proper *Urticaceæ* were published a few years ago in the "Annales des Sciences Naturelles." Since then, botanists, aware from this and his other works that the subject was in most able hands, have been anxiously waiting for his full monograph. This, we understand, is now completed, although the last fasciculus has not yet reached this country. The greater part is before us, and an admirable monograph [1] it is, worthy of a place in

[1] *Monographie de la Famille des Urticées.* Par H. A. Weddell (*Archives du Museum*, ix., livr. 1–4), 1856–57. (American Journal of Science and Arts, 2 ser., xxv. 109.)

the "Archives" which contain that model one on the *Malpighiaceæ* of his lamented botanical master. It illustrates in detail about 470 species, under 40 genera, and is accompanied by twenty well-filled plates, drawn by the author. It opens with a conspectus of the members of the great group to which the true *Urticaceæ* belong (which the author inclines to receive rather as the orders of a class than as suborders of an extensive order, fully admitting, however, their close affinity *inter se*), followed by a brief indication of the principal investigators of these plants, and of the resources at his own command. A general account of the organs of vegetation and reproduction, of the affinities, and the geographical distribution of the plants of the group, and of their properties and uses, conclude the preliminary matter. The body of the work is occupied by their systematic arrangement and description.

Apetalæ being viewed as degenerations of *Polypetalæ*, our author searches among the latter orders for the nearest relatives of the great Urticaceous order of alliance, and finds them in the *Tiliaceæ*, that is, in the group of orders of which the *Malvaceæ* are the highest development. According to Weddell's happy illustration, *Malvaceæ* crown the summit of a three-sided pyramid, with *Sterculiaceæ*, *Byttneriaceæ*, and *Tiliaceæ* just below them, one upon each face; under the *Byttneriaceæ* he ranks the *Euphorbiaceæ* with the *Antidesmeæ*, and under these, at the very base of the pyramid, the *Scepaceæ*, the lowest degradation in this direction of the Malvaceous type. On the adjacent face, under the *Tiliaceæ*, and on the same level with the *Euphorbiaceæ*, he inscribes the *Urticaceæ*, with the *Cupuliferæ* perhaps underneath them. Upon this ingenious plan of representation, the apetalous orders throughout may be most conveniently and instructively ranked under their superior types;—bearing in mind that some types degrade as much within an order (*e. g.*, *Euphorbiaceæ*, *Onagraceæ* inclusive of *Halorageæ*, *Caryophyllaceæ* including *Illecebreæ*), as others do through a series of two or three orders, or even as the same group does (*e. g.*, *Caryophyllaceæ*) through a series of orders on the other side of the pyramid.

The reason why this mode of representation will exhibit botanical affinities so well is, that (as we have elsewhere remarked) the vegetable kingdom does not culminate, — as the animal kingdom does, — and therefore offers no foundation in nature for a lineal arrangement even of its great groups. But it would appear that the Dicotyledonous orders might be arranged under a considerable number of short series, in groups converging upon the most fully developed or representative order of each type, so as to exhibit what we now know of the system of nature much better than in any other way.

We think that Dr. Weddell's idea of the affinity of *Urticaceæ* is a good one. The floral and seminal characters, the true *criteria* of affinity, are not abhorrent, but present some strong points of relationship, as do the organs of vegetation. These, once established, allow us to feel the force of the striking coincidence in the bast-tissue of the bark, so remarkable in all this alliance for the length, fineness, and toughness of the fibres, their union end to end, and their lateral independence, admirably adapting them for their use as textile materials, in which *Urticaceæ* vie with *Malvaceæ* and *Tiliaceæ*.

As to geographical distribution, Europe is very poor in *Urticaceæ*, poorer even than would at first view be supposed, as the author remarks. For as nettles like an enriched soil, the five or six European species of Urtica and Parietaria so abound around habitations that they make up in the multitude of individuals for the paucity of species, and perhaps cover nearly as much ground as the great number of intertropical species; two or three excepted, which are also weeds in the tropics. Temperate North America is not much richer in species than Europe. The greater part are found in the torrid zone, and in islands rather than continents; the Malay region, India, Mexico, and the West Indies together possess almost two thirds of the known species.

Our remaining remarks shall be restricted to one well-known plant described in the work, and to another of recent discovery, which unfortunately was not communicated in season to find a place in it.

The first is our common *Pilea pumila*. Dr. Weddell has overlooked the fact that Rafinesque had founded a genus (Adice or Adike) upon it, although the name is mentioned in the work cited by him, where the plant was first published as a Pilea, and although Dr. Torrey had adopted Rafinesque's genus, and figured the species, in an earlier and more considerable work ("Flora of the State of New York"), which, having unfortunately been published by the State, and in a large edition, has in consequence remained almost unknown to science. Considering that the three sepals of the fertile flower in this species are nearly equal and not gibbous, it may be doubted whether the single species of Blume's genus Achudemia, differing only in having five sepals, should not rather be appended to Pilea. We dare say that Dr. Weddell would have so arranged it, if Blume had not published the genus.

Since the appearance of the third part of Weddell's monograph, but before it had reached this country, Dr. Torrey has published, in the report on Dr. Bigelow's fine California collection made in Lieutenant Whipple's Railroad Survey to the Pacific, a new Nettle allied to Bœhmeria but with the penicillate stigma of Urtica, namely, his *Hesperoenide tenella* ("Pacific Railroad Reports," iv. p. 139). This little plant, it now appears, comes nearest to Wight's monotypic genus *Chamabaina* of India, of which better details than Wight's as to the female flowers and fruit are figured in the present monograph. The stigma is intermediate in character between that of Chamabaina and that of Urtica; and, moreover, as the sepals of the male flower want the pointed gibbous tips of the former, the stipules are inconspicuous, and the cotyledons are not only reniform but (which is unnoticed in the published description) pretty strongly emarginate at the summit also, the genus will probably be retained.

Great thanks are due to Dr. Weddell for his labors upon this family, which he found in a most unsatisfactory and difficult state, and has left in such condition that Nettles and their allies are easy and inviting objects of study.

RADLKOFER'S PROCESS OF FECUNDATION IN THE VEGETABLE KINGDOM.[1]

THIS gives in English, and in an accessible form, a systematic and historical survey of the whole subject of vegetable fecundation, including the recent discoveries of Pringsheim, Cohn, Braun, and Bary.

As to *Fungi* and *Lichens*, — thanks to the observations of Itzigsohn upon the latter, and the most useful and persevering investigations of Tulasne upon both families, — the analogues of male organs in all probability are discovered, and their general presence recognized; but the fact of fecundation is not made out.

In the lower or green *Algæ*, fecundation was first demonstrated by Pringsheim. The "horns" of Vaucheria which Vaucher half a century ago observed and conjectured to be male organs, Pringsheim proved to be so, having seen them open at the summit and emit a great number of free-moving corpuscles (spermatozoids), many of which found their way into the now open orifice of the protuberance, which contains the forming spore, and were seen crowding against it, after which a membrane of cellulose appears over the surface of the mass of protoplasm and completes the spore. Whether one or more of the spermatozoids actually penetrates the protoplasm and so is included within the cell-membrane is uncertain; but Pringsheim thought it was the case, from having detected a colorless corpuscle like one of the spermatozoids inside of the membrane. Next Pringsheim demonstrated a similar fecundation in Œdogonium. His results, briefly published in the proceedings of the Berlin Academy, and thence translated into French and English, are now given in detail in the first part of his "Jahrbücher," noticed above. Œdogonium consists of a row of cylindrical cells.

[1] *Der Befruchtungsprocess im Pflanzenreiche.* L. Radlkofer. Leipsic, 1857. (English translation by Arthur Henfrey in *Annales* and *Magazine of Natural History*, October and November, 1857.) (American Journal of Science and Arts, 2 ser., xxv. 112.)

Some of these cells, usually shorter than the rest, become tumid, and, without conjugation, have their whole green contents transformed into a large spore. Pringsheim has ascertained that other cells of the same individual plant have their green contents transformed into a multitude of active corpuscles or zoöspores, which, from their subsequent evolution and office, he names androspores; these escape by the opening of the mother cell moving about freely by the vibration of a crown of cilia attached near the smaller end. One or more of these androspores fix themselves by the smaller end upon the surface of the cell in which a large ordinary spore is forming, or in the vicinity, and germinate there, growing longer and narrower at the point of attachment, while near the free end a cross partition forms, and sometimes another, making one or two small cells; this is the true antheridium: for in it a crowd of spermatozoids are formed, also endowed with motivity by means of vibratile cilia. Now the top of the antheridium falls off as a lid, the spermatozoids escape; the spore-cells at this time open at the top; one of the spermatozoids enters the opening, its pointed end foremost; this becomes stationary upon or slightly penetrates the surface of the young spore, into which its contents are doubtless transferred, and a coat of cellulose is then, and not until then, deposited upon it, completing its organization as a spore, which in due time germinates, and grows directly into a plant like the parent.

But in Bulbochæte, and especially in Sphæroplea, so beautifully investigated by Cohn ("Annales des Sciences Naturelles," 4 ser., v.), the spore does not directly develop into the normal or fruit-bearing plant. Instead of this, by an alternation of generations (to adopt that well understood phrase), the spore proceeds to convert its contents by successive division into a large number of zoöspores, different from the androspores, namely, small oval or oblong bodies, furnished with two long cilia on a short beak at one end, and for a time moving actively about by their vibration. Coming to rest, these zoöspores germinate, by elongation and the formation of transverse partitions, into adult thread-like plants, consisting of a row of cells. In Sphæroplea the whole contents of the cells of some adult individuals condense into large green spores, as yet without a coat; while those of different individuals give rise to myriads of slender spermatozoids, moving by means of a pair of cilia fixed at the narrow end. The latter escape from the parent cell through a small perforation which now appears, enter the spore-bearing cells of the fertile plant through a similar perforation in them, play around the spores, and at length one or more of them drives its pointed extremity into their naked surface; after which, fertilization being accomplished, a thick coat of cellulose is deposited to complete the spore. "Cohn does not consider that observations justify his assuming a direct penetration of the spermatozoids into the primordial spore-cell. It rather seemed to him as if they attached themselves on the outside of the spore, and were finally converted into mucilaginous globules."

Reproduction by conjugation of course had long been familiarly known in the lower *Algæ*. But it was questioned whether this was really analogous to sexual reproduction, since what appeared to be similar spores are often formed of the contents of a single cell without conjugation. Areschoug shows that these are abortive spores, incapable of germination; while those which result from actual conjugation will grow into new plants, without further metamorphosis; Vaucher's old observations to this effect having been confirmed by Braun and Pringsheim.

That in the *Fucaceæ* or olive-green *Algæ* the large spores are fecundated by spermatozoids, produced in antheridia, was demonstrated by Thuret in the year 1850. And in more recent memoirs he has shown that the fertilization takes place through direct contact of the spermatozoids with the naked surface of the unimpregnated spore, then having only a protoplasmic coating; and that these spores will not develop nor hardly acquire a cell-wall unless so fertilized. His experiments upon diœcious species are perfectly decisive upon these points. He observed the lively spermatozoids playing over the surface of the still naked spore, fix themselves to it by the ciliated end, apparently by one of the cilia, and at length come to rest in contact with it; but he could not detect any material penetration of them into the body of the spore. Pringsheim, confirming all Thuret's observations, thinks also that the spermatozoids actually penetrate the sperm-mass; but there is no direct proof of it. Indeed Thuret, in a very recent article ("Annales des Sciences Naturelles," 4 ser., vii., 1857), indicates the grounds of Pringsheim's probable mistake. The most interesting point in this last article by Thuret relates to the suddenness with which the cell-membrane is formed on the spore of Fucus after the access of the spermatozoids and the accomplishment of the act of fecundation. In six or eight minutes traces of the formation of the membrane are recognizable upon a considerable number of the spores. In ten minutes the presence of a membrane may be clearly made manifest by the application of chloride of zinc. In an hour the membrane has acquired considerable firmness and thickness, and the presence of cellulose is revealed by the action of sulphuric acid and iodine; an hour later and the blue coloration under the test is decided.

In the higher *Cryptogamia* and in the *Phanerogamia*, Radlkofer's treatise, though interesting for the history, offers nothing new to our readers. In fact, its date precluded it from containing much of what is referred to in the preceding paragraphs. But the subject is still to be continued.

DR. HOOKER ON THE BALANOPHOREÆ.

ALTHOUGH read before the Linnæan Society nearly three years ago, this fine memoir [1] was published only last summer. The delay has probably been owing, in a great part, to the time requisite for the engraving of the very beautiful and

[1] *On the Structure and Affinities of Balanophoreæ*, by J. D. Hooker, (separately issued from the *Transactions of the Linnæan Society* of London, xxii.) London, 1857. (American Journal of Science and Arts, 2 ser., xxv. 116.)

elaborate plates which illustrate the memoir. It is a clear, patient, and philosophical elucidation of an extremely anomalous group of plants, and a succinct exposition of the principal lessons to be learned from their study, both organographically and systematically; and it bears the impress throughout of the spirit, freshness, and independence which so distinguish this author, and make all his writings so attractive and instructive. While the whole subject is developed in proper order, the divisions are not quite clearly marked out in the essay. The first sectional heading is: "1. Parasitism and structure of the Rhizome." But there is no section 2 answering to the first, which moreover continues, without a break, to treat of the general anatomy, organography, and morphology of these plants, the structure of the flowers, ovules, and seeds, and of the diverse doctrines which have been propounded respecting them. The affinities of *Balanophoreæ* are then considered under a special heading; their Classification is then the subject of a few general remarks; also their Geographical Distribution and Variation. Then a Synoptical Table of the genera is given; and the fourteen genera with their known species (28 in all) are finally described and illustrated.

As to the structure and affinities of *Balanophoreæ*, and the curious questions that have arisen about their place in the natural system, Dr. Hooker, in the first place, affirms them to be truly phænogamous. It now seems strange that this should ever have been doubted. The arguments to the contrary, says our author, "all appear to have originated, on the one hand, in mistaking feeble analogies between the forms of organs that are not homologous, for affinities; and, on the other, in overlooking a multitude of positive characters. These arguments may be summed up as: — 1. An erroneous view of the nature of the seeds, by Endlicher, Martius, Blume, and others, who describe them as a sporuliferous mass, — a term which, even if it were applicable, has no meaning. 2. An erroneous view of their origin being in a diseased state of the plants they grow upon, adopted by Junghuhn and Trattinick. 3. A supposed similarity in appearance to *Fungi*, and an erroneous idea that their appearance is meteoric and their growth rapid; — a theory advanced by Endlicher, who says of the horizontal rhizome of Helosis and Langsdorffia, 'mycelio Fungorum quam maxime analogum.' 4. The resemblance between the articulated filaments on the capitula of the *Helosideæ* and the paraphyses of *Musci;* and between the pistils of *Balanophoreæ* and the pistillidia of Mosses; strongly advocated by Griffith and Lindley. 5. The resemblance of the cellular and vascular tissues in some of their characters to some of those of Ferns, as indicated by Unger and Gœppert. 6. A very peculiar view of the nature and relations of the parts of the female flower entertained by Weddell; who hence considers *Balanophoreæ* (together with *Rafflesiaceæ*) to approach nearer to Gymnosperms than to any other group of plants." Instead of discussing at length opinions which "had the authors who advocate them been sufficiently furnished with specimens and facts they would never have entertained," Dr. Hooker merely recalls attention to the essential facts that these plants exhibit true flowers with stamens and pistils, genuine ovules, and even embryo, and so accord in no one particular with Cryptogams. He shows moreover that the embryo is dicotyledonous in the few cases where it is sufficiently developed to manifest the character, and that the stem is constructed upon the exogenous plan. Even with these facts before him, Lindley has retained his Rhizogens, as "logically a class"; as an intermediate form of organization between Endogens and Thallogens, and characterized by vegetation rather than fructification. But there is little or nothing really peculiar in their vegetation; and, as Lindley himself reduces the differences to questions of degree, it suffices to say that the classes are not founded upon degradation of type, but upon change of type.

Viewing *Balanophoreæ*, then, as degraded members of the Dicotyledonous class, Dr. Hooker follows Brown and Griffith in regarding *Rafflesiaceæ* as near to *Aristolochiaceæ*, and in denying all affinity between these and *Balanophoreæ*. In searching for the affinities of the latter, Dr. Hooker is guided by the sound rule of disregarding "the negative characters, as those may be termed which are founded on the imperfections of organs;" and he takes the most perfectly developed species as the best exponents of the typical structure of any group, — a principle laid down, we believe, by Mr. Brown. This gives a substantial scientific basis for the estimation of affinity. Agreement in plan of structure is just what constitutes affinity; agreement in grade of evolution may indicate only distant analogy, can indicate only collateral relationship, — not to be neglected, indeed, but in itself of no account in assigning a family to its true position in a system. The principle as applied in the present case leads Dr. Hooker to the conclusion that the nearest relatives of *Balanophoreæ* are the *Halorageæ*, a group itself, "consisting for the most part of reduced forms of *Onagrarieæ*," or, more strictly speaking, that the link which connects these plants with the higher forms of vegetation is furnished by Gunnera. The qualifying phrase above is appropriate; for it is hard to conceive of Gunnera with its minute embryo as a reduced *Onagracea*, while it is impossible to sever the chain of evidence which binds the genus to Loudonia and Haloragis. Be this as it may, Dr. Hooker has surely made a happy hit in seizing upon Gunnera as the key to the true affinities of *Balanophoreæ*. Of all the objections that may be urged against this approximation not the strongest, but rather the least valid, in our opinion (so long as the question is one of alliance and not of co-ordination), is that to be derived from the habit and the imperfection of the foliar organs. Any type is liable to have its parasitic phase, and this is generally a degraded one in these respects; the Gesneriaceous has it in *Orbancheæ*, which it might with the greatest propriety include; the Scrophulariaceous graduates insensibly into similar parasitic forms; the Ericaceous has them in *Monotropeæ;* and the Cornaceous or Olacaceous degrades through *Santalaceæ* into *Loranthaceæ*.

It is quite probable that our author would deny the degradation in the latter case, judging from some points which he makes when considering whether the group of *Balanophoreæ*, "putting aside any consideration of its relationship with other orders, and regarding it *per se*, . . . should abstractedly be considered as ranking high, or the contrary." This is an abstraction of which we are hardly capable,—that of determining the rank of an order *per se*. Still our author's ideas are clear and clearly expressed; the comparison is really between these plants and the ideal plant-type. And what is wanting to make the comparison practical is a settled idea as to what constitutes the highest style of plant, and what is the relative importance of deviations from it; questions too large to be entered upon here, if indeed the science is yet ready for their discussion, but which underlie the most important inquiries which good systematic botanists are everywhere tentatively prosecuting. Assuming that the conventional definition of perfection in use among zoölogists is applicable to the vegetable kingdom, and which argues that a high degree of specification of organs and morphological differentiation of them for the performance of the highest functions indicate a high rank, Dr. Hooker ingeniously argues that "*Balanophoreæ* may in some respects be considered to hold a very high one;" and the points are presented under seven heads. Now we will not deny that the principles are logically applied in the present case, nor that the considerations of the kind are perhaps as applicable to the vegetable as to the animal kingdom. But we should *a priori* expect that principles of fundamental importance in the latter could have no sound application to the former; that even such as relate to functions common to the two, or to structures analogous, would require to be based each upon its own ground. As to morphology, and as to what constitutes perfection of type, we should look to the fundamental differences rather than to the resemblances of the two for our starting-point.

Plants for obvious reasons are constructed on the principle of extension of surface. Concentration or consolidation, wherever it occurs in the vegetable kingdom, is a special provision against some peculiar danger. Animals, on the contrary, are formed on the principle of restriction of surface. As if to withdraw them as much as practicable from the direct action of the external world, their shape is compact, their extent as individuals strictly limited, the external organs by which they take their sustenance comparatively few and small, while the most essential organs are safely sheltered within. Consolidation of organs and even their restriction in number, accordingly, are not likely to be indications of high rank in the vegetable kingdom. Not the latter, because the object of the plant in vegetation is attained by the indefinite repetition of the same organs; nor the former, for the type of the plant is realized only in the distinct elimination of leaves from the axis. A Melon-Cactus and a Cuscuta are low forms of plants as to vegetation. As it is a fundamental character of plants that their organs of reproduction are only specialized organs of vegetation; as the higher great divisions of plants are those in which the leaf-type is most apparent throughout; as the perfect accomplishment of the end in view — the production, protection, and nourishment of the embryo even of the highest or most developed kind — does not require the confluence of homogeneous parts, why should such confluence be regarded as indicating higher rank, merely because the type is more disguised in such cases? We see no sufficient ground for ranking a monopetalous plant higher than a polypetalous one on that account; and still less for regarding a Loranthus or a Viscum as the highest style of plant. On the contrary, we incline to look upon the consolidation of heterogeneous parts in the blossom not as high specialization at all, but as want of development, *i. e.* imperfect elimination; and in this light those who maintain an inferior ovary to be one immersed in a receptacle, must needs regard it.

Again, suppression or abortion of organs that belong to the type of the blossom cannot be considered as other than an imperfection, although the loss of the corolla is no great matter, and the abortion of one of the sexes little more. Still hermaphroditism is plainly in the type of the highest style of plant; while the opposite is the case in the animal kingdom. But we cannot here enter further into the discussion of this class of questions. No one feels more deeply than our author the want of fixed and philosophical principles for the subordination of characters and the study of affinities in plants;

and no botanist of his age is more competent, or so well placed and furnished for the investigation of this problem, to which we invite him as to a task worthy of his powers.

As to the rank of *Balanophoreæ*, if our author has demonstrated anything, it is that they belong to the highest class of plants, but that they are probably the most degraded members of it.

BOUSSINGAULT ON THE INFLUENCE OF NITRATES ON THE PRODUCTION OF VEGETABLE MATTER.

SEVERAL years ago Boussingault demonstrated, in the clearest way, that plants are incapable of assimilating the free nitrogen of the atmosphere. Two years ago, in a paper communicated to the French Academy of Sciences, he showed that nitrates eminently favor vegetation. He now shows,[1] by decisive experiments, —

(1) That the amount even of ternary vegetable matter produced by a plant depends absolutely upon the supply of assimilable nitrogen (ammonia and nitrates). A plant, such as a sunflower, with a rather large seed, may grow in a soil of recently calcined brick, watered with pure water, so far as even to complete itself with a blossom; but it will only have trebled or quadrupled the amount of vegetable matter it had to begin with in the seed. In the experiments, the seeds weighing 0.107 grams, in three months of vegetation formed plants which when dried weighed only 0.392 grams, — a little more than trebling their weight. The carbon they had acquired from the decomposition of carbonic acid of the air was only 0.114 grams; the nitrogen they had assimilated from the air in three months was only 0.0025 grams.

(2) Phosphate of lime, alkaline salts, and earthy matters

[1] *Researches upon the Influence which assimilable Nitrogen in manures exerts upon the production of Vegetable Matter; and* (2) *Upon the Quantity of Nitrates contained in the Soil and in Water of various kinds.* J. B. J. D. Boussingault. *Annales des Sciences Naturelles,* 4 ser., vii., No. 1, 1857. (American Journal of Science and Arts, 2 ser., xxv. 120.)

indispensable to the constitution of plants exert no appreciable action upon vegetation, except when accompanied by matters capable of furnishing assimilable nitrogen. Two plants of the same kind, grown under the same conditions as above, but with the perfectly sterile soil adequately supplied with phosphate of lime, alkali in the form of bicarbonate of potash, and silex from the ashes of grasses, resulted in only 0.498 grams of dried vegetable matter, from seeds weighing 0.107 grams; and had acquired only 0.0027 grams of nitrogen beyond what was in the seeds.

(3) But nitrate of potash furnishing assimilable nitrogen, associated with phosphate of lime and silicate of potash, forms a complete manure, and suffices for the full development of vegetation. Parallel experiments with nitrate in place of bicarbonate of potash resulted in the vigorous growth of the Sunflower plants, and the formation of 21.248 grams of organic matter, from seeds weighing as before only 0.107. This 21.111 grams of new vegetable matter, produced in three months of vegetation, contained 8.444 of carbon derived from the carbonic acid of the air and 0.1666 grams of nitrogen. The 1.4 grams of nitrate of potash supplied to the soil contained 0.1969 grams of nitrogen, leaving a balance of 0.0303, nearly all of which was found unappropriated in the soil.

Finally Boussingault made a neat series of comparative experiments, introducing into calcined sand the same amount of phosphate of lime and carbonate of potash, but different proportions of nitrate of soda, or in other words of assimilable nitrogen, and watering with water free from ammonia but containing a quarter of its volume of carbonic acid. The soil was divided among four pots, each having two seeds of Sunflower (*H. argophyllus* was the species used in all the experiments); the pot

No. 1	received of nitrate of soda			0.00	grams.
No. 2	"	"	"	0.02	"
No. 3	"	"	"	0.04	"
No. 4	"	"	"	0.16	"

The results of fifty days' vegetation are given in the rate

of growth, size and number of leaves, weight of the product, etc. : —

No. 1 made of new vegetable matter	. . .	0.397 grams.
No. 2 " " "	. . .	0.720 "
No. 3 " " "	. . .	1.130 "
No. 4 " " "	. . .	3.280 "

In No. 2 so little as three milligrams of assimilable nitrogen introduced into the soil enabled the plant to double the amount of organic matter. The proportion of the weight of the seeds to that of the plant formed was in

No. 1, as	1 : 4.6 grams.
No. 2, as	1 : 7.6
No. 3, as	1 : 11.3
No. 4, as	1 : 30.8

In no case did the nitrogen acquired by the plant exceed that of the nitrate added to the soil.

In the experiments where no nitrate was added to the soil the two or three milligrams of nitrogen acquired by the plants during three months of vegetation came in all probability from ammoniacal vapors and nitrates existing or formed in the atmosphere. To establish their presence, Boussingault arranged an apparatus which detected the production of some nitrates. And, in exposing to the air 500 grams of calcined sand, which had 10 grams of oxalic acid mixed with it, in a glass vessel with an open surface equal to that of one of the flower-pots used in the above experiments, the sand took 0.0013 grams of nitrogen from the air, of which a part was certainly ammonia.

The object of the researches of which a summary is given in the second paper was, to determine the quantity of nitrates contained, at a given moment, in one *hectare* of cultivated ground, one of meadow, one of the forest soil, and in one *metre* of river or spring water. The quantity in the soil was of course found to vary extremely with the extremes of wet or dry weather. Garden soil, highly manured every autumn, contained on the 9th of August, 1856, after fourteen dry and warm days, 316.5 grams of nitre in a cubic litre of soil. On the 29th of the month, after twenty rainy days, the same quantity of the same soil contained only 13 grams of nitre. The greater part had been dissolved out of the superficial soil.

Some specimens of forest-soil, in a state of nature, furnished no indication of nitrates; others gave 0.7 and 3.27 grams of nitre to the cubic metre.

The soil of meadows and pastures afforded from 1 to 11 grams of nitre to the cubic metre. Nineteen specimens of good cultivated land gave, four of them none; others from 0.8 to 1.33; the richer ones from 10.4 to 14.4, and one fallow, of exceptional richness, as much as 108 grams of nitre to the cubic metre. To the latter much calcareous matter has been added.

The soil of the conservatory, from which the nitrates would not be washed away by rains, contained 89, or 161, and some rather deep soil 185 grams of nitre to the cubic metre.

The sources of the nitre are not difficult to understand when we reflect that a manured soil, especially a calcareous one, is just in the condition of an artificial nitre-bed. The ultimate result of the decomposition of ordinary manure is a residuum of alkaline and earthy salts, phosphates, and nitrates, the latter, with the ammonia furnishing the assimilable nitrogen, all-essential to productive vegetation. In incorporating with the soil undecomposed manure, instead of the ultimate results of decomposition, less loss is suffered from prolonged rains wasting out the formed nitrates.

The soluble matters washed out of the soil are to be sought in the water. River and spring waters therefore act as manure by the silex and alkali, the organic matter, and the nitrates which they hold. The spring waters, poorest in nitre of those examined, contained from 0.03 to 0.14 milligrams of nitre to the *litre;* the richer ones from 11 to 14 grams in the cubic metre.

As to the river-water, the Vesle in Champagne held 12 grams, the Seine at Paris 9 grams the cubic metre. These were the richest. The Seine at Paris carries to the sea, in times of low water, 58,000 kilograms, in times of high water, 194,000 kilograms, of nitre every twenty-four hours. What enormous amounts of nitre must be carried into the sea by the

Mississippi, the Amazon, and by every great continental river; and how active, beyond all ordinary conception, must the process of nitrification be all over the land; and how vast the supply of assimilable nitrogen for the use of vegetation!

BENTHAM'S HAND-BOOK OF THE BRITISH FLORA.

ONE of the best systematic botanists — of the soundest judgment and the largest experience, both in European and exotic botany — has deemed it no unfit employment of a portion of his valuable time to prepare a volume[1] by which beginners, having no previous acquaintance with the science, may learn to know most advantageously and readily, the wild flowers and plants of his native land. The result is a genuine popular Flora, and a clear proof that the plants of a limited country may be described, by one who understands them thoroughly, in comparatively simple language, without any sacrifice of scientific accuracy, or of scientific interest. No really good work of this kind was ever made by a compiler; and no one who has not essayed the task, can comprehend how thoroughly faithful writing for beginners brings one's knowledge to the proof.

The characteristic features of the work before us are: 1. The full use of analytical keys, after the mode of De Candolle's "Flora Française," leading easily not only to the order and the genus, but also to the species of the plant in hand. These keys, or analyses, are here made to supersede specific characters as such, neat and free in descriptions, longer or shorter according to circumstances, occupying their place. But generic characters are given with considerable fullness. 2. The exclusion of all technical terms which were not required for the purpose in view, and "the omission, in numerous instances, of microscopical, anatomical, or theoretical

[1] *Hand-book of the British Flora; a Description of the Flowering Plants and Ferns indigenous to, or naturalized in, the British Isles: for the use of Beginners and Amateurs.* By George Bentham. London, 1858. (American Journal of Science and Arts, 2 ser., xxvi. 413.)

characters, often of the greatest importance in scientific botany, but useless to the mere amateur." 3. The descriptions are original, and have been drawn up from British specimens in the first instance, and afterwards compared with the characters given in the standard Floras, and verified upon continental specimens from various parts of the geographical range of the species. As a describer of species (which is something very different from a describer of specimens), Mr. Bentham has no superior. 4. The geographical range of each species, at least its European range, is carefully specified; then the British stations are given in general terms, the object being to state where the plant is likely to be found, rather than to indicate the precise spot where it has been gathered. 5. The judicious limitation of species, and the reduction of a crowd of nominal or "critical" species to their supposed types, with a thoroughness which only a botanist of Mr. Bentham's great experience and authority could well venture upon. The following extract from the preface will explain his views:—

"Taking into account the omission of all plants erroneously indicated as British, it will still, no doubt, be a matter of astonishment that, whilst the last edition of Hooker and Arnott's Flora contains 1571 species, and that of Babington's Manual as many as 1708 (exclusive of Chara), the number, in the present work, is reduced to 1285. This is not owing to any real difference of opinion as to the richness and diversity of our vegetable productions, but is occasioned by a different appreciation of the value of the species themselves. The author has long been persuaded that the views originally entertained by Linnæus, of what really constitutes a species, were far more correct than the limited sense to which many modern botanists seem inclined to restrict the term; and that in most cases where that great master had good means of observation, he succeeded admirably in the practical application of his principles. At any rate, if those minute distinctions by which the innumerable varieties of Brambles, of Roses, of Hawk-weeds, or of Willows, have of late years been characterized, are really more constant and more important than the author's experience has led him to conclude, they cannot be understood without a more complete acquaintance with trifling, vague, and sometimes theoretical characters, than he has himself been able to attain, or than can ever be expected from the mere amateur. . . . The species are limited according to what are conceived to have been the original principles of Linnæus: and the author, in submitting his views to the judgment of the scientific world, trusts that they will not be attributed to hasty generalizations, or conjectural theories, but that they will be generally recognized as founded on personal observation of living plants, made during many years' residence on the continent, as well as in this country, and on repeated comparison of specimens collected from the most varied and distant points of the geographical areas of the several species."

6. Popular names are employed and reduced to a system in accordance with the principles of botanical nomenclature. "An attempt has, on the present occasion, been made to give prominence to a series of English names to the British plants, rendering them as far as possible consistent with the recognized principles of systematic nomenclature, so essential for the study of plants. It was at first intended merely to have adopted those which are appended to all the genera and species in Hooker and Arnott's Flora; but the first attempts to apply them practically, gave evidence that they had never been framed with a view to being used by botanists, or amateurs, in the place of the Latin ones. It will be observed that there is among them a continual confusion between popular, trivial, and generic names; between epithets and specific names; between substantives and adjectives; that on frequent occasions one name is applied to several genera, or several names to one genus; that the number of words forming the name of a plant varies from one to five, instead of being constantly two; and that some of the names put forward as English, are very local, almost unknown or obsolete, and no easier to learn than the more useful Latin ones they represent. It became necessary, therefore, thoroughly to revise the whole system, and to recast it upon the Linnæan principles universally adopted for the Latin botanical names. . . . The full

statement of the principles which have induced the rejection of certain names, and the substitution of others, and the details of their application to individual cases, . . . are given at length in a paper prepared by the author, to be laid before the Linnæan Society on the publication of this Flora."

Criticism may well be deferred until this paper comes to hand. Of the propriety of an English nomenclature of some kind in a Flora where a great part of the plants have well-known vernacular names, there is no room for doubt; and if used at all, it is desirable that these names should be reduced to a systematic form. This is readily done for perhaps half of the common plants of the British Flora; but for the rest, the difficulties are various and much greater than one would imagine before making the attempt. The present undertaking must be deemed a decided success. What imperfections it has, are on the safer side. We should have inclined to a larger use of the vernacular for the generic names; and where they were inapplicable to the whole genera, to apply them to subgenera, *e. g.*, Apple and Pear, Gooseberry and Currant. As these are real and universal English generic names, they ought, if possible, to be given as such. Still we appreciate the reasons which appear to have compelled the adoption of Pyrus and Ribes as English names, though English they never can become. Our author is strongly disposed throughout to make the Latin name do duty as an English one, doubtless supposing that they may become popular appellations in time, as Geranium and Aster have done. Sometimes he adopts the Latin word entire; sometimes he truncates or anglicizes the termination. Happy instances of the latter sort are: —

Trigonel, from Trigonella.

Limosel, from Limosella.

Corydal, from Corydalis.

Corrigiole, from Corrigiola.

Chrysosplene, from Chrysosplenium; but why not Goldenspleen?

Samole, from Samolus; but why not Brookweed?

Limnanth, from Limnanthemum.

Scleranth, from Scleranthus; but why not Knawel?

Osmund, from Osmunda.

Myriophyll, from Myriophyllum; but why not Milfoil?

Matricary, for Matricaria.

Eupatory, for Eupatorium, etc.

Those names which are not at all to our tastes are: —

Cerast, for Cerastium; but if such a word must be coined, why not Holost, for Holosteum on the preceding page?

Doronic, for Doronicum.

Onopord, for Onopordon; why not Cotton-Thistle?

Polycarp (newly martyred), for Polycarpon.

Myosote, for Myosotis; in place of Forget-me-not.

Capsell, for Capsella; in the place of the vernacular Shepherd's-purse.

The best coinage of an English name is Rockcist for Helianthemum.

An Introduction of thirty-six pages teaches the elements of botany to beginners, and explains the technical terms used in the flora, and many besides. The definitions of perigynous and epigynous, however conformable to etymology, are not the quite usual ones, and are not adhered to in the work itself. We were not aware that "in general the word ovary is used to designate all the ovaries of a flower," unless when united into one body, and are glad to observe that the author does not use the word in this way in the body of the work, one or two instances excepted. We always supposed the word to be an exact synonym of the Linnæan germen. And if we may not use it, as botanists always have done, for the ovule-bearing portion of the pistil, whether simple or compound (reserving carpel for the simple of elementary pistil, whether separate or combined), then a new word must needs be coined for this very purpose. To mistake the radicle of the embryo for the root, is common to all English botanists. The short sections upon classification and the examination and determination of plants, are full of practical wisdom.

VILMORIN'S IMPROVEMENT OF CULTIVATED PLANTS.

THIS very interesting pamphlet[1] is a collection and reprint of several of Louis Vilmorin's important communications to the Central Agricultural Society of France and to the Academy of Sciences: to which is prefixed a French translation of a memoir upon the Amelioration of the Wild Carrot, contributed by his venerable father to the Transactions of the London Horticultural Society (but not before published in the vernacular of the author), which memoir, as the younger Vilmorin informs us, was the point of departure for his own investigations in this field, and even contains the germ of most of the ideas which he has since developed upon the theory of the amelioration of the plants from the seed. These papers claim the attention of the philosophical naturalist, no less than of the practical horticulturalist.

Most of our esculents are deviations from the natural state of the species, which have arisen under the care and labor of man in very early times. New varieties of these cultivated races are originated almost every year, indeed; but between these particular varieties, the differences, however well marked, are not to be compared for importance with those changes which the wild plant has generally undergone in assuming the esculent state. In this amelioration or alteration, as in other cases, *c'est la première pas qui coûte.* For the altered race, once originated, has more stability than the wild stock; it accordingly tends not only to degenerate (as the cultivator would term it) towards its original and less useful state, but also to sport into new deviations, in various directions, with a freedom and facility not manifested by its wild ancestors. This explains the readiness with which we continually obtain new varieties of those esculent plants which have been a long time in cultivation, while a newly-introduced

[1] *Notice sur l'Amelioration des Plantes par le Semis et Considerations sur l'Hérédité dans les Vegetaux.* Par M. Louis Vilmorin. Paris, 1859. (American Journal of Science and Arts, 2 ser., xxvii. 440.)

plant exhibits little flexibility. To detect the earliest indications of sporting, and to select for the parents of the new race those individuals which begin to vary in the requisite direction, is the part of the scientific cultivator. In this way, the elder Vilmorin succeeded in producing the esculent carrot from the wild stock in the course of three generations, — no addition to our resources, indeed, but significant of what may be done by art directed by science. By adopting and skillfully applying these principles, the younger Vilmorin has conferred a benefit upon France which (if she will continue to make sugar from the beet) may almost be compared with that of causing two blades of grass to grow where only one grew before, having, so to say, created a race of beets containing twice as much sugar as their ancestors, and indicated the practicability of its perpetuation. The mode of procedure, and the ingenious methods he contrived for rapidly selecting the most saccharine out of a whole crop of beets, as seed-bearers for the next season, are detailed in these papers.

Once originated, and established by selection and segregation for a few generations, the race becomes fixed and perpetuable in cultivation, with proper care against intermixture, in virtue of the most fundamental of organic laws, namely, that the offspring shall inherit the characteristics of the parent, — of which law that of the general permanence of species is one of the consequences. The desideratum in the production of a race is, how to initiate the deviation. The divellent force, or idiosyncrasy, the source of that "infinite variety in unity which characterizes the works of the Creator," though ever active in all organisms, is commonly limited in its practical results to the production of those slighter differences which ensure that no two descendants of the same parent shall be just alike, being overborne by that opposite or centripetal force, whatever it be, which ensures the particular resemblance of offspring to parents. Now the latter force, as Mr. Louis Vilmorin has well remarked, is really an aggregation of forces, composed of the individual attraction of a series of ancestors, which we may regard as the attraction of the type of the species, and which we perceive is generally all-powerful. There is also the attraction or influence of the immediate parent, less powerful than the aggregate of the ancestry, but more close, which ever tends to impress upon the offspring all the parental peculiarities. So, when the parent has no salient individual characteristics, both the longer and the shorter lines of force are parallel, and combine to produce the same result. But whenever the immediate parent deviates from the type, its influence upon its offspring is no longer parallel with that of the ancestry; so the tendency of the offspring to vary no longer radiates around the type of the species as a centre, but around some point upon the line which represents the amount of its deviation from the type. Left to themselves, as Mr. Vilmorin proceeds to remark, such varieties mostly perish in the vast number of individuals which annually disappear, — or else, we may add, are obliterated in the next generation through cross-fertilization by pollen of the surrounding individuals of the typical sort, — whence results the general fixity of species in Nature. But under man's protecting care they are preserved and multiplied, perhaps still further modified, and the better sorts fixed by selection and segregation.

Keeping these principles in view, Mr. Vilmorin concluded that, in order to obtain varieties of any particular sort, his first endeavor should be to elicit variation in any direction whatever; that is, he selected his seed simply from those individuals which differed most from the type of the species, however unlike the state it was desired to originate. Repeating this in the second, third, and the succeeding generations, the resulting plants were found to have a tendency to vary widely, as was anticipated; being loosed, as it were, from the ancestral influence, which no longer acted upon a straight and continuous line, but upon one broken and interrupted by the opposing action of the immediate parents and grandparents. Thus confused by the contrariety of its inherited tendencies, it is the more free to sport in various ways; and we have only to select those variations which manifest the qualities desired, as the progenitors of the new race, and to develop and fix the product by selection upon the same principle continued for several generations.

It is in this way that Mr. Vilmorin supposes cross-fertilization to operate in the production of new varieties; and even in the crossing of two distinct species, the result, he thinks, is rarely, if ever, the production of a fertile hybrid, but of an offspring which, thus powerfully impressed by the strange fertilization and rendered productive by the pollen of its own female parent, is then most likely to give origin to a new race.

We cannot follow out this interesting but rather recondite subject in a brief article like this. But we are naturally led to inquire whether the history of those plants with which man has had most to do, and the study of the laws which regulate the production and perpetuation of domesticated races, may not throw some light upon the production of varieties in Nature; and whether races may not have naturally originated, occasionally, under circumstances equivalent to artificial selection and segregation. Some recent attempts which have been made in this direction we may hope to notice upon another occasion.

THE BUFFALO-GRASS.

THE Buffalo-Grass,[1] so abundant and so widely diffused over the broad, arid region which separates our Pacific from our Atlantic possessions, is one of the humblest plants of its order, rising only a few inches above the surface of the soil; but at the same time it is one of the most important and useful, since it forms the principal subsistence of the buffalo for a part of the year, and no less so of the cattle of the emigrant. The botanical history of this little grass, now happily completed by Dr. Engelmann, is remarkable. Nuttall first named and described it nearly thirty years ago; and, while he referred it to Sesleria, suspected it to be *sui generis*, and threw

[1] *Two new Genera of Diœcious Grasses of the United States.* By George Engelmann, M. D. Extr. from the Transactions of the Academy of Natural Sciences of St. Louis, i. p. 431; with three plates. 1859. (American Journal of Science and Arts, 2 ser., xxviii. 439.)

out a happy conjecture as to its natural relationship. Torrey figured it twelve years ago, and also announced its affinity to the *Chlorideæ;* he at the same time discovered its diœcious character, and showed that only the male plant was known. At length Dr. Engelmann has detected the female plant in a rather rare grass, the *Anthephora axilliflora* of Steudel, which is so unlike the common Buffalo-Grass that it naturally had been referred to a widely different tribe. Struck by the similarity of their foliage and stoloniferous growth, as they occurred together in a collection made by his brother, Dr. Engelmann shrewdly suspected the relationship, and finally set the question at rest by finding a male Buffalo-Grass which happened to bear a stalk of female flowers from the same rootstock; and these flowers were those of the so-called Anthephora. So different are the two that nothing short of this ocular proof would have been convincing. It hardly need be said that the male plant is not a Sesleria, nor the female an Anthephora; although they severally resemble these genera, or at least the female spikelets have a very great external resemblance to the Paniceous genus Anthephora. So that Dr. Engelmann, having to characterize this new generic type, very naturally named it Búchloë (shorter and more euphonious than Bubalóchloë), *i. e.* Buffalo-Grass; and he retained the specific appellation of Dactyloides, although the male plant is not much like a Dactylis, and the female wholly unlike. Very glad we are to see the genus established under so appropriate a name, — the more so as it has narrowly escaped a different fate. That is to say, two inchoate attempts seem to have been made to found a genus upon the male sex. First, in Sir William Hooker's enumeration of the plants of Geyer's western collection we find "*Calanthera dactyloides.* Kth. — Nutt. Sesleria, Nutt. Gen. i. p. 65." But neither Kunth nor any other author has described a genus Calanthera. We have a suspicion that the "Kth." is a slip of the pen, and that the name is really Nuttall's, given by him to a specimen in the Hookerian herbarium. But if this be so, the manuscript name (which, moreover, is destitute of any particular significance) can by no means now supersede Engelmann's published one; though we might have been constrained by courtesy to adopt it, if this suspicion had occurred to him, and he had been able to confirm it. Again, in the corrections at the close of the "Plantæ Hartwegianæ," Mr. Bentham applies the name of "*Lasiostega humilis*, Rupprecht (ined.)" to No. 250, which he had before called a Triodia. The plant is undoubtedly a male Buffalo-Grass. But no genus Lasiostega is found to be published, nor has this name any appropriateness as applied to the plant in question.

It is curious to remark that the male plant, being more proliferous by stolons than the female, has nearly displaced the latter, or has (so far as known) attained a wider geographical range as well as a far greater abundance. Probably, in accordance with a general law, the tendency to barrenness from seed which accompanies copious multiplication by offshoots, has also assisted in the production of this result, — a state of things quite contrary to the genius of that polygamous community which has effected a lodgment in the region of Buffalo-Grass.

Dr. Engelmann's second genus, Monanthochloë, is founded upon a singular, exceedingly stoloniferous, littoral grass with leaves scarcely half an inch long, with solitary sessile spikelets, which has long been known to occur on the coast of Texas and Florida (collected by Berlandier, Drummond, and Blodgett), but has never been studied until now. In fact, it has been thought to be something abnormal, on account of its showing as its most interesting feature, a regular transition from the foliage to the paleæ of the flowers. Dr. Engelmann notes that the glumes are wanting (perhaps represented by ordinary leaves of the axis of which the spikelet is a direct continuation), the uppermost leaf representing the lowest palea of the spikelet. The latter consists of from three to five flowers, of which the lowest flower and sometimes the next are neutral or rudimentary, from one to three succeeding ones are staminiferous or pistilliferous, according to the sex, and the uppermost is also reduced to a rudiment. In the hands of agrostologists such a grass as this will be likely further to elucidate the floral structure of the order, the theory of which is by no means settled yet. Dr. Engelmann's three excellent plates, displaying all the details of the flowers, will facilitate this investigation.

The youthful Academy of Natural Sciences of St. Louis is well inaugurating its public career by publications of such character as this paper, and the more elaborate "Monograph of Cuscuta" by the same author, which is now in press.

THE TREES OF NORTH CAROLINA.

We have turned over the pages of this popular exposition[1] with much interest, and gleaned some valuable information. "Botanists will of course find fault with it," says the author, who we well know could write scientifically and profoundly enough, if he so pleased, but who has here come down to the level of his most unlearned readers, discoursed separately of trees, shrubs, and vines, and classified these in a fashion which might well shock the susceptibilities of a stickler for technical nomenclature and natural system in botany. Now, we are not shocked at all; indeed we quite enjoy a glimpse of Flora en deshabille and slip-shod, and are well aware how much easier it is, and how much better in such cases, to fit your book to its proper readers than to fit the readers to it. The fault we should find is not with the plan of this Report but with the quantity. We could wish for more of it, for a volume as large at least as Mr. Emerson's Report on the Trees and Shrubs of Massachusetts. We quite like to see the popular names put foremost, but would suggest that the botanist who does this should lead as well as follow the indigenous nomenclature, so far as to correct absurd or incongruous local names and introduce right or fitting ones as far as practicable. For instance, "Virgin's Bower" is not a proper name for *Wistaria frutescens*, and is rightly applied to

[1] *Geological and Natural History Survey of North Carolina.* Part III. Botany. The Woody Plants of North Carolina. By M. A. Curtis. Raleigh, 1860. (American Journal of Science and Arts, 2 ser., xxx. 275.)

Clematis Virginiana over the leaf. (We venture to add, in passing, *C. Viorna* to the list, having gathered it in Ashe County.) And although the people alongshore call Baccharis by the name of the English annual weed, "Groundsel," it were better to write it "Groundsel-tree." "Yellow-wood" is the name of Cladrastis, rather than of Symplocos, which the Carolinians call "Horse-Sugar." Dr. Curtis can coin a name upon occasion; for surely nobody in Carolina knows *Menziesia globularis* as False Heath, nor has it any scientific claim to this appellation. While in critical mood we may express a strong dissent from the proposition that *Rhododendron punctatum* is too inferior to the other two species "to attract or deserve much attention." With us, it is surpassingly beautiful in cultivation, none the less so because its habit is so different, having light and pendent branches, when well grown forming broad and thick masses, and loaded with its handsome rose-colored blossoms. While *Leucothoë Catesbœi* is called "a very pretty shrub," the far handsomer *Andromeda floribunda*, so much prized by our nurserymen, gets no commendation. *Magnolia Fraseri* may not only be "cultivated in the open air near Philadelphia," but is perfectly hardy near Boston, and the earliest to blossom; but we never noticed the fragrance of the flowers. On the other hand, as it is a native so far south as Florida, it might thrive in plantations anywhere in North Carolina. The flowers of *M. cordata* are described as if larger than those of *M. Fraseri*, instead of the contrary; we could hardly say much for their beauty, except in comparison to those of the common Cucumber-tree. *Prunus Virginiana* is omitted; yet surely it is not wanting in North Carolina. And it is almost an excess of conscientiousness to leave out Cladrastis, the handsomest tree of the country, all things considered, when it is known to grow only a few rods over the Tennessee line.

On the other hand, we are disposed to doubt if the genuine White Spruce (*Abies alba*) occurs in North Carolina. At length we know this tree, but only in Canada and parts adjacent. It is more, instead of less, northern in its range than *A. nigra*. But since President Wheeler has pretty nearly determined the existence of *A. Fraseri* on the Green Mountains in Vermont, we could not deny that *A. alba* grows with the latter on the high mountains of North Carolina. We make our little criticism freely, — as we know the excellent author would wish, — for we think it likely that this part of the Report will pass to a second edition, — when we hope it will be largely augmented.

BENTHAM'S FLORA OF HONGKONG.

THE present work[1] is the third of the series of British Colonial Floras, upon a new and simple plan, compact in form, written in English throughout, authorized and supported by the British Government. The Colonial department pays a very moderate recompense to the authors, and turns the work over to a publisher upon such terms as to render the volume generally accessible to working botanists and colonists. This is a much wiser as well as vastly more economical plan of government patronage to scientific publication than that adopted in this country, one which secures that the publications are just what is wanted and that they reach the hands which are to use them, and not others, — one which, when our present task is done and we again cultivate the arts of peace, we might profitably adopt. The present work is by a master-hand; for Mr. Bentham is one of the most experienced, industrious, and judicious of systematic botanists. The island of Hongkong has an area of scarcely thirty square miles, its general aspect is bleak and barren; yet it has already yielded about a thousand phænogamous species. "At a first glance," as the author observes, "one is struck with the very large total amount of species crowded upon so small an island, which all navigators depict as apparently so bleak and bare;

[1] *Flora Hongkongensis*; a Description of the Flowering Plants and Ferns of the Island of Hongkong, by George Bentham. With a Map of the Island. Published under the Authority of Her Majesty's Secretary of State for the Colonies. London, 1861. (American Journal of Science and Arts, 2 ser., xxxii. 124.)

— with the tropical character of the great majority of species, when botanists agree in representing the general aspect (derived from the majority of individuals) to present the features of a much more northern latitude; — with the large proportion of arborescent and shrubby species, on a rocky mass where the woods are limited to a few ravines, or short narrow valleys half monopolized by cultivation; — and with the very great diversity in the species themselves, the proportion of orders and genera to species, and the comparative number of monotypic genera, being far greater in the Hongkong Flora than in any other Flora of similar extent known to me. The very large number of endemic species — of species known to us only from the island — is probably occasioned by cur ignorance, already alluded to, of the vegetation of continental south China."

A fitting acknowledgment is given for the important contribution to this Flora furnished by the botanical collection (of above 500 species) made by Charles Wright, as botanist of the U. S. North Pacific Exploring Expedition under Captains Ringgold and Rodgers, duplicates of which were obligingly and most properly furnished by direction of the Commander and the enlightened Secretary of the Smithsonian Institution.

In aid of the colonial botanists or amateurs who may use this Flora, the author has prefixed (with some minor alterations) the admirable brief outlines of Botany and Glossary prepared for his popular British Flora.

In these Outlines the subject is regarded, not from the morphological or the physiological, but from the descriptive point of view. It opens with a statement of the nature and design of a Flora, and of what a botanical description ought to be.

"These descriptions should be clear, concise, accurate, and characteristic, so that each one should be readily adapted to the plant it relates to, and to no other; they should be as nearly as possible arranged under natural divisions, so as to facilitate the comparison of each plant with those nearest allied to it; and they should be accompanied by an artificial key or index, by means of which the student may be guided step by step in the observation of such peculiarities, or characters, in his plant as may lead him, with the least delay, to the individual description belonging to it.

"For descriptions to be clear and readily intelligible, they should be expressed as much as possible in ordinary well-established language. But, for the purpose of accuracy, it is necessary not only to give a more precise technical meaning to many terms used more or less vaguely in common conversation, but also to introduce purely technical names for such parts of plants or forms as are of little importance except to the botanist. In the present chapter it is proposed to define such technical or technically limited terms as are made use of in these Floras.

"At the same time mathematical accuracy must not be expected. The forms and appearances assumed by plants and their parts are infinite. Names cannot be invented for all; those even that have been proposed are too numerous for ordinary memories. Many are derived from supposed resemblances to well-known forms and objects. These resemblances are differently appreciated by different persons; and the same term is not only differently applied by two different botanists, but it frequently happens that the same writer is led on different occasions to give somewhat different meanings to the same word. The botanist's endeavors should always be, on the one hand to make as near an approach to precision as circumstances will allow, and on the other hand to avoid that prolixity of detail and overloading with technical terms which tends rather to confusion than to clearness. In this he will be more or less successful. The aptness of a botanical description, like the beauty of a work of imagination, will always vary with the style and genius of the author."

These Outlines are throughout so well sketched, and so worthy to be regarded as of standard authority, that we must still venture a criticism or two, looking to their possible improvement.

In the first place, referring to paragraphs 8 and 88, we must dissent from the proposition that the subject of *homology* does not belong to "*morphology* in the proper sense of the term;" — unless, indeed, morphology relates simply to form in the lowest sense, to mere shape, arbitrarily viewed, — which would belittle the subject down to mere terminology, and empty that of all scientific interest. If the comparison even of a perfoliate or clasping with a cordate leaf, or of membranaceous or coriaceous with thickened leaves, such as those of a Houseleek, a Mesembryanthemum, and an Aloë, falls within the province of morphology, surely so also must the comparison of an ordinary leaf with a cotyledon, with a bulb-scale, a bud-scale, and no less with a sepal, a petal, a carpel, etc. In the latter we merely trace morphological relations of the very same kind somewhat further and higher. The relation of a leaf as foliage to the scale of a bud, or to the thorn of a Barberry, is clearly of the same category as its relation to a sepal or a petal, — the latter, as we regard it, bringing in no new idea, and requiring no new point of view.

Next, *Quincuncial* imbrication is defined by Mr. Bentham to be that arrangement in which "one petal is outside, an adjoining one wholly inside, the three others intermediate and overlapping on one side." But why give this name to a mixed form, to that which is merely convolute æstivation deranged by one of the five petals getting both edges under? And why change the uniform usage from De Candolle's "Théorie Elémentaire," if not earlier, down to the present time, which defines the quincuncial mode as having two members exterior, two interior, and one with one edge overlapping its neighbor and the other overlapped; an arrangement which especially merits a distinguishing name, since it is the normal imbrication in a pentamerous perianth, answering as it does to two fifths phyllotaxis. So that current usage and reason tell against the innovation.

In the third place, we are equally inclined to demur to the proposed modifications of the sense of the terms *perigynous* and *epigynous* (paragraph 140), Mr. Bentham restricting the former to those cases in which the petals, etc., are adnate to a perfectly free calyx, as in the Cherry, and applying the latter in cases where the calyx, equally bearing the petals, etc., is adnate even merely to the base of the ovary, if only the adhesion reaches above the level of the insertion of the lowest ovule; which would make most Saxifrages epigynous. Besides the etymological objections, and the inconvenience of a change, the new definitions seem to us to be at least as ambiguous as the old in practice; and it is not surprising that they are not uniformly adopted in the Hongkong Flora itself.

Finally, as to paragraph 166, we are not much better satisfied with the definition that the radicle is the "base of the future root," than with the original statement that it is "the future root." To us nothing in botany is clearer, or more patent to observation during germination, than that while the radicle is, if you please, "the base of the future root" inasmuch as it is that from which the root proceeds, it is itself the first internode of the stem. This view, to which morphological considerations and observation of the development long since brought us, appears to be generally adopted by the French and German botanists, but not by the English. If the radicle universally failed to elongate, as in Monocotyledons, and in the Pea, Oak and others with hypogæous germination, this organ might be deemed to be merely the base of the future root; but its more usual elongation, in the manner of any other internode, plainly reveals the cauline nature which analogy would also assign to it.

The chapter on Vegetable Anatomy and Physiology is new, is very condensed, and considering that it deals with matters to which Mr. Bentham has never specially attended, is remarkably good and accurate. We merely observe in passing, of paragraphs 195, 197, that the distinction between exogenous and endogenous stems is as obvious during the first season, and even at its beginning, as ever afterward, and it is then that the purely systematic botanist will more commonly have occasion to examine the structure in this regard; of § 198[5], that "the liber or inner bark" is by no means always "formed of bast cells;" of § 200, that we cannot accept the statement that "in the leaf the structure of the petioles and principal ribs or veins is the same as that of the young branches of which they are ramifications," at least in any sense in which the sentence would be understood by the learner. Paragraph 207, that roots grow in length at the extremities, "in proportion as they find the requisite nutriment," might imply the popular fallacy that they grow directly by means of what they take in from the soil, which surely they do not, unless they live in the manner of *Fungi*. To say that the starch, etc. in a tuber or in a seed "appears to be a store of nourishment" for the early growth of the buds or the embryo, is a remarkably over-cautious statement (how could these grow without some store of elaborated matter to feed upon?); nor does the consideration that similar accumulations in the pericarps of many fruits "perish long before germination," and so do not nourish the embryo, afford to us any presumption to the contrary, even if we could not conceive — as we readily can — of other final causes, some of them important to the continuance of the species thereby subserved.

The fourth chapter, on the Collection, Preservation, and Determination of Plants, and upon Aberrations from the ordinary type or appearance, is most excellent.

DR. HOOKER'S DISTRIBUTION OF ARCTIC PLANTS.

THE immediate subjects of the treatise[1] are the Arctic plants, of every phænogamous species known to occur spontaneously anywhere within the Arctic circle; the geographical distribution of which, so far as known, is carefully indicated: 1. Within the Arctic region, under the several divisions — Europe, Asia, western America (Behring's Straits to the Mackenzie River), eastern America (Mackenzie River to Baffin's Bay), and arctic Greenland. 2. Without this circle, and under the general divisions of north and central European and north Asiatic Distribution, with three longitudinal subdivisions; American Distribution, with appropriate subdivisions;

[1] *Outlines of the Distribution of Arctic Plants.* By Joseph D. Hooker. Extr. *Transactions Linnæan Society*, of London. Vol. xxiii. pp. 251–348. 1861. (American Journal of Science and Arts, 2 ser., xxxiv. 144.)

south European and African Distribution; central and south Asiatic Distribution. The theory upon which the facts are collocated and discussed, and which they are thought strongly to confirm, is that of Edward Forbes, which was completed, if not indeed originated, by Darwin: — "first, that the existing Scandinavian flora is of great antiquity, and that previous to the glacial epoch it was more uniformly distributed over the Polar Zone than it is now; secondly, that during the advent of the glacial period this Scandinavian vegetation was driven southward in every longitude, and even across the tropics into the south temperate zone; and that, on the succeeding warmth of the present epoch, those species that survived both ascended the mountains of the warmer zones, and also returned northward, accompanied by aborigines of the countries they had invaded during their southern migration. Mr. Darwin shows how aptly such an explanation meets the difficulty of accounting for the restriction of so many American and Asiatic arctic types to their own peculiar longitudinal zones, and for what is a far greater difficulty, the representation of the same arctic genera by closely allied species in different longitudes. . . . Mr. Darwin's hypothesis accounts for many varieties of one plant being found in various alpine and arctic regions of the globe, by the competition into which their common ancestor was brought with the aborigines of the countries it invaded. Different races survived the struggle for life in different longitudes; and these races again, afterwards converging on the zone from which their ancestor started, present there a plexus of closely allied but more or less distinct varieties, or even species, whose geographical limits overlap, and whose members, very probably, occasionally breed together." A further advantage claimed for this hypothesis is, that it explains a fact brought out by Dr. Hooker in a former publication, namely: "that the Scandinavian flora is present in every latitude of the globe, and is the only one that is so."

Moreover, Dr. Hooker discovers in the flora of Greenland a state of things explicable upon this hypothesis, but hardly by any other, namely: its almost complete identity with that of

Lapland; its general paucity, as well as its poverty in peculiar species; the rarity of American species there; the fewness of temperate plants in temperate Greenland; and the presence of a few of the rarest Greenland and Scandinavian species in enormously remote alpine localities of west America and the United States. Our author reasons thus: "If it be granted that the polar area was once occupied by the Scandinavian flora, and that the cold of the glacial epoch did drive this vegetation southwards, it is evident that the Greenland individuals, from being confined to a peninsula, would have been exposed to very different conditions from those of the great continents. In Greenland many species would, as it were, be driven into the sea, that is, exterminated; and the survivors would be confined to the southern portion of the peninsula, and, not being there brought into competition with other types, there could be no struggle for life amongst their progeny, and, consequently, no selection of better adapted varieties. On the return of heat, survivors would simply travel northwards, unaccompanied by the plants of any other country."

The rustic denizens of Greenland, huddled upon the point of the peninsula during the long glacial cold, have never enjoyed the advantages of foreign travel; those of the adjacent continents on either side have "seen the world," and gained much improvement and diversity thereby. Considering the present frigid climate of Greenland, the isotherm of 32° just impinging upon its southern point, its moderate summer and low autumnal temperature, we should rather have supposed the complete extermination of the Greenland ante-glacial flora; and have referred the Scandinavian character of the existing flora (all but eleven of the 207 arctic species, and almost all those of temperate Greenland, being European plants) directly to subsequent immigration from the eastern continent. Several geographical considerations, and the course of the currents, which Dr. Hooker brings to view on p. 270, would go far towards explaining why Greenland should have been re-peopled from the Old rather than from the New World; while the list (on pp. 272, 273) of upwards of 230

Arctic-European species, which are all likewise American plants, but are remarkable for their absence from Greenland, would indicate no small difficulty in the westward migration, and render it most probable that the diffusion of species from the Old World to the New was eastward through Asia, for the arctic no less than (as has elsewhere been shown) for the temperate plants. Was it that Greenland and the adjacent part of the American continent remained glacial longer than the rest of the zone? And if our northern regions were thus colonized by an ancient Scandinavian flora, this seems to have been in return for a still earlier donation of American plants to Europe, to which a very few existing but numerous fossil remains bear testimony. Speculative inquiries of this sort are enticing, and the time is approaching in which they may be fruitful.

Indeed, the characteristic features and the immediate interest and importance of the present memoir, as of others of the same general scope and interest, are found in this: 1. That the actual geographical distribution of species is something to be accounted for; 2. That our existing species, or their originals, are far more ancient than was formerly thought, mainly if not wholly antedating the glacial period; and, 3. That they have therefore been subject to grave climatic vicissitudes and changes. There may be many naturalists who still hesitate to accept these propositions, as there are one or two who deny them; but these or similar conclusions have evidently been reached by those botanists, paleontologists, and geologists in general who have most turned their thoughts to such inquiries, and who march foremost in the advancing movement of these sciences. In this position, the author of the present memoir, — prepossessed with Darwin's theory of the diversification of species through natural selection, — having occasion to revise systematically the materials of the arctic flora, is naturally led to compare the new theory with the facts of the case in this regard; to see how far the vicissitudes to which it is all but demonstrated that the plants of the northern hemisphere have long been subjected, and the modifications and extinctions which he thinks must have ensued under such grave changes and perils, during such lapse of time, may serve to explain the actual distribution of arctic species and the remarkable dispersion of many of them. That the enquiry is a legitimate and a hopeful one we must all agree, whether we favor Darwinian hypotheses or not. How well it works in the present trial we could not venture to pronounce without a far more critical examination than could now be undertaken. But there are good reasons for the opinion that this is just the ground upon which the elements of the new hypothesis figure to the best advantage.

The mass of facts, so patiently and skillfully collected and digested in this essay, have a high and positive value, irrespective of all theoretical views. We cannot undertake to offer an abstract, but may note here and there a point of interest. The flowering plants which have been collected within the arctic circle number 762, namely, 214 Monocotyledons and 548 Dicotyledons. They occupy a circumpolar belt of 10° to 14° of latitude. The only abrupt change in the vegetation anywhere along this belt is at Baffin's Bay, the opposite shores of which present, as has been already intimated, an almost purely European flora on the east coast, but a large admixture of purely American species on the west.

"Regarded as a whole, the arctic flora is decidedly Scandinavian; for Arctic Scandinavia, or Lapland, though a very small tract of land, contains by far the richest arctic flora, amounting to three fourths of the whole." This would not be very surprising, since this is much the least frigid portion of the zone, and has the highest summer temperature; but "upwards of three-fifths of the species, and almost all the genera of Arctic Asia and America, are likewise Lapponian;" so that the Scandinavian character pervades the whole.

In the section on the local distribution of plants within the arctic circle, Dr. Hooker shows that there is no close relation discoverable between the isothermal lines (whether annual or monthly) and the amount of vegetation, beyond the general fact that the scantiness of the Siberian flora is associated with a great southern bend in Asia, and its richness in Lapland, with an equally great northern bend there, of the annual isotherm of 32°. Yet "the same isotherm bends northwards in passing from eastern America to Greenland, the vegetation of which is the scantier of the two; and it passed to the northward of Iceland, which is much poorer in species than those parts of Lapland to the southward of which it passes." A glance at the supposed former state of things would suggest the explanation of all that is anomalous here.

"The June isothermals, as indicating the most effective temperatures in the arctic regions (when all vegetation is torpid for nine months, and excessively stimulated during the three others), might have been expected to indicate better the positions of the most luxuriant vegetation. But neither is this the case; for the June isothermal of 41°, which lies within the arctic zone in Asia, where the vegetation is scanty in the extreme, descends to lat. 54° in the meridian of Behring's Straits, where the flora is comparatively luxuriant." The aridity of the former and the humidity of the latter district here offers an obvious explanation; also the great severity of the winter in the former, and its mildness in the latter. And Great Britain, in which a far greater diversity of species are capable of surviving without protection than in the eastern United States under the same annual isotherms, indicates the advantage of a mean over an extreme climate in this respect, if only there be a certain amount of summer heat. For lack of that, doubtless, very many of the introduced denizens of Britain would soon disappear, if deprived of human care.

"The northern limit to which vegetation extends varies in every longitude; the extreme is still unknown; it may, indeed, reach to the pole itself. Phænogamic plants, however, are probably nowhere found far north of lat. 81°. Seventy flowering plants are found in Spitzbergen; and Sabine and Ross collected nine on Walden Island, towards its northern extreme, but none on Ross's Islet, fifteen miles further to the north.

"*Saxifraga oppositifolia* is probably the most ubiquitous, and may be considered the commonest and most arctic flowering plant." There are only eight or nine phænogamous species peculiar to the arctic zone, and only one peculiar genus, namely, the grass Pleuropogon.[1] Of the 762 found south of the circle, all but 150 have advanced beyond lat. 40° N. in some part of the world; about 50 of them are identified as natives of the mountainous regions of the tropics, and 105 as inhabiting the south temperate zone.

"The proportion of species which have migrated southward in the Old and New World also bear a fair relation to the facilities for migration presented by the different continents." The tables given to illustrate this "present in a very striking point of view the fact of the Scandinavian flora being the most widely distributed over the world. The Mediterranean, south African, Malayan, Australian, and all the floras of the New World, have narrow ranges compared with the Scandinavian, and none of them form a prominent feature in any other continent than their own. But the Scandinavian not only girdles the globe in the arctic circle, and dominates over all others in the north temperate zone of the Old World, but intrudes conspicuously into every other temperate flora, whether in the northern or southern hemisphere, or on the Alps of tropical countries. . . . In one respect this migration is most direct in the American meridian, where more arctic species reach the highest southern latitudes. This I have accounted for ('Flora Antarctica,' p. 230) by the continuous chain of the Andes having favored their southern dispersion."

In presenting the actual number of arctic species, and in delineating their geographical ranges, the question, what are to be regarded as species, becomes all-important. As to this, it does not so much matter what scale is adopted, as to know clearly what the adopted scale is. Here we are not left in

[1] *Douglasia* is mentioned in another place (p. 269) as an absolutely peculiar arctic or arctic-alpine genus of eastern America. But we have considered this genus as identical with Gregoria, of Duby. It would appear as if these two genera were established in the same year, since Lindley himself, in the "Botanical Register," refers to Brande's Journal for January, 1828, for his original article. But this article will be found in the volume of that Journal for 1827 ; so that the name Douglasia is to be adopted, if the genus is sufficiently distinct from Androsace.

doubt. Taking European botanists by number, we are confident that nine out of ten would have enlarged the list of 762 phænogamous arctic species to 800 or more, and would not have recognized a goodly number of the synonyms adduced, thereby considerably affecting the assigned ranges, especially into temperate and austral latitudes. In this regard we should side with Dr. Hooker on the whole, but with differences and with questionings — with halting steps following his bold and free movement, but probably arriving at the same goal at length. Indeed, we freely receive the view which Dr. Hooker presents as appropriate to his particular purpose, and as the most useful expression of our knowledge of the relationships of the plants in question, when collocated in reference to the ideas upon which this memoir is based. That is: "if, with many botanists, we consider these closely allied varieties and species as derived by variation and natural selection from one parent form at a comparatively modern epoch, we may with advantage, for certain purposes, regard the aggregate distribution of such very closely allied species as that of one plant." "An empirical grouping of allied plants, for the purposes of distribution, may thus lead to a practical solution of difficulties in the classification and synonymy of species. My thus grouping names must not be regarded as a committal of myself to the opinion that the plants thus grouped are not to be held as distinct species. . . . My main object is to show the affinities of the polar plants, and I can best do this by keeping the specific idea comprehensive." And further: "I wish it then to be clearly understood, that the catalogue here appended is intended to include every species hitherto found within the arctic circle, together with those most closely allied forms which I believe to have branched off from one common parent within a comparatively recent geological epoch, and that immediately previous to the glacial period or since then" (p. 279). All we could ask more would be some distinction (typographical or other), to mark, 1, undoubted and complete synonyms; 2, mere variations or states, local or otherwise, or undoubted varieties; 3, such as, theory apart, would claim to be regarded as distinct but closely related species. For example: to take one order, while *Rhinanthus minor* may well be considered as "not a sufficiently constant form to rank as a race even," while *Limosella tenuifolia* could rank for no more than a race, and while *Castilleia septentrionalis* and *C. pallida*, we are now convinced, however distinct in this single character, differ only (and inconstantly) in the relative development of the galea, we think it likely that *Pedicularis lanata*, Willd., does not rightfully merge in *P. hirsuta* this side of the glacial period, although it perhaps may into *P. Langsdorffii*, and that into *P. Sudetica.* But this is no place for criticisms upon the limitation of species, upon which the opinions of botanists will so greatly depend upon the amount of their materials, and upon which the best considered opinions must be subject to frequent revisal. Nor does the value of the present memoir at all depend upon the settlement of such points. To the philosophical naturalist, as to the archæologist, just now the most interesting and pregnant epoch of the world's natural history is that immediately antecedent to the present, that near past from which the present has proceeded, and upon which so much light, from very diverse sources, is now being concentrated: towards its elucidation the memoir we have been considering is a very valuable contribution.

ALPHONSE DE CANDOLLE ON THE VARIATION AND DISTRIBUTION OF SPECIES.

THIS is the title of a paper [1] by Monsieur De Candolle growing out of his study of the Oaks. It was published in the November number of the "Bibliothèque Universelle," and separately issued as a pamphlet. A less inspiring task could hardly be assigned to a botanist than the systematic elaboration of the genus Quercus and its allies. The vast materials assembled under De Candolle's hands, while disheartening

[1] *Etude sur l'Espèce, à l'occasion d'une Revision de la Famille des Cupulifers.* Par M. Alphonse De Candolle. (American Journal of Science and Arts, 2 ser., xxxv. 431.)

for their bulk, offered small hope of novelty. The subject was both extremely trite and extremely difficult. Happily it occurred to De Candolle that an interest might be imparted to an onerous undertaking, and a work of necessity be turned to good account for science, by studying the Oaks in view of the question of Species.

What this term Species means, or should mean, in natural history, what the limits of species, *inter se* or chronologically, or in geographical distribution, their modifications, actual or probable, their origin, and their destiny, — these are questions which surge up from time to time; and now and then in the progress of science they come to assume a new and hopeful interest. Botany and Zoölogy, Geology, and what our author, feeling the want of a new term, proposes to name *Epiontology*,[1] all lead up to and converge into this class of questions, while recent theories shape and point the discussion. So we look with eager interest to see what light the study of Oaks, by a very careful, experienced, and conservative botanist, particularly conversant with the geographical relation of plants, may throw upon the subject.

The course of investigation in this instance does not differ from that ordinarily pursued by working botanists; nor, indeed, are the theoretical conclusions other than those to which a similar study of other orders might not have equally led. The Oaks afford a very good occasion for the discussion of questions which press upon our attention, and perhaps they offer peculiarly good materials on account of the number of fossil species.

Preconceived notions about species being laid aside, the

[1] A name which, at the close of his article, De Candolle proposes for *the study of the succession of organized beings*, to comprehend, therefore, palæontology and everything included under what is called geographical botany and geographical zoölogy, — the whole forming a science parallel to geology, — the latter devoted to the history of unorganized bodies, the former, to that of organized beings, as respects origin, distribution, and succession. We are not satisfied with the word, notwithstanding the precedent of *palæontology;* since *ontology*, the science of being, has an established meaning as referring to mental existence, — *i. e.*, is a synonym or a department of metaphysics.

specimens in hand were distributed, according to their obvious resemblances, into groups of apparently identical or nearly identical forms, which were severally examined and compared. Where specimens were few, as from countries little explored, the work was easy, but the conclusions, as will be seen, of small value. The fewer the materials, the smaller the likelihood of forms intermediate between any two, and — what does not appear being treated upon the old law-maxim as non-existent — species are readily enough defined. Where, however, specimens abound, as in the case of the Oaks of Europe, of the Orient, and of the United States, of which the specimens amounted to hundreds, collected at different ages, in varied localities, by botanists of all sorts of views and predilections, — here alone were data fit to draw useful conclusions from. Here, as De Candolle remarks, he had every advantage, being furnished with materials more complete than any one person could have procured from his own herborizations, more varied than if he had observed a hundred times over the same forms in the same district, and more impartial than if they had all been amassed by one person with his own ideas or predispositions. So that vast herbaria, into which contributions from every source have flowed for years, furnish the best possible data — at least are far better than any practicable amount of personal herborization — for the comparative study of related forms occurring over wide tracts of territory. But as the materials increase, so do the difficulties. Forms, which appeared totally distinct, approach or blend through intermediate gradations; characters, stable in a limited number of instances or in a limited district, prove unstable occasionally, or when observed over a wider area; and the practical question is forced upon the investigator, — what here is probably fixed and specific, and what is variant, pertaining to individual, variety, or race?

In the examination of these rich materials, certain characters were found to vary upon the same branch, or upon the same tree, sometimes according to age or development, sometimes irrespective of such relations or of any assignable reasons. Such characters, of course, are not specific, although many of them are such as would have been expected to be constant in the same species, and are such as generally enter into specific definitions. Variations of this sort, De Candolle, with his usual painstaking, classifies and tabulates, and even expresses numerically their frequency in certain species. The results are brought well to view in a systematic enumeration: —

(1) Of characters which frequently vary upon the same branch: over a dozen such are mentioned.

(2) Of those which sometimes vary upon the same branch: a smaller number of these are mentioned.

(3) Those so rare that they might be called monstrosities.

Then he enumerates characters, ten in number, which he has never found to vary on the same branch, and which therefore may better claim to be employed as specific. But, as among them he includes the duration of the leaves, the size of the cupule, and the form and size of its scales, which are by no means wholly uniform in different trees of the same species, even these characters must be taken with allowance. In fact, having first brought together, as groups of the lowest order, those forms which varied upon the same stock, he next had to combine similarly various forms which, though not found associated upon the same branch, were thoroughly blended by intermediate degrees.

"The lower groups (varieties or races) being thus constituted, I have given the rank of *species* to the groups next above these, which differ in other respects, *i. e.*, either in characters which were not found united upon certain individuals, or in those which do not show transitions from one individual to another. For the Oaks of regions sufficiently known, the species thus formed rest upon satisfactory bases, of which the proof can be furnished. It is quite otherwise with those which are represented in our herbaria by single or few specimens. These are *provisional species*, — species which may hereafter fall to the rank of simple varieties. I have not been inclined to prejudge such questions; indeed, in this regard, I am not disposed to follow those authors whose tendency is, as they say, to reunite species. I never reunite them without proof in each particular case; while the botanists to whom I refer do so on the ground of analogous variations or transitions occurring in the same genus or in the same family. For example, resting on the fact that *Quercus Ilex*, *Q. coccifera*, *Q. acutifolia*, etc., have the leaves sometimes entire and sometimes toothed upon the same branch, or present transitions from one tree to another, I might readily have united my *Q. Tlapuxahuensis* to *Q. Sartorii* of Liebmann, since these two differ only in their entire or their toothed leaves. From the fact that the length of the peduncle varies in *Q. Robur* and many other Oaks, I might have combined *Q. Seemannii*, Liebm., with *Q. salicifolia*, Née. I have not admitted these inductions, but have demanded visible proof in each particular case. Many species are thus left as provisional; but in proceeding thus, the progress of the science will be more regular, and the synonymy less dependent upon the caprice or the theoretical opinions of each author."

This is safe and to a certain degree judicious, no doubt, as respects published species. Once admitted, they may stand until they are put down by evidence, direct or circumstantial. Surely a species may rightfully be condemned on good circumstantial evidence. But what course does De Candolle pursue in the case — of every-day occurrences to most working botanists having to elaborate collections from countries not so well explored as Europe — when the forms in question, or one of the two, are as yet unnamed? Does he introduce as a new species every form which he cannot connect by ocular proof with a near relative, from which it differs only in particulars which he sees are inconstant in better known species of the same group? We suppose not. But if so, little improvement for the future upon the state of things revealed in the following paragraph can be expected.

"In the actual state of our knowledge, after having seen nearly all the original specimens, and in some species as many as 200 representatives from different localities, I estimate that, out of the 300 species of *Cupuliferæ* which will be enumerated in the 'Prodromus,' two thirds at least are *provisional* species. In general, when we consider what a multitude of species were described from a single specimen, or from the forms of a single locality, of a single country, or are badly described, it is difficult to believe that above one third of the actual species in botanical works will remain unchanged."

Such being the results of the want of adequate knowledge, how is it likely to be when our knowledge is largely increased? The judgment of so practised a botanist as De Candolle is important in this regard, and it accords with that of other botanists of equal experience.

"They are mistaken," he pointedly asserts, "who repeat that the greater part of our species are clearly limited, and that the doubtful species are in a feeble minority. This seemed to be true, so long as a genus was imperfectly known, and its species were founded upon few specimens, that is to say, were provisional. Just as we come to know them better, intermediate forms flow in, and doubts as to specific limits augment."

De Candolle insists, indeed, in this connection, that the higher the rank of the groups, the more definite their limitation, or, in other terms, the fewer the ambiguous or doubtful forms; that genera are more strictly limited than species, tribes than genera, orders than tribes, etc. We are not convinced of this. Often where it has appeared to be so, advancing discovery has brought intermediate forms to light, perplexing to the systematist. "They are mistaken," we think more than one systematic botanist will say, "who repeat that the greater part of our natural orders and tribes are absolutely limited," however we may agree that we will limit them. Provisional genera we suppose are proportionally hardly less common than provisional species; and hundreds of genera are kept up on considerations of general propriety or general convenience, although well known to shade off into adjacent ones by complete gradations. Somewhat of this greater fixity of higher groups, therefore, is rather apparent than real. On the other hand, that varieties should be less definite than species, follows from the very terms employed. They are ranked as varieties, rather than species, just because of their less definiteness.

Singular as it may appear, we have heard it denied that spontaneous varieties occur. De Candolle makes the important announcement that, in the Oak genus, the best known species are just those which present the greatest number of spontaneous varieties and sub-varieties. The maximum is found in *Q. Robur*, with twenty-eight varieties, all spontaneous. Of *Q. Lusitanica* eleven varieties are enumerated, of *Q. Calliprinos* ten, of *Q. coccifera* eight, etc. And he significantly adds that " these very species which offer such numerous modifications are themselves ordinarily surrounded by other forms, provisionally called species, because of the absence of known transitions or variations, but to which some of these will probably have to be joined hereafter." The inference is natural, if not inevitable, that the difference between such species and such varieties is only one of degree, either as to amount of divergence, or of hereditary fixity, or as to the frequency or rarity, at the present time, of intermediate forms.

This brings us to the second section of De Candolle's article, in which he passes on, from the observation of the present forms and affinities of Cupuliferous plants, to the consideration of their probable history and origin. Suffice it to say, that he frankly accepts the inferences derived from the whole course of observation, and even contemplates with satisfaction a probable historical connection between congeneric species. He accepts and, by various considerations drawn from the geographical distribution of European *Cupuliferæ*, fortifies the conclusion — long ago arrived at by Edward Forbes — that the present species, and even some of their varieties, date back to about the close of the Tertiary epoch, since which time they have been subject to frequent and great changes of habitation or limitation, but without appreciable change of specific form or character; that is, without profounder changes than those within which a species at the present time is known to vary. Moreover, he is careful to state that he is far from concluding that the time of the appearance of a species in Europe at all indicates the time of its origin. Looking back still further into the Tertiary epoch, of which the vegetable remains indicate many analogous, but few, if any, identical forms, he concludes, with Heer and others, that specific changes of form, as well as changes of station, are to be presumed. And finally, that " the theory of a succession of forms through the deviation of anterior forms is the most natural hypothesis, and the most accordant with the known facts in palæontology, geographical botany and zoölogy, of anatomical structure and classification: but direct proof of it is wanting, and moreover, if true, it must have taken place very slowly; so slowly indeed, that its effects are discernible only after a lapse of time far longer than our historic epoch."

In contemplating the present state of the species of *Cupuliferæ* in Europe, De Candolle comes to the conclusion that, while the Beech is increasing, and extending its limits southward and westward (at the expense of *Coniferæ* and Birches), the Common Oak, to some extent, and the Turkey Oak decidedly, are diminishing and retreating, and this wholly irrespective of man's agency. This is inferred of the Turkey Oak from the great gaps found in its present geographical area, which are otherwise inexplicable, and which he regards as plain indications of a partial extinction. Community of descent of all the individuals of species is of course implied in these and all similar reasonings.

An obvious result of such partial extinction is clearly enough brought to view. The European Oaks (like the American species) greatly tend to vary, — that is, they manifest an active disposition to produce new forms. Every form tends to become hereditary, and so to pass from the state of mere variation to that of race; and of these competing incipient races some only will survive. *Quercus Robur* offers a familiar illustration of the manner in which one form may in the course of time become separated into two or more distinct ones.

To Linnæus this Common Oak of Europe was all of one species. But of late years the greater number of European botanists have regarded it as including three species, *Q. pedunculata*, *Q. sessiliflora*, and *Q. pubescens*. De Candolle looks with satisfaction to the independent conclusion which he reached from a long and patient study of the forms (and which Webb, Gay, Bentham and others had equally reached), that the view of Linnæus was correct, inasmuch as it goes to show that the idea and the practical application of the term species have remained unchanged during the century which has elapsed since the publication of the "Species Plantarum." But the idea remaining unchanged, the facts might appear under a different aspect, and the conclusion be different, under a slight and very supposable change of circumstances. Of the twenty-eight spontaneous varieties of *Q. Robur*, which De Candolle recognizes, all but six, he remarks, fall naturally under the three sub-species, *pedunculata*, *sessiliflora*, and *pubescens*, and are therefore forms grouped around these as centres; and, moreover, the few connecting forms are by no means the most common. Were these to die out, it is clear that the three forms which have already been so frequently taken for species, would be what the group of four or five provisionally admitted species which closely surround *Q. Robur* (see p. 435) now are. The best example of such a case, as having in all probability occurred, through geographical segregation and partial extinction, is that of the Cedar, thus separated into the Deodar, the Lebanon, and the Atlantic Cedars, — a case admirably worked out by Dr. Hooker two or three years ago.[1]

A special advantage of the *Cupuliferæ* for determining the probable antiquity of existing species in Europe, De Candolle finds in the size and character of their fruits. However it may be with other plants (and he comes to the conclusion generally that marine currents and all other means of distant transport have played only a very small part in the actual dispersion of species), the transport of acorns and chestnuts by natural causes across an arm of the sea in a condition to germinate, and much more the spontaneous establishment of a forest of Oaks or Chestnuts in this way, De Candolle conceives to be fairly impossible in itself, and contrary to all experience. From such considerations, *i. e.*, from the actual dispersion of the existing species, with occasional aid from Post-tertiary deposits, it is thought to be shown that the principal *Cupuliferæ* of the Old World attained their actual extension before the present separation of Sicily, Sardinia and Corsica, or of Britain, from the European continent.

[1] Natural History Review, Jan., 1862.

This view once adopted, and this course once entered upon, has to be pursued farther. *Quercus Robur* of Europe with its bevy of admitted derivatives, and its attending species only provisionally admitted to that rank, is very closely related to certain species of eastern Asia, and of Oregon and California, — so closely that "a view of the specimens by no means forbids the idea that they have all originated from *Q. Robur*, or have originated, with the latter, from one or more preceding forms so like the present ones that a naturalist could hardly know whether to call them species or varieties." Moreover, there are fossil leaves from diluvian deposits in Italy, figured by Gaudin, which are hardly distinguishable from those of *Q. Robur* on the one hand, and from those of *Q. Douglasii*, etc., of California on the other. No such leaves are found in any Tertiary deposit in Europe; but such are found of that age, it appears, in northwest America, where their remote descendants still flourish. So that the probable genealogy of *Q. Robur*, traceable in Europe up to the commencement of the present epoch, looks eastward and far into the past on far distant shores.

Q. Ilex, the Evergreen Oak of southern Europe and northern Africa, reveals a similar archæology; but its presence in Algeria leads De Candolle to regard it as a much more ancient denizen of Europe than *Q. Robur;* and a Tertiary Oak, *Q. ilicoides*, from a very old Miocene bed in Switzerland, is thought to be one of its ancestral forms. This high antiquity once established, it follows almost of course that the very nearly related species in central Asia, in Japan, in California, and even our own Live Oak with its Mexican relatives, may probably enough be regarded as early offshoots from the same stock with *Q. Ilex*.

In brief, — not to continue these abstracts and remarks, and without reference to Darwin's particular theory (which De Candolle at the close very fairly considers), — if existing species, or many of them, are as ancient as they are now generally thought to be, and were subject to the physical and geographical changes (among them the coming and the going of the Glacial epoch) which this antiquity implies; if in former times they were as liable to variation as they now are; and if the individuals of the same species may claim a common local origin, — then we cannot wonder that "the theory of a succession of forms by deviations of anterior forms" should be regarded as "the most natural hypothesis," nor at the general advance made towards its acceptance in some form or other.

The question being, not, how plants and animals originated, but, how came the existing animals and plants to be just where they are and what they are; it is plain that naturalists interested in such inquiries are mostly looking for the answer in one direction. The general drift of opinion, or at least of expectation, is exemplified by this essay of De Candolle; and the set and force of the current are seen by noticing how it carries along naturalists of widely different views and prepossessions, — some faster and farther than others, — but all in one way. The tendency is, we may say, to extend the law of continuity, or something analogous to it, from inorganic to organic nature, and in the latter to connect the present with the past in some sort of material connection. The generalization may indeed be expressed so as not to assert that the connection is genetic, as in Mr. Wallace's formula: "Every species has come into existence coincident both in time and space with preëxisting closely allied species." Edward Forbes, who may be called the originator of this whole line of inquiry, long ago expressed a similar view. But the only material sequence we know, or can clearly conceive, in plants and animals, is that from parent to progeny; and, as De Candolle implies, the origin of species and that of races can hardly be much unlike, nor governed by other than the same laws, whatever these may be.

The progress of opinion upon this subject in one generation is not badly represented by that of De Candolle himself, who is by no means prone to adopt new views without much consideration. In an elementary treatise published in the year 1835, he adopted, and, if we rightly remember, vigorously maintained, Schouw's idea of the double or multiple origin of species, at least of some species, — a view which has been carried out to its ultimate development only perhaps by Agassiz, in the denial of any necessary genetic connection among the individuals of the same species, or of any original localization more restricted than the area now occupied by the species. But in 1855, in his "Géographie Botanique," the multiple hypothesis, although in principle not abandoned, is seen to lose its point, in view of the probable high antiquity of existing species. The actual vegetation of the world being now regarded as a continuation, through numerous geological, geographical, and more recently historical, changes, of anterior vegetations, the actual distribution of plants is seen to be a consequence of preceding conditions and geological considerations, and these alone may be expected to explain all the facts, many of them so curious and extraordinary, of the actual geographical distribution of the species. In the present essay, not only the distribution but the origin of congeneric species is regarded as something derivative; whether derived by slow and very gradual changes in the course of ages, according to Darwin, or by a sudden, inexplicable change of their Tertiary ancestors, as conceived by Heer, De Candolle hazards no opinion. It may, however, be inferred that he looks upon "natural selection" (which he rather underrates) as a real, but insufficient, cause; while some curious remarks (pp. 57, 58), upon the number of monstrosities annually produced, and the possibility of their enduring, may be regarded as favorable to Heer's view.

As an index to the progress of opinion in the direction referred to, it will be interesting to compare Sir Charles Lyell's well-known chapters of twenty or thirty years ago, in which the permanence of species was ably maintained, with his treatment of the same subject in a work just issued in England, which, however, has not yet reached us.

A belief in the derivation of species may be maintained along with a conviction of great persistence of specific characters. This is the idea of the excellent Swiss vegetable palæontologist Heer, who imagines a sudden change of specific type at certain periods, and perhaps is that of Pictet. Falconer adheres to somewhat similar views in his elaborate paper on Elephants, living and fossil, in the "Natural History Review" for January last. Noting that "there is clear evidence of the true Mammoth having existed in America long after the period of the northern drift, when the surface of the country had settled down into its present form," and also in Europe so late as to have been a contemporary of the Irish Elk, and on the other hand that it existed in England so far back as before the deposition of the boulder Clay; also that four well-defined species of fossil Elephant are known to have existed in Europe; that "a vast number of the remains of three of these species have been exhumed over a large area in Europe; and, even in the geological sense, an enormous interval of time has elapsed between the formation of the most ancient and the most recent of these deposits, quite sufficient to test the persistence of specific characters in an Elephant," he presents the question: "Do then the successive Elephants occurring in these strata show any signs of a passage from the older form into the newer?"

To which the reply is: "If there is one fact which is impressed on the conviction of the observer with more force than any other, it is the persistence and uniformity of the characters of the molar teeth in the earliest known Mammoth and his most modern successor. . . . Assuming the observation to be correct, what strong proof does it not afford of the persistence and constancy, throughout vast intervals of time, of the distinctive characters of those organs which are most concerned in the existence and habits of the species? If we cast a glance back on the long vista of physical changes which our planet has undergone since the Neozoic Epoch, we can nowhere detect signs of a revolution more sudden and pronounced, or more important in its results, than the intercalation and sudden disappearance of the glacial period. Yet the 'dicyclotherian' Mammoth lived before it, and passed through the ordeal of all the hard extremities it involved, bearing his organs of locomotion and digestion all but unchanged. Taking the group of four European fossil species above enumerated, do they show any signs in the successive deposits of a transition from the one form into the other? Here again the result of my observation, in so far as it has extended over the European area, is, that the specific characters of the molars are constant in each, within a moderate range of variation, and that we nowhere meet with intermediate forms." . . . Dr. Falconer continues (p. 80): —

"The inferences which I draw from these facts are not opposed to one of the leading propositions of Darwin's theory. With him, I have no faith in the opinion that the Mammoth and other extinct Elephants made their appearance suddenly, after the type in which their fossil remains are presented to us. The most rational view seems to be, that they are in some shape the modified descendants of earlier progenitors. But if the asserted facts be correct, they seem clearly to indicate that the older elephants of Europe, such as *E. meridionalis* and *E. antiquus*, were not the stocks from which the later species, *E. primigenius* and *E. Africanus*, sprung, and that we must look elsewhere for their origin. The nearest affinity, and that a very close one, of the European *E. meridionalis* is with the Miocene *E. planifrons* of India; and of *E. primigenius*, with the existing India species.

"Another reflexion is equally strong in my mind, — that the means which have been adduced to explain the origin of the species by 'Natural Selection,' or a process of variation from external influences, are inadequate to account for the phenomena. The law of phyllotaxis, which governs the evolution of leaves around the axis of a plant, is as nearly constant in its manifestation as any of the physical laws connected with the material world. Each instance, however different from another, can be shown to be a term of some series of continued fractions. When this is coupled with the geometrical law governing the evolution of form, so manifest in some departments of the animal kingdom, *e. g.*, the spiral shells of the Mollusca, it is difficult to believe that there is not, in nature, a deeper-seated and innate principle, to the operation of which Natural Selection is merely an adjunct. The whole range of the Mammalia, fossil and recent, cannot furnish a species which has had a wider geographical distribution, and passed through a longer term of time, and through more extreme changes of climatal conditions, than the Mammoth. If species are so unstable, and so susceptible of mutation through such influences, why does that extinct form stand out so signally a monument of stability? By his admirable researches and earnest writings, Darwin has, beyond all his cotemporaries, given an impulse to the philosophical investigation of the most backward and obscure branch of the biological sciences of his day; he has laid the foundations of a great edifice; but he need not be surprised, if, in the progress of erection, the superstructure is altered by his successors, like the Duomo of Milan from the Roman to a different style of architecture."

Entertaining ourselves the opinion that something more than natural selection is requisite to account for the orderly production and succession of species, we offer two incidental remarks upon the above extract.

First, we find in it — in the phrase "Natural Selection, or a process of variation from external influences" — an example of the very common confusion of two distinct things, namely, variation and natural selection. The former has never yet been shown to have its cause in "external influences," nor to occur at random. As we have elsewhere insisted, if not inexplicable, it has never been explained; all we can yet say is, that plants and animals are prone to vary, and that some conditions favor variation. Perhaps in this Dr. Falconer may yet find what he seeks: for "it is difficult to believe that there is not in [its] nature, a deeper-seated and innate principle, to the operation of which Natural Selection is merely an adjunct." The latter, which is the *ensemble* of the external influences, including the competition of the individuals themselves, picks out certain variations as they arise, but in no proper sense can be said to originate them.

Secondly, although we are not quite sure how Dr. Falconer intends to apply the law of phyllotaxis to illustrate his idea, we fancy that a pertinent illustration may be drawn from it, in this way. There are two species of phyllotaxis, perfectly distinct, and, we suppose, not mathematically reducible the one to the other, — namely, 1, that of alternate leaves, with its varieties; and 2, that of verticillate leaves, of which opposite leaves present the simplest case. That, although generally constant, a change from one variety of alternate phyllotaxis to another should occur on the same axis, or on successive axes, is not surprising, the different sorts being terms of a regular series, — although indeed we have not the least idea as to how the change from the one to the other comes to pass. But it is interesting, and in this connection perhaps instructive, to remark that, while some dicotyledonous plants hold to the verticillate, *i. e.*, opposite-leaved phyllotaxis throughout, a larger number — through the operation of some deep-seated and innate principle, which we cannot fathom — change abruptly into the other species at the second or third node, and change back again in the flower, or else effect a synthesis of the two species in a manner which is puzzling to understand. Here is a change from one fixed law to another, as unaccountable, if not as great, as from one specific form to another.

An elaborate paper on the vegetation of the Tertiary period in the southeast of France, by Count Gaston de Saporta, published in "Annales des Sciences Naturelles," xvi. pp. 309–344, — which we have not space to analyze, — is worthy of attention from the general inquirer, on account of its analysis of the Tertiary flora into its separate types, Cretaceous, Austral, Tropical, and Boreal, each of which has its separate and different history, — and for the announcement that "the *hiatus*, which, in the idea of most geologists, intervened between the close of the Cretaceous and the beginning of the Tertiary, appears to have had no existence, so far as concerns the vegetation; that in general it was not by means of a total overthrow, followed by a complete new emission of species, that the flora has been renewed at each successive period; and that while the plants of southern Europe inherited from the Cretaceous period more or less rapidly disappeared, as also the austral forms, and later the tropical types (except the Laurel, the Myrtle, and the *Chamærops humilis*), the boreal types, coming later, survived all the others, and now compose, either in Europe, or in the north of Asia, or in North America, the basis of the actual arborescent vegetation." Especially "a very considerable number of forms nearly identical with Tertiary forms now exist in America, where they have found, more easily than in our [European] soil, — less vast and less extended southward, — refuge from ulterior revolutions." The extinction of species is attributed to two kinds of causes: the one material or physical, whether slow or rapid; the other inherent in the nature of organic beings, incessant, but slow, in a manner latent, but somehow assigning to the species, as to the individuals, a limited period of existence, and, in some equally mysterious but wholly natural way, connected with the development of organic types: — "By type meaning a collection of vegetable forms constructed upon the same plan of organization, of which they reproduce the essential lineaments with certain secondary modifications, and which appear to run back to a common point of departure."

In this community of types, no less than in the community of certain existing species, Saporta recognizes a prolonged material union between North America and Europe in former times. Most naturalists and geologists reason in the same way, — some more cautiously than others, — yet perhaps most of them seem not to perceive how far such inferences imply the doctrine of the common origin of related species.

For obvious reasons such doctrines are likely to find more favor with botanists than with zoölogists. But with both the advance in this direction is seen to have been rapid and great; yet to us not unexpected. We note, also, an evident disposition, notwithstanding some endeavors to the contrary, to allow derivative hypotheses to stand or fall upon their own merits, — to have indeed upon philosophical grounds certain presumptions in their favor, — and to be, perhaps, quite as capable of being turned to good account as to bad account in natural theology.[1]

[1] What the Rev. Principal Tulloch remarks in respect to the philosophy of miracles has a pertinent application here. We quote at second hand: "The stoutest advocates of interference can mean nothing more than that the Supreme Will has so moved the hidden springs of nature that a new issue arises on given circumstances. The ordinary issue is supplanted by a higher issue. The essential facts before us are a certain set of phenomena, and a Higher Will moving them. How moving them? is a question for human definition; the answer to which does not and cannot affect the divine meaning of the change. Yet when we reflect that this Higher Will is everywhere reason and wisdom, it seems a juster as well as a more comprehensive view to regard it as operating by subordination and evolution, rather than by interference or violation."

Among the leading naturalists, indeed, such views — taken in the widest sense — have one and, so far as we are now aware, only one thorough-going and thoroughly consistent opponent, namely, Mr. Agassiz.

Most naturalists take into their very conception of a species, explicitly or by implication, the notion of a material connection resulting from the descent of the individuals composing it from a common stock, of local origin. Mr. Agassiz wholly eliminates community of descent from his idea of species, and even conceives a species to have been as numerous in individuals, and as widespread over space, or as segregated in discontinuous spaces, from the first as at a later period.

The station which it inhabits, therefore, is with other naturalists in nowise essential to the species, and may not have been the region of its origin. In Mr. Agassiz's view the habitat is supposed to mark the origin, and to be a part of the character of the species. The habitat is not merely the place where it is, but a part of what it is.

Most naturalists recognize varieties of species; and many, like De Candolle, have come to conclude that varieties of the highest grade, or races, so far partake of the characteristics of species, and are so far governed by the same laws, that it is often very difficult to draw a clear and certain distinction between the two. Mr. Agassiz will not allow that varieties or races exist in nature, apart from man's agency.

Most naturalists believe that the origin of species is supernatural, their dispersion or particular geographical area, natural, and their extinction, when they disappear, also the result of physical causes. In the view of Mr. Agassiz, if rightly understood, all three are equally independent of physical cause and effect, are equally supernatural.

In comparing preceding periods with the present and with each other, most naturalists and palæontologists now appear to recognize a certain number of species as having survived from one epoch to the next, or even through more than one formation, especially from the Tertiary into the Post-tertiary period, and from that to the present age. Mr. Agassiz is understood to believe in total extinctions and total new creations at each successive epoch, and even to recognize no existing species as ever contemporary with extinct ones, except in the case of recent exterminations.

These peculiar views, if sustained, will effectually dispose of every form of derivative hypothesis.

Returning for a moment to De Candolle's article, we are disposed to notice his criticism of Linnæus's "definition" of the term species ("Philosophia Botanica," No. 157): "Species tot numeramus quot diversæ formæ in principio sunt creatæ," — which he declares illogical, inapplicable, and the worst that has been propounded. "So, to determine if a form is specific, it is necessary to go back to its origin, which is impossible. A definition by a character which can never be verified is no definition at all."

Now, as Linnæus practically applied the idea of species with a sagacity which has never been surpassed, and rarely equalled, and indeed may be said to have fixed its received meaning in natural history, it may well be inferred that in the phrase above cited he did not so much undertake to frame a logical definition, as to set forth the idea which, in his opinion, lay at the foundation of species. On which basis A. L. Jussieu did construct a logical definition: "nunc rectius definitur perennis individuorum similium successio continuata generatione renascentium." The fundamental idea of species, we would still maintain, is that of a chain, of which genetically-connected individuals are the links. That, in the practical recognition of species, the essential characteristic has to be inferred, is no great objection, — the general fact that like engenders like being an induction from a vast number of instances, and the only assumption being that of the uniformity of nature. The idea of gravitation, that of the atomic constitution of matter, and the like, equally have to be verified inferentially. If we still hold to the idea of Linnæus, and of Agassiz, that existing species were created independently, and essentially all at once at the beginning of the present era, we could not better the propositions of Linnæus and of Jussieu. If, on the other hand, the time has come in which we may accept, with De Candolle, their successive origination, at the commencement of the present era or before, and even by derivation from other forms, then the "in principio" of Linnæus will refer to that time, whenever it was, and his proposition be as sound and wise as ever.

In his "Géographie Botanique" (ii. pp. 1068–1077) De Candolle discusses this subject at length, and in the same interest. Remarking that of the two great facts of species, namely, likeness among the individuals, and genealogical connection, zoölogists have generally preferred the latter,[1] while botanists have been divided in opinion, he pronounces for the former as the essential thing, in the following argumentative statement: —

"Quant à moi, j'ai été conduit, dans ma définition de l'espèce, à mettre décidément la ressemblance au-dessus des caractères de succession. Ce n'est pas seulement à cause des circonstances propres au règne végétal, dont je m'occupe exclusivement ; ce n'est pas non plus afin de sortir ma définition des théories et de la rendre le plus possible utile aux naturalistes descripteurs et nomenclateurs, c'est aussi par un motif philosophique. En toute chose il faut aller au fond des questions, quand on le peut. Or, pourquoi la reproduction est-elle possible, habituelle, féconde indéfiniment, entre des êtres organisés que nous dirons de la même espèce ? Parce qu'ils se ressemblent et uniquement à cause de cela. Lorsque deux espèces ne peuvent, ou, s'il s'agit d'animaux supérieurs, ne peuvent et ne veulent se croiser, c'est qu'elles sont très differentes. Si l'on obtient des croisements, c'est que les individus sont analogues ; si ces croisements donnent des produits féconds, c'est que les individus êtaient plus analogues ; si ces produits eux-mêmes sont féconds, c'est que la ressemblance était plus grande ; s'ils sont fécond habituellement et indéfiniment, c'est que la ressemblance intérieure et extérieure était très grande. Ainsi le degré de ressemblance est le fond ; la reproduction en est seulement la manifestation et la mesure, et il est logique de placer la cause au-dessus de l'effet."

We are not at all convinced. We still hold that genealogical connection, rather than mutual resemblance, is the fundamental thing, — first on the ground of fact, and then from the

[1] Particularly citing Flourens : "La ressemblance n'est qu'une condition secondaire ; la condition essentielle est la descendance : ce n'est pas la ressemblance, c'est la succession des individus, qui fait l'espèce."

philosophy of the case. Practically, no botanist can say what amount of dissimilarity is compatible with unity of species; in wild plants it is sometimes very great, in cultivated races, often enormous. De Candolle himself informs us that the different variations which the same Oak tree exhibits are significant indications of a disposition to set up separate varieties, which, becoming hereditary, may constitute a race; he evidently looks upon the extreme forms, say of *Quercus Robur*, as having thus originated; and on this ground, inferred from transitional forms, and not from their mutual resemblance, as we suppose, he includes them in that species. This will be more apparent should the discovery of the transitions, which he leads us to expect, hereafter cause the four provisional species which attend *Q. Robur* to be merged in that species. It may rightly be replied that this conclusion would be arrived at from the likeness step by step in the series of forms; but the cause of the likeness here is obvious. And this brings in our "motif philosophique."

Not to insist that the likeness is after all the variable, not the constant, element, — to learn which is the essential thing, resemblance among the individuals or their genetic connection, we have only to ask which can be the cause of the other.

In hermaphrodite plants (the normal case), and even as the question is ingeniously put by De Candolle in the above extract, the former surely cannot be the cause of the latter, though it may, in case of crossing, offer occasion. But, on the ground of the most fundamental of all things in the constitution of plants and animals, "the fact incapable of farther analysis, that individuals reproduce their like, that characteristics are inheritable," the likeness is a direct natural consequence of the genetic succession, — and it is logical to place the cause above the effect.

We are equally disposed to combat a proposition of De Candolle's about genera, elaborately argued in the "Géographie Botanique," and incidentally re-affirmed in his present article, namely, that genera are more natural than species, and are more correctly distinguished by people in general, as is shown by vernacular names. But we have no space left in which to present some evidence to the contrary.

Here we must abruptly close our long exposition of a paper which, from the scientific position, ability, and impartiality of its author, is likely at this time to produce a marked impression. We would also direct attention to an earlier article in the same important periodical (namely, in the "Bibliothèque Universelle" for May, 1862), on the European Flora and the Configuration of Continents in the Tertiary Epoch, a most interesting abstract of, and commentary on, the introductory part of Heer's "Flora Tertiaria Helvetiæ," as reëdited and translated into the French by Gaudin, with additions by the author.

DR. HOOKER ON WELWITSCHIA.

THIS is a separate issue, in folio form, of a memoir[1] in the current (24) volume of the Transactions of the Linnæan Society of London, illustrated by fourteen superb and elaborate plates, the expense of which has been mainly defrayed by the Royal Society, from a parliamentary fund placed at its disposal for the promotion of scientific research. By the co-operation of these two learned societies, the fruits of Dr. Hooker's admirable researches are given to the scientific world in a form and manner worthy of them and of the wonderful subject.

A good idea of the vegetable wonder in question is given in the following brief account of its appearance and prominent characters, drawn partly from the descriptions of its discoverer, and partly from specimens sent to England: —

"The Welwitschia is a woody plant, said to attain a century in duration, with an obconic trunk about two feet long, of which a few inches rise above the soil, presenting the appearance of a flat, two-lobed depressed mass, sometimes (according to Dr. Welwitsch) attaining fourteen feet in circumference (!), and looking like a round table. When full grown it is dark brown, hard, and cracked over the whole surface (much like

[1] *On the Welwitschia*, a new genus of *Gnetaceæ*. By Joseph Dalton Hooker. (American Journal of Science and Arts, 2 ser., xxxvi. 434.)

the burnt crust of a loaf of bread) ; the lower portion forms a stout tap-root, buried in the soil, and branching downward at the end. From deep grooves in the circumference of the depressed mass two enormous leaves are given off, each six feet long when full grown, one corresponding to each lobe; these are quite flat, linear, very leathery, and are split to the base into innumerable thongs that lie curling upon the surface of the soil. Its discoverer describes these same two leaves as being present from the earliest condition of the plant, and assures me that they are in fact developed from the two cotyledons of the seed, and are persistent, being replaced by no others. From the circumference of the tabular mass, above, but close to the insertion of the leaves, spring stout, dichotomously branched cymes, nearly a foot high, bearing small erect scarlet cones, which eventually become oblong, and attain the size of those of the common Spruce Fir. The scales of the cones are very closely imbricated, and contain, when young and still very small, solitary flowers, which in some cones are hermaphrodite (structurally but not functionally), in others female. The hermaphrodite flower consists of a perianth of four pieces, six monadelphous stamens, with globose trilocular anthers, surrounding a central ovule, the integument of which is produced into a styliform sigmoid tube, terminated by a discoid apex. The female flower consists of a solitary erect ovule, contained in a compressed utricular perianth. The mature cone is tetragonous, and contains a broadly winged fruit in each scale. Its discoverer observes that the whole plant exudes a resin, and that it is called *Tumbo* by the natives, — whence he suggests that it may bear the generic name of *Tumboa;* but this he withdrew at my suggestion, for reasons which I shall presently give. It inhabits the elevated sandy plateau near Cape Negro (lat. 15° 40′ S.) on the southwest coast of Africa."

Welwitschia mirabilis, Hook. fil., was also detected and made known — indeed the first actual materials, with a drawing of the plant, were sent to England — by Mr. Baines from the Damara country, in lat. 24° or 24° S., about 500 miles south of Cape Negro. Mr. Baines is an artist, and his original colored sketch of a plant in fruit is reproduced on the first plate of the memoir. It appears as if five-leaved; but probably one of the two original leaves is split in two, and the other into three segments. As might be inferred from the form and structure, the Welwitschia inhabits a dry region. Mr. Monteiro writes to Dr. Hooker: —

" . . . About thirty miles distant from the coast, I passed a plain about three miles across, on which this plant was growing abundantly; that is to say, I saw about thirty specimens on my line of march. The plain was perfectly dry, and bare of other vegetation than the Welwitschia and a little short grass. The ground was a hard quartzose schist. The Welwitschia was generally growing near the little ruts worn in the plain by running water during the rainy season."

And from Damara Land, Mr. Anderson writes that, —

" Rain rarely or never falls where this plant exists. (Yet the night dews are heavy, as other authorities mention.) I have crossed and recrossed Damara Land throughout its entire length and breadth, but only found the plant growing on that desperately arid flat, stretching far and wide about Walvisch Bay."

We are familiar with plants of very diverse orders of Dicotyledons and Monocotyledons which are adapted to arid regions by great restriction of surface. Here a similar plan is adopted by a Gymnosperm; for the resemblance to *Coniferæ* and *Casuarineæ* indicated by Dr. Welwitsch is shown by Dr. Hooker to import a close affinity, the author referring the plant to *Gnetaceæ* near to Ephedra. Its permanently abbreviated ascending axis — of which the greater part consists of the first internode, below the cotyledons — increases in thickness but hardly in length, develops no other than the seminal pair of leaves, above which the disciform bilobate axile portion or "crown," gradually produced, bears year by year only leafless inflorescence.

Hæmanthus equally bears a pair of leaves; but these die off as the season of drought advances, when the plant is reduced to a minimum of surface in its spherical bulb, — which outspreads a new pair of leaves when the rainy season returns. But in Welwitschia the two leaves are permanent. Wherefore they are firm and coriaceous, and, increasing by basal growth from year to year, the older parts doubtless become inactive at length, while fresh surface below is annually renewed, under the shelter (as Dr. Hooker describes) afforded by the deep grooves which in old plants separate the growth of the hypocotyledonary stock from that of the crown above, and is filled by the tender growing bases of the leaves.

Having given a detailed generic character of Welwitschia and a comparative view of the Gnetaceous genera, now three in number, Dr. Hooker proceeds to describe at length the trunk, leaves, inflorescence, flowers, fertilization, embryogeny, and seeds of this curious subject, — comparing it, in the latter respects, with *Coniferæ* and *Cycadaceæ* on the one hand, and with Santalum and Loranthus on the other, and closing with a general summary of the results.

An abstract or analysis of this most important paper is beyond our present reach and space. But we may refer to some of the special points.

The most obvious peculiarity of Welwitschia is, that "it appears to be the only perennial flowering plant which at no period has other vegetative organs than those proper to the embryo itself, — the main axis being represented by the radicle, which becomes a gigantic caulicle, and develops a root from its base and inflorescences from its plumulary end, and the leaves being the two cotyledons in a very highly developed and specialized condition." It is an excellent case, accordingly, if any such were still needed for showing the nature of the radicle as stem, or ascending axis (not root), — a view which we supposed observation had long ago demonstrated. Dr. Hooker, in a note, refers to this view as expressed by Adr. de Jussieu in his "Cours Elémentaire" (which appeared in 1843 and 1844), and in Gray's Introduction to Botany ("Botanical Text Book"), 1858. But the same view is taken in all the earlier editions of the latter work; even in the first (1842) the radicle is spoken of as the first internode of the stem (p. 28, note); and probably the idea will be found distinctly expressed in works of an earlier date. Dr. Hooker, in the note referred to, assents to the proposition that "the radicle is rightly regarded as an axis," *i. e.*, an ascending axis, "and not a root," but does not agree that it is an internode. To us, the one implies the other. Conceiving, as we do, the fundamental idea of the morphology of the phænogamous plant to be, that the ascending axis consists of a series of superposed internodes, each crowned by a leaf-bearing point or ring (the node), the first internode must needs be that which is crowned by the first leaf or pair of leaves, the cotyledons; and its whole development confirms this view.

Dr. Hooker notes the curious fact that in Welwitschia flower-buds are occasionally produced on the stock below the insertion of the leaves, that is on the radicle or caulicle itself; and Dr. Masters pointed out to him analogous cases of shoots thus originating, one of which was described by Bernhardi thirty years ago. It is simply the case of adventitious buds; these might seem as likely to occur on the first internode as on any later one.

Welwitschia, having a dicotyledonous embryo, has also essentially an exogenous stem, *i. e.*, "the vascular system is referable to the exogenous plan, but its arrangement into concentric wood wedges is very rude." But the superadded isolated and closed vascular bundles of the stock and root, and especially their losing themselves in the periphery of the stock, are endogenous analogies. So also is the strictly parallel and free venation of the leaves; yet, as there are no cross veinlets, thus favoring the splitting up of the leaf into laciniæ, this looks as much or more towards *Cycadaceæ* and broad-leaved *Coniferæ*.

The total absence of anastomosing veinlets in the leaf, each nerve representing a single and independent vascular axis, extending, in Welwitschia, from the axis of the trunk to the apex of the leaf, causes such leaves as these and those of Dammara, etc., to "resemble more closely a series of parallel uninerved leaves united by cellular tissue, than a foliar expansion of parenchyma traversed by one system of inosculating vessels, and the frequent presence of many linear cotyledons in these plants seems to favor this view, as does the mixed character of the foliage of Podocarpus, of which some species have uninerved and others many-nerved leaves. The numerous flower-buds along the periphery of the crown also further favor this view." That is, in Welwitschia, where this ingenious surmise carries a plausibility, which it does not when applied to Podocarpus.

The binary symmetry of Welwitschia, beginning with the cotyledons, is carried through the inflorescence up to the decussating pairs of bracts of the cones and the two leaflets in each whorl of the hermaphrodite perianth. But the stamens are six, at first sight a monocotyledonous analogy; yet they may be regarded three sets of two, notwithstanding their monadelphy. The flowers are diœcio-polygamous, *i. e.*, of two sorts, one female, the other structurally hermaphrodite, but the gynæcium sterile, though well-developed, except that no embryo-sac appears. The hermaphrodite cones and their flowers accord in many respects strikingly with the male cones of Ephedra; but the anthers are trilocular, which is remarkable. The simple ellipsoidal pollen is the same in both. In Ephedra the stamens vary from two to eight, and the column is solid, there being no rudiment of a gynæcium.

The female fruitful cones are about three inches long, and bright red when fresh.

The integument of the ovule, as in Gnetum, is prolonged at the summit into a style-like body, thus closely simulating a pistil; and the apex of this styliform tube, which is thin and merely erose in the fertile flower, in the structurally hermaphrodite flower is dilated into a broad papillose disk, exactly imitating a highly developed stigma — a marked instance of a highly developed organ which is functionless; for no pollen has been detected upon it, and no embryo-sac in the nucleus. Here Dr. Hooker speculates upon "the possibility of Welwitschia being the only known representative of an existing or extinct race of plants, in which such a stigma-like organ was really capable of performing the function of a stigma. And, when we see this organ occurring in a hermaphrodite flower, it is easy to suppose that we have in Welwitschia a transition in function, as well as in structure, between the gymnospermous and angiospermous Dicotyledons; and that the ideal race consisted of hermaphrodite plants, in which the office of the stigma of the carpellary leaf was performed by a stigmatic dilatation of the ovular coat itself."

This assumes that the gymnospermous theory established by Brown is correct (whatever be the nature of the cone-scales, rameal or carpellary, simple or compound), and applicable to the Gnetaceous as well as to the Coniferous type. This view, lately much questioned, Dr. Hooker maintains, and enforces, as respects *Gnetaceæ*, by very convincing and in part wholly original arguments, drawn from his own researches upon the present plant and its allies. We refer to pp. 28 to 31, which we could not readily condense, and have not room to copy. The same is to be said in regard to the resemblances or analogies in gynæcial structure between *Gnetaceæ* and *Loranthaceæ*, etc., — a subject upon which we await expectantly Professor Oliver's investigations. Moreover, as Dr. Hooker remarks upon another page, the decisive or final comparative view of the structures in question cannot be had until the homology of the ovule itself is settled. In cases where the flower is so simplified that the nucleus of an ovule directly terminates the floriferous axis, and is surrounded by few and simple, or peculiarly specialized, investments, the discrimination of these must be difficult enough, and must ultimately depend upon the theory adopted as to the nature of the ovular coats. If these be regarded as foliar (as a rigid application of adopted morphological principles will require), then a complete transition from gymnospermy to angiospermy is probable, and may be expected to be demonstrable.

The fertilization and embryology of Welwitschia have been wonderfully worked out, considering the materials, by Dr. Hooker, and the two most elaborate and valuable plates of the memoir are filled with the details. Suffice it to say, that it appears that the pollen must be brought by insects to the ovule of the female flowers, at an early period, before the nucleus is covered by the ovular coat or by the perianth, and before the former has produced its styliform apex, down which it would be nearly impossible to convey the grains of pollen which were bodily found on the nucleus, with their tubes there produced. So that, notwithstanding the carpel-like form of the ovule, the impregnation is absolutely gymnospermous. As to embryo-formation also, "there is a general agreement in many most essential particulars with *Cycadaceæ* and *Coniferæ*," especially, and beyond what has already been adverted to, in "the free embryo-sac being filled with endosperm-cells previous to fertilization, the numerous secondary embryo-sacs, the position of the germinal vesicle at the base of these sacs, and in the high development of the long tortuous suspensor." There is an agreement with Angiosperms, however, in several particulars, especially in that of "the germinal vesicle giving rise to one embryo only." And it is concluded that, in special reference to Sentalum and Loranthus, "Welwitschia presents an embryogenic process intermediate between that of Gymnosperms and Angiosperms."

And here we should not omit to mention that its wood differs from that of all known Gymnosperms in wanting the disc-bearing wood-cells!

It will be conceded that Welwitschia is "the most wonderful discovery, in a botanical point of view, that has been brought to light during the present century." Also, that Dr. Hooker has enjoyed (and improved) an opportunity unequalled by any botanist since that which placed the Rafflesia in Mr. Brown's hands.

DARWIN'S MOVEMENTS AND HABITS OF CLIMBING PLANTS.

THIS is a long paper [1] read before the Linnæan Society in February last. The investigations which it records were made, we believe, during a period when the author's ordinary

[1] *On the Movements and Habits of Climbing Plants.* By Charles Darwin. Journal Linnæan Society of London, ix. London, 1867. (American Journal of Science and Arts, 2 ser., xl. 273 ; xli. 125.)

scientific labors were interrupted by illness, — as was no less the case with respect to his former papers on Dimorphous and Trimorphous Flowers and his volume on the Fertilization of Orchids by the aid of Insects. Of these works and of the present, — side issues as they are, — it may fairly be said, that they show a genius for biological investigation, and a power of turning common materials and ordinary observations to high scientific account, which, if equalled, have not been surpassed since the days of Hunter and Charles Bell. This will be the opinion equally, we suppose, of those who favor and of those who dislike Mr. Darwin's theory of the gradual transformation of specific forms through natural selection, upon which, indeed, all these collateral researches have a bearing, direct or incidental. In the present case the bearing is obvious. The gradual acquisition by certain plants of advantageous peculiarities is inferred from the gradation of forms and functions. Properties and powers which are latent or feebly developed in most plants are taken advantage of by some, made specially useful, and enhanced from generation to generation. Tendril-bearing plants, the most specialized in structure and the most exquisitely adapted to the end in view, are supposed to have been derived from leaf-climbers, and these in turn from simple twiners.

The author states that he was led to this subject by a brief note, communicated to the American Academy in the summer of 1858 (and reprinted in this Journal), in which the writer of the present notice recorded his observation of the coiling of certain tendrils by a visible movement promptly following an extraneous irritation. Mr. Darwin's observations were more than half completed before he became aware that the spontaneous revolution of the stems and of some tendrils of climbing plants had been observed and recorded almost forty years ago, and nearly at the same time, by Palm and by Von Mohl, and had been the subject of two memoirs by Dutrochet, published more than twenty years ago. But the mode in which the free and growing end of a stem sweeps around seems not to have been previously well made out, having been more or less confounded with the torsion of the axis which many twining stems, such as those of the Hop and the Morning Glory, are apt to undergo. It is plain to see, however, that many stems which revolve do not twist at all; and those that do never could twist on their axis at every revolution without speedy destruction, — indeed usually do not twist until they have ceased revolving. Every one must have noticed that the growing extremity of a Hop, Convolvulus, or other twiner, when unsupported, hangs over or stretches out horizontally to one side. But it is not so well known that this outstretched portion, while at the proper age, is continually sweeping round, in circles widening as it grows, and always in the same direction, in search of some object round which to twine. The Hop revolves with the sun; the Convolvulus, Bean (Phaseolus), etc., against the sun, that is, in the same directions that they twine. Two or three internodes are usually revolving at the same time. Mr. Darwin observed thirty-seven revolutions in one internode of a Hop, — the first revolution made in about twenty-four hours, the second in nine hours, the third and the following ones up to the eighth in a little over three hours each. "The shoot had now grown $3\frac{1}{2}$ inches in length, and carried at its extremity a young internode an inch in length, which showed slight changes in its curvature. The next or ninth revolution was effected in two hours and thirty minutes. From this time forward the revolutions were easily observed. The thirty-sixth revolution was performed at the usual rate; so was the last or thirty-seventh, but it was not quite completed; for the internode abruptly became upright, and after moving to the centre became motionless. I tied a weight to its upper end, so as slightly to bow it, and thus to detect any movement; but there was none. Some time before the last revolution the lower part of the internode had ceased to move. . . . It moved during five days; but the more rapid movement after the third revolution lasted during three days and twenty hours. The regular revolutions from the 9th to the 36th inclusive, were performed at the average rate of 2 h. 31 m. The weather was cold, and this affected the temperature of the room, especially during the night, and consequently retarded a little the rate of movement. . . . After the seventeenth revolution the internode had grown from $1\frac{3}{4}$ to 6 inches in length, and carried an internode $1\frac{7}{8}$ inch long, which was just perceptibly moving; and this carried a very minute ultimate internode. After the 21st revolution the penultimate internode was $2\frac{1}{2}$ inches long, and probably revolved in a period of about three hours. At the 27th revolution our lower internode was $8\frac{5}{8}$, the penultimate $3\frac{1}{2}$, and the ultimate $2\frac{1}{2}$ inches in length; and the inclination of the whole shoot was such that a circle 19 inches in diameter was swept by it. When the movement ceased the lower internode was 9 and the penultimate 6 inches in length; so that, from the 27th to the 37th revolutions inclusive, three internodes were at the same time revolving." — (pp. 3, 4.)

The shoots of many climbers sweep their circles more rapidly than the Hop, the common Pole Bean (*Phaseolus vulgaris*) in rather less than two hours, Convolvuluses of various species in the same time or rather less; while more woody stems naturally move more slowly, some requiring from 24 to 50 hours for each revolution. But the thickness or texture of the shoot does not govern the rate, many slender shoots moving slower than some stout ones, and some lignescent quicker than other purely herbaceous ones. The movement appears to be accelerated, up to a certain point, by raising the temperature, or rather is retarded by lowering it; but while the conditions are nearly the same, the rate is often remarkably uniform. The quickest rate of revolution of a proper stem observed by Mr. Darwin was that of a Scyphranthus, in 77 minutes. When the light comes from one side, the semicircle towards the light is usually described in less time, often in less than half the time, of that from the light. The tendency of young stems to turn toward the light is here active as usual, but is overcome by a superior force. The end of the shoot describes circles or broad ellipses, or else, from insufficient power or mechanical disadvantages, narrow ellipses, semicircles, or irregular figures. A horizontal shoot of considerable length will thus be found, not unfrequently, to sway from side to side in a semicircular course, while the extreme internodes are making complete revolutions.

A striking illustration of the amount of space that may be swept over is afforded by a case in which Mr. Darwin allowed the top of a Ceropegia to grow out almost horizontally to the length of 31 inches, — three long internodes terminated by two short ones. The whole revolved at rates between $5\frac{1}{4}$ and $6\frac{3}{4}$ hours for each revolution, the tip sweeping a circle of above five feet in diameter and 16 feet in circumference, traveling therefore at the rate of at least 32 inches per hour. "It was an interesting spectacle to watch the long shoot sweeping, night and day, this grand circle, in search of some object around which to twine."

As to the nature of this revolving movement, Mr. Darwin clearly shows that it is not a torsion of the axis, but a successive bending (similar to that by which ordinary stems bend toward the light), the direction of which is constantly and uniformly changing. "If a colored streak be painted (this was done with a large number of twining plants) along, we will say, the convex line of surface, this colored streak will, after a time depending on the rate of revolution, be found to lie along one side of the bow, then along the concave side, then on the opposite side, and, lastly, again on the opposite convex surface. This clearly proves that the internodes, during the revolving movement, become bowed in every direction. The movement is, in fact, a continuous self-bowing of the whole shoot, successively directed to all points of the compass." It is an automatic movement, of the same character as those which these and other parts of plants effect in changing position or direction, sometimes slowly and sometimes with a visible motion. The movement may be likened in one case to that of the hour-hand or the minute-hand of a clock, in the other to the second-hand, but in both is as truly a vital movement as is the contraction of an involuntary muscle. It must be effected — as Mr. Darwin recognizes — either by the contraction of the cells on the concave side, or by the turgescence and elongation of those on the convex side of the internode, or by both, — probably the former, as various facts go to show; but questions of that kind are not investigated in the present essay.

No differences in this regard are observable in the behavior of exogenous or endogenous stems, or even of those of climbing Ferns. *Lygodium scandens*, according to Darwin, revolves like other twiners; it completes its revolutions in six hours, or on a very hot day in five (moving against the sun, which is much the commoner case); this is about the average rate of Phænogamous twiners, like which it comports itself in all respects. Our own *L. palmatum*, we find, revolves in the same way, in about four hours, the temperature being 75° Fahr.

The power of revolving depends, of course, upon the general health and vigor of the plant, and upon the age of the shoot, is retarded by lowering the temperature, is interrupted by any considerable disturbance, such as exposure to cold or to much jarring; carrying the plant from one place to another, or cutting off a shoot and placing it in water, stops the movement for a time, just as it does the more vivid automatic movement of *Desmodium gyrans*. But each internode is so independent that cutting off an upper one does not affect the revolutions of the one beneath. Twining stems are far from being insensible to the action of light (as Mohl supposed), the half-revolution toward the light being not uncommonly twice faster than that from it; but as the rate of revolution by day and by night is nearly the same, the one half of the circuit is accelerated just as much as the other is retarded. This influence of the light is quite remarkable when we consider the slenderness of most revolving internodes, the small surfaces they expose, and that their leaves are little developed.

The design, as we must term it, of this revolving of the end of twining stems is obvious, and usually effectual. Such stems, even when no supporting object is within their reach, will reach each other, and by twining together make a mutual support, from which, as they lengthen, they may reach yet farther. The connection of the revolving with twining is obvious, though the latter is not a necessary consequence; for many stems revolve which do not twine, but climb in some other way.

"When at last the [revolving] shoot meets with a support,

the motion at the point of contact is necessarily arrested, but the free projecting part goes on revolving. Almost immediately another and upper point of the shoot is brought into contact with the support and is arrested; and so onward to the extremity of the shoot; and thus it winds round its support. When the shoot follows the sun in its revolving course, it winds itself round the support from right to left, the support being supposed to stand in front of the beholder; when the shoot revolves in an opposite direction, the line of winding is reversed. As each internode loses from age its power of revolving, it loses its power of spirally twining round a support. If a man swings a rope round his head, and the end hits a stick, it will coil round the stick according to the direction of the swinging rope; so it is with twining plants, the continued contraction or turgescence of the cells along the free part of the shoot replacing the momentum of each atom of the free end of the rope.

"All the authors, except Von Mohl, who have discussed the spiral twining of plants maintain that such plants have a natural tendency to grow spirally. Mohl believes (S. 112) that twining stems have a dull kind of irritability, so that they bend toward any object which they touch. Even before reading Mohl's interesting treatise, this view seemed to me so probable that I tested it in every way that I could, but always with negative results. I rubbed many shoots much harder than is necessary to excite movement in any tendril or in any foot-stalk of a leaf-climber, but without result. I then tied a very light forked twig to a shoot of a Hop, a Ceropegia, Sphærostema, and Adhatoda, so that the fork pressed on one side alone of the shoot and revolved with it; I purposely selected some very slow revolvers, as it seemed most likely that these would profit from possessing irritability; but in no case was any effect produced. Moreover, when a shoot winds round a support, the movement is always slower, as we shall immediately see, than whilst it revolves freely and touches nothing. Hence I conclude that twining stems are not irritable; and indeed it is not probable that it should be so, as nature always economizes her means, and irritability would be superfluous. Nevertheless I do not wish to assert that they are never irritable; for the growing axis of the leaf-climbing, but not spirally twining, *Lophospermum scandens* is, as we shall hereafter see, certainly irritable; but this case gives me confidence that ordinary twiners do not possess this quality, for directly after putting a stick to the Lophospermum, I saw that it behaved differently from any true twiner or any other leaf-climber.

"The belief that twiners have a natural tendency to grow spirally probably arose from their assuming this form when wound round a support, and from the extremity, even whilst remaining free, sometimes assuming this same form. The free internodes of vigorously growing plants, when they cease to revolve, become straight, and show no tendency to be spiral; but when any shoot has nearly ceased to grow, or when the plant is unhealthy, the extremity does occasionally become spiral. I have seen this in a remarkable degree with the ends of the shoots of the Stauntonia and of the allied Akebia, which became closely wound up spirally, just like a tendril, especially after the small, ill-formed leaves had perished. The explanation of this fact is, I believe, that the lower parts of such terminal internodes very gradually and successively lose their power of movement, whilst the portions just above move onward, and in their turn become motionless; and this ends in forming an irregular spire.

"When a revolving shoot strikes a stick, it winds round it rather more slowly than it revolves. For instance, a shoot of the Ceropegia took 9 hours and 30 minutes to make one complete spire round a stick, whilst it revolved in 6 hours; *Aristolochia gigas* revolved in about 5 hours, but took 9 hours and 15 minutes to complete its spire. This, I presume, is due to the continued disturbance of the moving force by its arrestment at each successive point; we shall hereafter see that even shaking a plant retards the revolving movement. The terminal internodes of a long, much-inclined, revolving shoot of the Ceropegia, after they had wound round a stick, always slipped up it, so as to render the spire more open than it was at first; and this was evidently due to the force which caused the revolutions being now almost freed from the constraint of gravity, and allowed to act freely. With the Wistaria, on the other hand, a long, horizontal shoot wound itself at first, in a very close spire, which remained unchanged; but subsequently, as the shoot grew, it made a much more open spire. With all the many plants which were allowed freely to ascend a support, the terminal internodes made at first a close spire; and this, during windy weather, well served to keep the shoots in contact with their support; but as the penultimate internodes grew in length, they pushed themselves up for a considerable space (ascertained by colored marks on the shoot and on the support) round the stick, and the spire became more open.

"If a stick which has arrested a revolving shoot, but has not as yet been wound round, be suddenly taken away, the shoot generally springs forward, showing that it has continued to press against the stick. If the stick, shortly after having been wound round, be withdrawn, the shoot retains for a time its spiral form, then straightens itself, and again commences to revolve. The long, much-inclined shoot of the Ceropegia previously alluded to offered some curious peculiarities. The lower and older internodes, which continued to revolve, had become so stiff that they were incapable, on repeated trials, of twining round a thin stick, showing that the power of movement was retained after flexibility had been lost. I then moved the stick to a greater distance, so that it was struck by a point 2½ inches from the extremity of the penultimate internode; and it was then neatly wound round by this part and by the ultimate internode. After leaving the spirally wound shoot for eleven hours, I quietly withdrew the stick, and in the course of the day the curled part straightened itself and re-commenced revolving; but the lower and not curled portion of the penultimate internode did not move, a sort of hinge separating the moving and the motionless part of the same internode. After a few days, however, I found that the lower part of this internode had likewise recovered its revolving power. These several facts show, that, in the arrested portion of a revolving shoot, the power of movement is not immediately lost, and that when temporarily lost it can be recovered. When a shoot has remained for a considerable time wound round its support, it permanently retains its spiral form even when the support is removed.

"When a stick was placed so as to arrest the lower and rigid internodes of the Ceropegia at the distance at first of 15 and then of 21 inches from the centre of revolution, the shoot slowly and gradually slid up the stick, so as to become more and more highly inclined; and then, after an interval sufficient to have allowed of a semi-revolution, it suddenly bounded from the stick and fell over to the opposite side, to its ordinary slight inclination. It now recommenced revolving in its usual course, so that after a semi-revolution it again came into contact with the stick, again slid up it, and again bounded from it. This movement of the shoot had a very odd appearance, as if it were disgusted with its failure but resolved to try again. We shall, I think, understand this movement by considering the former illustration of the sapling, in which the contracting surface was supposed to creep from the southern, by the eastern, to the northern, and thence back again by the western side to the southern face, successively bowing the sapling in all directions. Now with the Ceropegia, the stick being placed a very little to the east of due south of the plant, the eastern contraction could produce no effect beyond pressing the rigid internode against the stick; but as soon as the contraction of the northern face began, it would slowly drag the shoot up the stick; and then, as soon as the western contraction had well begun, the shoot would be drawn from the stick, and its weight coinciding with the northwestern contraction, would cause it suddenly to fall to the opposite side with its proper slightly inclined positions; and the ordinary revolving movement would go on. I have described this case because it first made me understand the order in which the contracting or turgescent cells of revolving shoots must act.

"The view just given further explains, as I believe, a fact observed by Von Mohl (S. 135), namely, that a revolving shoot, though it will twine round an object as thin as a thread, cannot do so round a thick support. I placed some long revolving shoots of a Wistaria close to a post between 5 and

6 inches in diameter, but they could not, though aided by me in many ways, wind round it. This apparently is owing to the flexure of the shoot, when winding round an object so gently curved as this post, not being sufficient to hold the shoot to its place when the contracting force creeps round to the opposite surface of the shoot; so that it is at each revolution withdrawn from its support." — (pp. 9–13, *passim.*)

The successive shifting of the contracting side of the shoot, which explains the revolution or bowing in turn in every direction, no less explains the twining round a proper support, leaving however some idiosyncrasies unexplained. Some tendrils and some petioles of leaf-climbing plants equally possess this revolving power; but their usefulness depends mainly upon additional and more special endowments, — mainly upon the power of directly responding by curvature to the contact, more or less prolonged, of an extraneous body.

Of Leaf-climbers, no instance is more familiar than that of Clematis or Virgin's Bower. Little more was known of them than that they climbed by curling their petioles (common or partial) around neighboring objects. Mr. Darwin made observations upon eight species of Clematis, seven of Tropæolum, the common species of Maurandia, Lophospermum, Fumaria, etc., as also upon Gloriosa and Flagellaria, which climb by a tendril-like production of the tip of the leaf. From the summary it appears that plants which belong to eight families are known to have clasping petioles, and those of four families climb by the tips of their leaves. In almost all of them the young internodes revolve, in some of them as extensively as in twining plants, — the movement being plainly serviceable in bringing the petioles or the tips of the leaves into contact with surrounding objects. Those whose shoots revolve most freely are also capable of twining spirally around a support; but when the stem twines (as in *Clematis Sieboldii* and *C. calycina*, but not in *C. Viticella*), it has the peculiarity of winding first in one direction for two or three turns, and then in the opposite direction. The petioles are principally efficient in these plants, and that by means of an endowment which is not shown to belong to twining stems, with one or two exceptions. That is, the petioles or their divisions are sensitive to the contact of an extraneous body, contracting on the side touched so as to curve or coil around it. That the footstalk is directly sensitive to the touch, just as tendrils are, Mr. Darwin proved by lightly rubbing them with a twig for a few times, when in the course of some hours it bends to the rubbed side, afterwards becoming straight again; or by leaving the body in contact it is permanently clasped by the footstalk. So sensitive are some footstalks that "a loop of thread weighing a quarter of a grain caused them to bend; a loop weighing one-eighth of a grain sometimes acted, and sometimes not." In one instance, in *Clematis Flammula*, even the sixteenth part of a grain caused a petiole to bend through nearly 90 degrees. With rare exceptions only the young petioles are sensitive. Take the cultivated *Clematis Viticella* for an illustration of the mode in which the leaves do the work of climbing.

"The leaves are of large size. There are three pairs of lateral leaflets and a terminal one, all borne by rather long petioles. The main petiole bends a little, angularly, downward at each point where a pair of leaflets arises, and the petiole of the terminal leaflet is bent downward at right angles; hence the whole petiole, with its rectangularly bent extremity, acts as a hook. This, with the lateral petioles directed a little upward, forms an excellent grappling apparatus by which the leaves readily become entangled with surrounding objects. If they catch nothing, the whole petiole ultimately grows straight. Both the medial and lateral petioles are sensitive; and the three branches, into which the basi-lateral petioles are generally subdivided, likewise are sensitive. The basal portion of the main petiole, between the stem and the first pair of leaflets, is less sensitive than the remainder, but it will clasp a stick when in contact. On the other hand, the inferior surface of the rectangularly bent terminal portion (carrying the terminal leaflet), which forms the inner side of the end of the hook, is the most sensitive part; and this portion is manifestly best adapted to catch distant supports. To show the difference in sensibility, I gently placed loops of string of the same weight (in one instance weighing .82 of a grain) on the several lateral and on the terminal sub-petioles; in a few hours the latter were bent, but after 24 hours no effect was produced on any of the lateral petioles. Again, a terminal sub-petiole placed in contact with a thin stick became sensibly curved in 45 minutes, and in 1 hour 10 minutes had moved through ninety degrees, whereas a lateral petiole did not become sensibly curved until 3 hours and 30 minutes had elapsed. In this latter case, and in all other such cases, if the sticks be taken away, the petioles continue to move during many hours afterward; so they do after a slight rubbing; but ultimately, if the flexure has not been very great or long-continued, they become, after about a day's interval, straight again." — (p. 31.)

In numerous cases, notably in *Solanum jasminoides*, the petiole when clasped increases very greatly in thickness and rigidity, undergoing a change in its woody structure by which the fibro-vascular bundles, originally semi-lunar in cross-section, develop into a closed ring, like that of an exogenous stem.

Lophospermum scandens of the gardens climbs, like its allies Maurandia and Rhodochiton, by clasping petioles; but in this plant, alone, the young internodes are also sensitive to the touch.

"When a petiole clasps a stick, it draws the base of the internode against it; and then the internode itself bends toward the stick, which is thus caught between the stem and the petiole as by a pair of pincers. The internode straightens itself again, excepting the part in contact with the stick. Young internodes alone are sensitive, and these are sensitive on all sides along their whole length. I made fifteen trials by lightly rubbing two or three times with a thin twig several internodes; and in about 2 hours, but in one case in 3 hours, all became bent; they became straight again in about 4 hours, subsequently. An internode, which was rubbed as much as six or seven times with a twig, became just perceptibly curved in 1 hour 15 minutes, and subsequently in 3 hours the curvature increased much; the internode became straight again in the course of the night. I rubbed some internodes one day on one side, and the next day on the opposite side or at right angles; and the curvature was always toward the rubbed side."

Here, then, is one case in which the sensibility of a stem is manifest, and is turned to useful account. The peduncles of the allied *Maurandia semperflorens* are also sensitive and flexuous, although Mr. Darwin insists that they are useless for climbing. That some stems should be sensitive might have been expected; for tendrils of axial nature (*e. qr.* of *Passiflora gracilis*) are not less sensitive than those of foliar nature, as of *Leguminosæ*, *Cucurbitaceæ*, and Cobæa. And if twining stems in general are not endowed with "a dull kind of irritability," as Mohl conjectured, it may well be because the equally wonderful automatic revolving movement leaves no need for it. In general, the most striking cases of automatic movement belong to leaves or their homologues.

The distinction can be only somewhat arbitrarily drawn between Leaf-climbers — especially those with small or undeveloped leaflets, or where the tip of the leaf forms a hook or tendril-like projection — and Tendril-climbers. The tendril, however, whether answering to leaf or stem, is the more specialized organ, adapted only for climbing, and endowed in different plants with very various and some highly remarkable powers. To this subject Mr. Darwin has devoted more than half of his essay. An analysis of it must be deferred, for want of space.

Near the close of the essay, under Hook-climbers, Mr. Darwin remarks that: —

"Even some of the climbing Roses will ascend the walls of a tall house, if covered with a trellis; how this is effected I know not; for the young shoots of one such Rose, when placed in a pot in a window, bent irregularly toward the light during the day and from it during the night, like any other plant; so that it is not easy to understand how the shoots can get under a trellis close to a wall."

Now we have had occasion to observe that the strong summer-shoots of the Michigan Rose (*Rosa setigera*, Mx., *R. rubifolia*, R. Br.), trained on a latticed wall, are strongly disposed to push into dark crevices and away from the light; they would, many of them, pretty surely place themselves under the trellis, and the lateral shoots of the next spring would emerge as they seek the light. We suspect this is also true of the Sweet Brier.

Twiners and Leaf-climbers having been considered, Tendril-bearers, which are the highest style of climbing plants, next demand our attention. But our analysis of this important part of Mr. Darwin's treatise must be disproportionably brief.

There are two kinds of movement exhibited by plants, which should be distinguished. 1st, Automatic, usually continued movements, not set in action by extraneous invitation. The gyratory movement of the small leaflets of *Desmodium gyrans* is an exalted instance of this. 2d, Movements in consequence of the contact or action of an extraneous body, — of which those of the leaves of the Sensitive Plant may be taken as the type. Twining stems, as has been seen, strikingly exhibit the first, and their coiling around a support is a consequence of it.

Tendrils for the most part execute both kinds of movement. They revolve, with some exceptions, like twining stems; and they are all more or less sensitive to contact, — usually more so than the petioles of leaf-climbers, — bending toward the impinging body so as to hook or clasp around it, if the size will allow. Different tendrils act differently in some respects, some revolving freely, and sweeping wide circuits, some less evidently, and some, like those of the Virginia Creeper, do not revolve at all, but turn from the light to the dark. But whether the tendril is the homologue of a leaf, or of a stem (or of a peduncle, which is the same thing), appears to make no difference in its action. On the other hand their diversity of gifts in one and the same family, or even in species of the same genus, is very remarkable, as may be seen especially in the Bignonia Family, the Grape Family, etc. So, also, the tendrils are commonly aided in their endeavors by the revolving of the internodes of the stem, but sometimes not, even in plants of the same genus or family. Mr. Darwin takes up tendril-bearing plants by natural families, beginning with *Bignoniaceæ*, which order contains tendril-bearers, leaf-climbers, twiners, root-climbers, and various combinations of these diverse modes. We, however, will first consider the tendrils of the Gourd, and Passion-flower families, regarding them as typical and simple representatives of tendril-climbers.

Passiflora gracilis, a delicate annual species, lately introduced into the gardens, of the easiest cultivation, one which differs from most of its relatives in the young internodes having the power of revolving, is said by Mr. Darwin to exceed all other climbing plants in the rapidity of its movements, and all tendril-bearers in the sensitiveness of its tendrils. In the latter respect it decidedly surpasses our Echinocystis; but it is nearly if not quite equalled by Sicyos, in which the coiling upon contact was first noticed as a visible movement. The revolving internodes, when in the best condition, make almost hourly revolutions, and the long, delicate, straight tendrils revolve nearly in the same manner and at the same rate. The sensitiveness of the tendril, when full-grown, is correspondingly great, a single light touch on the concave surface of the tip causing a considerable curvature. "A loop of soft thread weighing $\frac{1}{32}$d of a grain, placed most gently on the tip, thrice plainly caused it to curve, as twice did a bent bit of thin platinum wire weighing $\frac{1}{50}$th of a grain; but this latter weight, when left suspended, did not suffice to cause permanent curvature." After touch with the twig, the tip begins to bend in from 25 to 39 seconds. After coiling into an open helix upon transient irritation, they soon straighten again, recovering their sensibility; but if left in contact, the action continues. We found it a pretty experiment, last summer, during the warmest days, to bring the upper part of an outstretched tendril by its inner or concave side against a twig or cord, and to see how promptly it would clasp it, revolving its free apex round and round it. A curious discrimination in the sensibility of such tendrils is mentioned by Mr. Darwin, as follows: —

"I repeated the experiment made on the Echinocystis, and placed several plants of this Passiflora so close together that the tendrils were repeatedly dragged over each other; but no curvature ensued. I likewise repeatedly flirted small drops of water from a brush on many tendrils, and syringed others so violently that the whole tendril was dashed about, but they never became curved. The impact from the drops of water on my hand was felt far more plainly than that from the loops of thread (weighing $\frac{1}{32}$d of a grain) when allowed to fall upon it, and these loops, which caused the tendrils to become curved, had been placed most gently on them. Hence it is clear, either that the tendrils are habituated to the touch of other tendrils and to that of drops of rain, or that they are sensitive only to prolonged though excessively slight pressure. To show the difference of the kind of sensitiveness in different plants, and likewise to show the force of the syringe used, I may add that the lightest jet from it instantly caused the leaves of the Mimosa to close; whereas the loop of thread weighing $\frac{1}{32}$d of a grain, when rolled into a ball and gently placed in the glands at the base of the leaflets of the Mimosa, caused no action." — (p. 90.)

Of Cucurbitaceous tendrils, the most active, after those of Sicyos (which Mr. Darwin has not observed), are those of *Echinocystis lobata.* The internodes and tendrils revolve in about an hour and three quarters, the former sweeping a circle or ellipse of two or three inches in diameter, the latter often one of 15 or 16 inches in diameter. Perhaps the most remarkable appearance of discrimination in tendrils is that which Mr. Darwin first noticed in this plant, but which may be seen in others, — and which he thus describes: —

"I repeatedly saw that the revolving tendril, though inclined during the greater part of its course at an angle of about 45° (in one case of only 37°) above the horizon, in one part of its course stiffened and straightened itself from tip to base, and became nearly or quite vertical. . . . The tendril forms a very acute angle with the extremity of the shoot, which projects above the point where the tendril arises; and the stiffening always occurred as the tendril approached and had to pass, in its revolving course, the point of difficulty, — that is, the projecting extremity of the shoot. Unless the tendril had the power of thus acting, it would strike against the extremity of the shoot, and be arrested by it. As soon as all these branches of the tendrils begin to stiffen themselves in this remarkable manner, as if by a process of turgescence, and to rise from an inclined into a vertical position, the revolving movement becomes more rapid; and as soon as the tendril has succeeded in passing the extremity of the shoot, its revolving motion, coinciding with that from gravity, often causes it to fall into its previous inclined position so quickly, that the end of the tendril could distinctly be seen travelling like the minute-hand of a gigantic clock." — (p. 75.)

Cucurbitaceous tendrils are mostly compound, in this case three-forked. When one of the lateral branches has firmly clasped any object, the middle branch continues to revolve. If a full-grown tendril fails to reach and lay hold of any object, it soon ceases to revolve, bends downwards, and coils up spirally from the apex. Indeed it often coils while still outstretched and revolving, the tendency to shorten (as we presume) on the inner side from the tip downward, which is usually brought into action by contact with an extraneous body, at length operating spontaneously. Uncaught tendrils when they thus coil up throw themselves of course into a simple helix or spire. One end being free, this is the simple and necessary consequence of the relative shortening of the concave side, sufficiently continued.

In a caught tendril, the relative shortening of one side, (through which the tip hooks round and fixes itself to the supporting object), being propagated downwards, the whole now throws itself into a spiral form, — with more or less promptitude according to the species, — thus pulling the free portion of the tendril-bearing shoot nearer to the support, and within easier reach of the next tendril above. Both ends of the tendril being fixed, and the winding round an axis (real or imaginary) necessarily involving or being a twist, it is certain that the caught tendril cannot now coil into a simple spiral, but that the spire will at least be double, a coil near one end of the tendril in one direction requiring the other to twist in the opposite direction, unless indeed it undergoes torsion. So, as is familiarly known, there is at least one neutral point in a caught and coiled-up tendril, usually in the middle, the turns on one side of it running from right to left, on the other side from left to right. That the coils, whether simple or double and reversed (as the case may be), are not determined by any peculiarity in the tendril, but merely by the relative shortening of one side, may be readily shown by a thread cut from a piece of india-rubber, of unequal tension of the two sides; this, when stretched and allowed to shorten while the two ends are held fast in the same plane, forms at once a pair of reverse coils, or three or four such coils, just as caught tendrils do.

Mr. Darwin explains the point by analogous practical illustrations. He shows, moreover, that an important service rendered by the coiling or spiral contraction "is that the tendrils are thus made highly elastic." In the Virginia Creeper, where the ends of the compound tendrils are peculiarly attached, "the strain is thus equally distributed to the several attached branches of a branched tendril; and this must render the whole tendril far stronger, as branch after branch cannot separately break. It is this elasticity which saves both simple and branched tendrils from being torn away during stormy weather. I have more than once gone on purpose, during a gale, to watch a Bryony growing in an exposed hedge, with its tendrils attached to the surrounding bushes; and as the thick or thin branches were tossed to and fro by the wind, the attached tendrils, had they not been excessively elastic, would have been instantly torn off and the plant thrown prostrate. But as it was, the Bryony safely rode out the gale, like a ship with two anchors down and a long range of cable ahead, to serve as a spring as she surges to the storm."

Moreover, while unattached tendrils soon shrink up or wither and fall off, as we observe in the Grapevine, Virginia Creeper, etc., these same plants show how an attached tendril thickens and hardens, gaining wonderfully in strength and durability. In a Virginia Creeper, "one single lateral branchlet of a (dead) tendril, estimated to be at least ten years old, was still elastic and supported a weight of exactly two pounds.

This tendril had five disk-bearing branches, of equal thickness and apparently equal strength; so that this one tendril, after having been exposed during ten years to the weather, would have resisted a strain of ten pounds."

Our space will not allow even an abstract of Darwin's account of the admirable adaptations and curious behavior of various tendrils, even of some very common plants; as for instance of the familiar *Cobœa scandens*, in which (the stem and the petioles being motionless) the great compound tendril borne at the summit of the leaf executes large circular sweeps with remarkable rapidity, carrying round an elaborate flexible grapnel, consisting of its five subdivisions, from 50 to 100 in number, which are very sensitive even to a slight touch, bending in a few minutes toward the touched side, so that they clasp twigs very promptly, and all tipped with minute, double or sometimes single, sharp hooks, which catch in little inequalities, and may prevent the tendril branchlets from being dragged away by the rapid revolving movement before their irritability has time to act, while the still free ones proceed to arrange themselves, by various queer and complicated movements, so as to secure the most advantageous hold; then contracting spirally so as to bring other portions up within reach of the support, until all are inextricably knotted and fastened, and finally growing stouter, rigid and strong, binding the plant firmly to its support.

We cannot omit all mention of *Bignonia capreolata*, a not uncommon climber of our Southern States, of which we especially wish to obtain fresh seeds or young plants, that we may ourselves observe the remarkable behavior of its tendrils which Mr. Darwin describes. These are said to turn from the light, as in many other cases; they will clasp smooth sticks, but soon lose their hold and straighten themselves again. A rough, fissured, or porous surface alone satisfies them; their young tips seek and crawl into dark holes and crevices, in the manner of roots; then they develop their hooked extremity, and, especially when they meet with any fibrous matter, the hook swells into irregular balls of cellular tissue, which first adhere to the fibres by a viscid cement, and then grow so as to develop them. This tendril can do nothing with a smooth post, fails to attach itself to a brick wall, but is well adapted to climb trees with rough and mossy bark.

The Virginia Creeper also turns its tendrils from the light, and, although they will occasionally clasp a slender support, in the manner of its relative the Grapevine, they uniformly seek dark crevices, or especially broad flat surfaces, as a wall, a rock, or the trunk of a tree. Having brought their curved tips into contact with such a surface, these swell and form, in the course of a few days, the well-known disks or cushions by which they firmly adhere. Here is a tendril-climber which emulates a root-climber, such as the Ivy, in the facility with which it ascends smooth trunks, rocks, or walls.

A very short chapter is devoted to Hook-climbers and Root-climbers. The stems of the latter are said to "have usually no power of movement, not even from the light to the dark. But *Hoya carnosa*, which twines, also climbs by rootlets spreading over the face of a damp wall; and *Tecoma radicans* (our Trumpet Creeper) exhibits in its young shoots some vestiges of the revolving power with which its twining relatives are endowed."

In a dozen pages of *Concluding Remarks*, Mr. Darwin gives much interesting matter in the way of deduction and speculation, which it would be difficult to condense into an abstract.

Plants become climbers, he remarks, in order to reach the light, and expose a large surface of leaves to its action and that of the free air. Their advantage is, that they do this with wonderfully little expenditure of organized matter in comparison with trees, which have to support a heavy load of branches by a massive trunk. Of the different sorts of climbers hook-climbers are the least efficient, at least in temperate countries, as they climb only in the midst of an entangled mass of vegetation. Next root-climbers, which are admirably adapted to ascend naked faces of rock; but when they climb trees they must keep much in the shade, and follow the trunk; for their rootlets can adhere only by long-continued and close contact with a steady surface. Thirdly, spiral twiners, with leaf-climbers and tendril-bearers, which agree in their power of spontaneously revolving and of grasping objects which they reach, are the most numerous in kinds, and most perfect in mechanism; they can easily pass from branch to branch, and securely ramble over a wide and sunlit surface.

After adducing some considerations in support of his opinion that both leaf-climbers and tendril-bearers "were primordially twiners, that is, are the descendants of plants having this power and habit," Mr. Darwin asks: "Why have nearly all the plants in so many aboriginally twining groups been converted into leaf-climbers or tendril-bearers? Of what advantage could this have been to them? Why did they not remain simple twiners? We can see several reasons. It might be an advantage to a plant to acquire a thicker stem, with short internodes, bearing many or large leaves; and such stems as are ill fitted for twining. Any one who will look during windy weather at twining plants will see that they are easily blown from their support; not so with tendril-bearers or leaf-climbers, for they quickly and firmly grasp their support by a much more efficient kind of movement. In those plants which still twine, but at the same time possess tendrils or sensitive petioles, as some species of Bignonia, Clematis, and Tropæolum, we can readily observe how incomparably more securely they grasp an upright stick than do simple twiners. From possessing the power of movement on contact, a tendril can be made very long and thin; so that little organic matter is expended in their development, and yet a wide circle is swept. Tendril-bearers can, from their first growth, ascend along the outer branches of any neighboring bush, and thus always keep in the full light; twiners, on the contrary, are best fitted to ascend bare stems, and generally have to start in the shade. . . .

"The object of all climbing plants is to reach the light and free air with as little expenditure of organic matter as possible; now, with spirally-ascending plants the stem is much longer than is absolutely necessary; for instance, I measured the stem of a Kidney-bean which had ascended exactly two feet in height, and it was three feet in length. The stem of a Pea, ascending by its tendrils would, on the other hand, have been but little longer than the height gained. That this saving of stem is really an advantage to climbing plants I infer from observing that those that still twine, but are aided by clasping petioles or tendrils, generally make more open spires than those made by simple twiners." — (p. 110.)

The gradations between one organ and another, and their special endowments, and the great diversity of their movements, are illustrated at length; and the very large number of natural families which exhibit these endowments, in some of their members, is indicated; and it is noted that two or three genera alone have those powers in some of the largest and best defined natural orders, such as *Compositæ*, *Rubiaceæ*, *Liliaceæ*, Ferns, etc.; from which he infers "that the capacity for acquiring the revolving power, on which most climbers depend, is inherent, though undeveloped, in almost every plant in the vegetable kingdom" (p. 117).

Mr. Darwin somewhere throws out the remark that the larger number, and the most perfectly organized climbing plants, as of the scandent animals, belong to one country, tropical America.

In abruptly closing these extracts and brief commentaries, we would add, that the Linnæan Society has issued a separate reprint of this charming treatise, thus opening to it a wider circle of readers.

WATSON'S BOTANY OF THE 40TH PARALLEL.

WE propose to notice this volume[1] particularly: indeed it well deserves a more thorough examination and more extended review than our time and space will now allow us to devote to it. It is published "by order of the Secretary of War, under the authority of Congress," as one of the Engi-

[1] *United States Geological Survey of the 40th Parallel.* V. Botany. By Sereno Watson. Washington, 1871. (American Journal of Science and Arts, 3 ser., iii. pp. 62 and 148.)

neer Corps series, has been carefully edited and beautifully printed, so that the volume is every way an attractive one. Errors of the press are to be found, but they are apparently few, and the whole typography is remarkably excellent for the Government Printing Office. Our comparison is naturally with corresponding volumes of the Pacific Railroad Survey, and of the Mexican Boundary Survey, upon which the present volume is a notable improvement. The forty plates, filled with well chosen subjects, if not of the very highest style, are so well done and of such excellent promise that the name of the draughtsman (who is new to this class of work, we believe), Mr. J. H. Emerton, of Salem, Massachusetts, should properly have been appended to them.

The General Report, of 53 pages, forms a separately paged introduction to the "Catalogue," as it is termed with excessive modesty, *i. e.*, the systematic account of the plants collected, which makes up the principal bulk of the volume. This General Report will naturally be most interesting to general readers and naturalists, but no less so to special botanists. It is thoroughly readable matter, and we expect to see it reproduced in the scientific journals. Four or five pages sketch the geographical features of the region, tersely and clearly. But, when a stream of water is said to "become demoralized with alkali and is lost," we could wish that this popularized use of the word were buried with it. The meteorological notes, with tabulated observations by thermometer, evaporator, etc., are equally interesting, displaying the dryness of the Great Basin, its cold winters and hot summers. The notes on the general character of the vegetation picture to us the botanical aspect of the region, the relative prevalence of the predominant species, the slow and cross-grained growth of what timber there is in the cañons, etc. A dead branch, apparently of *Pinus monophylla*, 8 inches in diameter, had the fibres so twisted that in 7 feet they made four complete circuits. A saw-mill in Ruby Valley offered the opportunity of ascertaining the age and dimensions of several specimens of *Pinus flexilis* from the upper cañons of the Humboldt Mountains; sections from 22 to 30 inches in diameter showed from 400 to 486 annual rings. The "everlasting sage-bush," *Artemisia tridentata*, displayed 65 rings on a section 8 inches in diameter, 37 upon 4 inches, etc. A *Juniperus occidentalis*, 12 inches in diameter, showed 250 rings. *Cercocarpus ledifolius*, it appears, may form a trunk of 2 feet in diameter, with 160 rings. The alkaline species, aquatic and meadow species, those of the drier valleys and foot-hills, the mountain species, etc., are separately enumerated; the introduced species, about 30 in number, are recorded, and finally the number of indigenous genera and species is given under their orders, and their distribution in the basin or over the borders on either side is tabulated. Of the 1141 species of the basin and of the Wahsatch and Uintas, 60 per cent. appear to inhabit also the Pacific slope, about 60 per cent. are not found east of the Rocky Mountains, 15 per cent. only approach the Mississippi or the Saskatchewan, 25 per cent. approach the Atlantic, 17 per cent. are Mexican or southern, and nearly 15 per cent. are Arctic.

A few pages at the close are devoted to the consideration of the agricultural resources of the basin, the limit to which is fixed by the deficiency of water. "The most fertile localities lie at the base of the Sierras; but, as a rule, there is an apparent absence everywhere of a true soil or mould resulting from the decomposition of vegetable matter." A moderate amount of alkali in the soil appears not to be detrimental to culture. The soil which produces "sage bush" seems to be always cultivable when it can be irrigated. With the present supply of water, most economically used, it is thought that only one thousand out of 34,000 square miles of northern Nevada could be brought under cultivation; of the southerly portion and of western Utah much less. Eastern Utah, with larger and more constant supplies of water from the Wahsatch and the Uinta Mountains, is much more favorably situated. The absence of graminivorous animals, excepting rabbits in the valleys and rarely a few mountain sheep or antelopes in the higher ranges, shows that the country is ill adapted for grazing. *Eurotia lanata* and a few other Chenopodiaceous plants are eaten by sheep as a substitute for grass.

Mr. Watson raises the question whether — considering the amount of low shrubby and perennial vegetation which inhabits the plains and thrives without irrigation — these plants themselves, or some more serviceable substitutes equally adapted to the climatic conditions, may not be turned to some profitable account under the necessities of a future population; and whether, in time to come, some forms of orchard, vineyard, or tree-culture may not possibly be made to thrive in that region. He finds that the present plants on the whole are not lacking in expansion of foliage or succulence, at least that the more prevalent plants had an average of from 55 to 80 per cent. of foliage or working surface; and a series of rough but seemingly well-devised experiments demonstrated that they give off by evaporation daily an amount equal to three eighths of the weight of their available material. Dry as the soil appears to be, it is this, and not the atmosphere, that must furnish the supply to make good this loss. Yet water is rarely to be had under a depth of 100 to 300 feet, often not even at that depth. The porous soil must allow of the free upward diffusion of moisture, also of deep penetration of the roots from above.

An excellent map is given, exhibiting the district from above the 42d parallel to below the 39th, on which the routes of the three several years are traced in colored lines, and the mountain ranges with the general configuration of the surface represented. We will endeavor hereafter to review the systematic part of this work.

(Second Notice.) — Under the modest name of a catalogue of the known plants of Nevada and Utah, Mr. Watson has given us a treatise, not to say a Flora, of a wide stretch of country between the Sierra Nevada and the Rocky Mountains, which is invaluable to the botanist studying the plants of that region in herbaria, and still more to explorers on the ground, — of which we hope there may be many. For not only are new or revised species described, but all species not contained in the common eastern manuals, etc., which every collector is supposed to possess; the characters of western genera are appended in foot-notes, and synopses of recently elaborated genera — some of them reprints or translations of scattered papers with corrections or additions, others original revisions by the editor himself — are added in an appendix, so as to afford every possible help to the student or collector who has not access to a full botanical library, and indeed most acceptable facilities to those few who have.

After thus calling attention to a volume of so much importance, we propose to restrict our comments to sundry details of criticism, or points of information, where opportunity occurs.

Under *Thalictrum Fendleri* some synonyms are adduced which are not all certain; as there is another Oregon species which has been confounded with *T. dioicum*, but is distinct from both in the fruit, which was sparingly collected in the British Boundary expedition, and lately by Mr. Hall.

Ranunculus alismæfolius var. *montanus* is essentially equivalent to the variety *alismellus* Gray; although the specimens from the "head of Provo River in the Uintas" are a stouter and larger-flowered form, identical with Parry's No. 79, which we had wrongly named when distributed, and which may be rightly characterized as merely a dwarf mountain state of Geyer's *R. alismæfolius*. We may now add that there is a much older name for this species, especially for this mountain form of it, namely, *R. Pseudo-Hirculus* of Schrank, 1842, a Songarian plant. It may also be noted that, while this species in eastern America takes the place of the European *R. Flammula*, both occur on the western side of the continent (as also in Siberia), and in forms so much alike that only the character of the style and that of the petal and its scale (so well indicated by Mr. Watson) will serve to distinguish them. Of amply developed *R. Flammula* — as large as any European form — copious specimens have been collected in Oregon last year, by Mr. Elihu Hall, and are soon to be distributed.

As to *R. fascicularis*, there is no clear evidence that this species extends to California, Nevada, or even to Oregon. The plant referred to and so named in Lyall's collection, though not in fruit, is apparently *R. orthorhynchus*, a plant most rare in collections, but now, thanks to E. Hall's collection of last summer, likely to be supplied to botanists. As to Mr. Watson's *R. orthorhynchus* var. *alpinus*, that is certainly not of this species, but a wholly new one, unless it be the rare, and to us obscure, *R. pedatifidus* of Smith, or at least of Hooker. For since Schlechtendal's plant of that name has been referred to *R. affinis*, the Siberian one of Smith may also be of that species.

A yellow-flowered Aquilegia, with flowers rather smaller and sometimes much smaller than those of *A. Canadensis*, and with spurs shorter than the widely spreading sepals, after the manner of *A. formosa*, and more or less curved (thus approaching the European type), which has been collected by Lyall, Bourgeau, and others, is now characterized as a new species, under the name of *A. flavescens*, Watson. It should be noted that this has been cultivated in European gardens, from seeds collected by Rœzl, under the name of *A. aurea*, but it is doubtful if it is yet published under that name.

The *Cruciferæ* constitute an important order in the interior basin and its borders. One of Mr. Watson's most notable discoveries is that of Brown's *Parrya macrocarpa*, hitherto found only on the Arctic coast. It was detected on the highest peak of the Uintas, at an altitude of 12,000 feet. The next point of interest is found in our author's discoveries and views of plants of the Streptanthus and Thelypodium type. Two or three well-marked new species are introduced, and Nuttall's obscure *Streptanthus cordatus* is confidently identified. In the present view this is the only Streptanthus of the collection; Mr. Watson, having ascertained that several species, such as *S. procerus*, the curious *S. crassicaulis*, and two new species, have oblong seeds in a terete elongated pod, and cotyledons inclining to be incumbent in the manner of Thelypodium, combines them into his new genus, Caulanthus. And Iodanthus, with a few other species, some of which had already been excluded from Streptanthus, are referred, as had also been tentatively suggested, to Thelypodium. Which is all to be highly approved, except, perhaps, the expediency of the new genus, when all could be disposed in the two genera: Streptanthus for the species with flat or flattish pods, flat seeds, and truly accumbent cotyledons; Thelypodium, for those with more or less terete pods, narrow seeds, and more or less incumbent cotyledons.

We are bound, moreover, to take steps for the suppression of a nominal species which is here introduced in consequence of our own short-sightedness. In an evil moment we gave the name of *Smelowskia ? Californica* to a plant of Professor Brewer's collections, thought to be perennial, with exceedingly short few-seeded pods. This Mr. Watson identified with a common Sisymbrium of the region, distinguished from *S. canescens* by its seeds, strictly in a single series, and transferring the name, calls it *S. Californicum.* He had overlooked an article in this Journal (for September, 1866) upon this Sophia group of Sisymbrium, from which it would have been seen that the plant in question is *Sisymbrium incisum* of Engelmann, and the later *S. longipedicellatum* of Fournier, besides one or two other names of the same author more or less strictly referable to it.

DECAISNE'S MONOGRAPH OF THE GENUS PYRUS.[1]

A VOLUME of Decaisne's great work — or rather of one of his great works — "Le Jardin Fruitier du Muséum, un Iconographie de touts les Espèces et Varietes d'Arbres Fruitiers cultives dans cet Etablissement," etc. (produced in first-rate style by Firmin Didot Frères), devoted to the genus Pyrus, is now before us. It is a complete monograph of the species of this genus, taken in its restricted sense, illustrated by figures of the wild types, and also of the cultivated races of those cider-pears known in France under the name of *Sauger.* There is a list of the cider-pears cultivated in the different provinces of France, a general alphabetical catalogue of all the published varieties of pears, and a table in which the synonyms are referred to the names severally adopted. The other volumes, and the illustrations of the edible varieties of

[1] American Journal of Science and Arts, 3 ser., iv. 489; x. 481.

pears, may have more interest for the horticulturist. But the present attracts the special attention of the scientific botanist.

As stated in the Introduction, Professor Decaisne entered upon his great undertaking more than twenty years ago, when, in the year 1850, he became the Professor of Culture. He cites the instructions under which the separate collection of fruit-trees was constituted, and the professor of culture was charged with its management, and was directed to bring together all the known varieties, with all their names, "afin d'établir une uniformité de nomenclature necessaire pour toutes les parties de la République." This is a decree of the National Convention, June 10, 1793. The collection which Decaisne has so diligently and acutely studied actually dates from the year 1792, when the fruit-garden of the Chartreux of Paris was broken up, and two trees of each variety transported to the Jardin des Plantes. In 1793 it contained 185 varieties. In 1824, when Thouin died, there were in it 265 varieties of pears alone; it has now more than 1400 varieties of this fruit. It is interesting and important to know that the collection still preserves the greater portion of the very types described a century ago by Duhamel. For seven years Professor Decaisne studied the incomparable collection under his charge, making drawings and analyses, in which he is so skilful, and an herbarium of their flowers and foliage, before he commenced the publication of the "Jardin Fruitier du Muséum," which he is now bringing to a close.

As to giving a correct nomenclature and available characters, this is difficult enough, as all botanists know, for the species themselves (which must needs have, or be assumed to have, real distinctions) in any large genus, such as Quercus, Rosa, Rubus, and the like; how much more difficult, even to impossibility, it must be in the case of cultivated varieties, of ever increasing numbers, usually named without system, sometimes of mixed origin, and often too like each other to be distinguished by any available descriptions. Here colored plates are a necessity; and those of this great standard work, upon which no pains have been spared, leave little to be desired that art can supply.

In France alone they count about 800 sorts of pears; the origin of most of them is unknown, and many are undoubtedly very ancient. Indeed, according to Jordan and his school these differences are primitive, and the so-called races and varieties, both of wild and cultivated plants, represent so many closely related species. But M. Decaisne, not content with the *reductio ad absurdum* of having about 2000 species of pears to be dealt with, proceeded to an experimental demonstration of the variability of the cultivated races. He sowed the seeds from four very distinct varieties in 1853, the Poire d'Angleterre, the Bosc, the Belle Alliance, and the Cirole. Of the last the four trees raised bore fruit of four different forms. From the Belle Alliance he obtained, in this first generation, nine new varieties, none of them representing the parent, neither in the form, size, color, nor even the time of ripening of the fruit. The Bosc equally produced new varieties. Of the Angleterre nine trees produced as many new forms, one of them a winter-pear similar to the Saint Germain, another apple-shaped fruit identical with one which was raised from the Belle Alliance. On plate 33, Decaisne gives figures of six different pears raised from the Angleterre. These results even led him to doubt the cases cited by Darwin of the reproduction of certain pears from seed. He insists, moreover, that very bad fruits may be raised from choice cultivated pears, and that good varieties may be obtained from the seeds of wild pears. The latter is not what one would expect in the first generation.

Our author proceeds to state that the trees raised from seed taken from the same fruit differed, not merely in their fruits and in the time of ripening, but no less in their flowers and in the form of the leaves. Some were thorny, others thornless; some produced slender shoots, others thick and stout shoots, etc. It is worth noticing, however, that no mention is made of any precautions to prevent cross-fertilization of the flowers from which the seeds planted were derived, which might have influenced the product through the now well-ascertained influence of the pollen upon the pericarp. We perceive, however, that he would regard this as unimportant, since pear varieties are of the lowest grade, incapable of propagating fruit by close-fertilization, and therefore wholly unlikely to impress by their pollen any characteristic upon the pericarp of another variety.[1] A large part of the Introduction is occupied with further evidence that the Pear-trees of cultivation are all of one species, from which have proceeded six races, completely fertile *inter se*, and varieties *ad infinitum*. In this respect the Pear-tree has but followed the example of most fruit and fruit-trees, and of the Grains, etc., which had apparently diverged into races, or distinct but closely related types, in very early times, and those under cultivation have themselves varied and subdivided more and more. Finally, M. Decaisne maintains, seemingly with good reason, that to combine into one genus the Apple, Pear, Quince, Sorb, and Mountain Ash, as done by Linnæus and followed by the latest authorities, is to misconceive the laws of the natural system; that "to unite generically these plants, which differ in the character of their wood, the vernation of their leaves, their inflorescence, the æstivation of the corolla, and the structure of their fruit," logically leads to the combination of all *Pomaceæ* into one genus. He accordingly restricts the genus Pyrus, or (restoring the classical orthography) Pirus, as did Tournefort and Jussieu, to the Pear proper. To the organography of this restricted genus, from the wood to the embryo, a full chapter is devoted. In the course of this the relative systematic value of characters observed is brought out. He

[1] Yet the Apple, which is in the same case, does so. An interesting instance of this kind lately came under our notice, an apple from a Spitzenberg tree, one half (at least as to the surface) *Spitzenberg*, the other half *Russet*. A tree of the latter fruit stood about 200 yards off. Several cases of this sort are known, in which, as in this, the division is into two exactly equal parts of the circumference, and the line of demarcation abrupt. This is quite unexpected, as the Secretary of the Smithsonian Institution, who sent us the fruit, remarked; for as the styles and carpels were five, we should have expected the division to be into fifths, and according to the number of the stigmas which were acted upon by the foreign pollen. It is, moreover, to be noticed that the action of the pollen in this case is manifest upon what is morphologically the calyx, not upon the pericarp. The apple we refer to was grown in the orchard of William Wicksham, of Washington County, Pennsylvania. — A. G.

notes that the vernation of the leaves is involute in Pyrus, but not in Cydonia, Mespilus, and Aria; that the cottony-leaved varieties, no less than the smooth ones, are glabrous in the seedling stage; that all varieties of the common Pear blossom at Paris whenever, in the month of April, the mean temperature reaches about 10° Centigrade, without perceptible difference between the earliest and the latest-ripening varieties; that the æstivation of the corolla is convolute in Cydonia, but imbricate in the Pear, although ordinarily quincuncial in other *Pomaceæ* (but in the two diagrams of Pear-flowers on Plate A, one has the quincuncial, *i. e.*, in our view typically imbricative æstivation of the corolla; in the other, there is only one wholly outer and one inner petal, — a combination of the quincuncial and the convolute modes which often occurs, but which need not be taken as the type of imbrication); that there are two types as to size of the corolla in the common Pear, the smaller flowered type comprehending most of the cultivated varieties; that the odor of Pear-blossoms is rather disagreeable than otherwise, in contrast with those of Malus, which are sweet-scented. Moreover, the anthers in the Pear genus are tinged with violet; those of the Apple genus are yellow.

As to the morphology and development of the gynæcium, Decaisne reproduces in full the note which he published in the "Bulletin of the Botanical Society of France" in 1857. From his investigation it appears that the five carpels in their early development are free and distinct in the concave centre of the flower; that at a later stage, when the concave receptacle has become much deeper, a cellular tissue develops from its base and inner face, moulds itself around and over the carpels, so as separately to envelop them, except at their inner angle, while it carries up the petals and stamens, and forms the perigynous disk upon which they are inserted; this forms the *core* or central part of the flesh of the fruit, which we have always regarded as receptacle, never ceasing to protest against the still prevalent notion (continued in the latest general works), that the cartilaginous or bony "cells" are "endocarp." But, while we were disposed to regard the whole exterior flesh as calyx, Professor Decaisne (no doubt correctly) regards it as mainly receptacle or axis, an *hypanthium* which in common pears is largely a hypertrophy of the peduncle, after the fashion of Anacardium.

In the proper Pear genus, the ovules never exceed a single pair; this should therefore enter into the generic character.

"Theophrastus had already remarked that the older the Pear-tree, the more prolific, and every day's experience confirms the justice of this observation." The gritty grains or lignified cells which are so abundant in the flesh of many sorts of pears are not wholly absent from any of them. To them is due the roughish surface of the skin, as contrasted with the smooth skin of apples. It is curious to remark that Meyen, in his "Pflanzen-Pathologie," considered the gritty grains to be a disease which attacked pears and quinces.

It appears that pear-growers are able to produce fruits of abnormal size by supporting the growing pear from underneath, instead of allowing it to hang on the peduncle. M. Decaisne has seen Poires de Livre of a kilogram, Gloumorceau of 600 grams, and a Chaumontel of 700 grams weight, produced in this way.

The testa of all Pomaceous seeds is smooth and more or less mucilaginous, except of a Photinia, in which it is reticulated. The cotyledons are accumbent relative to the rhaphe, except in a Photinia, Cotoneaster, Pyracantha (*Cratægus Pyracantha*, Pers.), and Eriobotrya, in which they are incumbent. At first there is a thin layer of albumen, which disappears at maturity of the seed.

Pears are commonly grafted upon a Quince stock. But it is confidently asserted, and generally supposed, that there are more than forty varieties which absolutely refuse this union, and which are therefore managed by subgrafting upon a Pear stock of a proper sort which has itself been engrafted upon the Quince. But, as Professor Decaisne remarks, horticulturists are too apt to generalize their impressions and to limit nature to the narrow horizon of their own practice. Upon the first trial of the experiment under his own observation, he succeeded with twenty of these antipathetic varieties without the least difficulty; but some (among which are the Clairgeau and the Bosc) obstinately refuse to unite with the Quince stock. He naturally discredits the assertion made by Cabanis and by Downing (cited by Darwin), that when certain pears are grafted on the Quince, their seeds produce trees of types different from those which they do when they are raised upon a Pear stock. Decaisne found, as already stated, that Pear-seeds produce indifferently new varieties in any case; that these varieties are not at all fixed into races. He regards as wholly unproven all the assertions that the fruit is ameliorated or in any degree altered by grafting upon a Quince or any other stock. He records a very exceptional instance in which the antipathy of the Pear to the Apple as a stock was so far overcome that the graft survived at least six years, but without vigor, and bore fruit; still this antipathy confirms the generic difference between Pyrus and Malus.

We must pass over the sections on the diseases of the Pear, and the parasitic plants and insects hurtful to it; while as to that on the classification of the Pears of cultivation, we may mention merely the conclusion, which is, that a natural classification of Pears is thus far an impossibility; and that in practice nothing better can be done than to follow the example of the older pomologists, who arranged them according to the period of ripening. A general list of the adopted names of the published varieties of cultivated Pears, alphabetically arranged, fills four pages of the volume. A list of their synonyms, in which each is referred to the adopted name, fills over 12 pages! Then follows a list of pears classed according to the period of maturing, and in which the best varieties are designated.

Finally comes a botanical monograph of the genus Pyrus, with a full generic character, and descriptions and figures of the *races*, as he would term them, considering as he does all known forms of the restricted genus as a single and very polymorphous species.

The six races are: 1. The Celtic, *Proles Armoricana*, of three quasi-species, *P. cordata*, *Boissieriana*, and *longipes*. 2. The Germanic, *Proles Germanica*, or *Pyrus communis*, including our common pears, both pear-shaped and apple-shaped, "both forms being often met with upon the same tree." Under this head Professor Decaisne gives some interesting pages upon the history of the cultivation of pears in France, which cannot be ancient, and of cider (perry) as a drink. It appears that it took the place of beer in the north of France in the fifteenth century or later, and is now giving way to wine and perhaps beer again; and that pears would have disappeared before this from a part of Normandy, were it not that they are carried in immense quantities to Epernay, where they are used in the manufacture of champagne. 3. The Hellenic Race, which comprises *P. parviflora* and three other subspecies. 4. The Pontic Race, *P. salicifolia* and its allies. 5. The Indian Race, *P. Pashia* and its relatives. 6. The Mongolian Race, *P. Sinensis* and its varieties. As one turns over the excellent plates one can hardly be persuaded that such extremely diverse forms can practically be regarded as of one species.

A list of the species remanded from Pyrus to other genera shows that the result of our author's prolonged and sagacious study is to increase the genera about as much as he diminishes the species of the Linnæan Pyrus.

A detailed analysis of Decaisne's monograph of the genus Pyrus was given in this Journal (3 ser., iv. 489, Dec., 1872). Some of the views taken in that work are fully expounded in the present paper,[1] which embracing the results of a prolonged study of an important group, by a botanist of great experience and ability, is worthy of particular attention. As the veteran author states it: —

"My principal object is here to call the attention of botanists to certain characters which have been neglected in systematic works, by the aid of which the ancient genera merged in Pyrus by most of the recent systematists may be neatly circumscribed. Such is the constancy and the value of these characters that the details of organization peculiar to each generic group may be expressed by very general propositions,

[1] *Memoire sur la Famille des Pomacées.* Par J. Decaisne (*Nouvelles Archives du Muséum*, x. pp. 113, 192). Paris, 1875.

which is the very object of a good method. Indeed, when a special organization is common to a large number of different plants, it is evident that comparatively slight but constant modifications of this structure ought to be particularly attended to; and this proposition seems to be especially true of the *Pomaceæ.*" M. Decaisne puts foremost his strongest point when he declares of the Quince, that "the nature of its bark and wood, its prefoliation, inflorescence, the æstivation of the corolla, the structure of the ovary and of the fruit differ essentially from that of the Pears, among which certain botanists still class it." Rather than combine the Quince and the Japan Quince with Pyrus, we are confident that botanists will generally accept his Docynia, along with Chænomeles Lindl. and Cydonia, as independent genera. The same may be said of Mespilus; and it must be allowed that the character which Kunth had noticed and which Decaisne has turned to account, that of the deformation of one of the ovules which becomes a kind of stipitate hood for the other, being common to it and to Cratægus, indicates a relationship to the latter genus rather than to Pyrus. Much nicer and more questionable characters are assigned to the genera here re-established from Pyrus in the Candollean sense, to which we are in this generation accustomed. These are, Aronia, Pers., our Chokeberry (in which eight species are set up from what we take to be a single polymorphous one); Sorbus, Tourn., the Mountain Ash (the synonym *S. microcarpa* omitted from *S. Americana*, and *S. sambucifolia* is still taken to belong only to our western coast, whereas it extends across the continent); Aria, Host., the Beam Trees, all of the Old World; Torminaria, Roem., for *Pyrus torminalis;* also Cormus, Spach, for *Sorbus domestica*, L., the Service-tree of Europe, with *Pyrus trilobata*, DC., and an allied species; Micromeles, a new genus for four Himalayan species thus far little known; lastly, Malus, Tourn. Here it is to be observed that *M. diversifolia* is held to be distinct from *M. rivularis;* and that a subgenus, Chloromeles, proposed for *M. angustifolia*, our narrow-leaved Crab-Apple, thus widely separated from *M. coronaria*, on account, as is stated, of its reddish anthers and the structure of the disk. Pyrus, Tourn., is thus brought down to the Pear; and this, as Decaisne had formerly announced, to a single collective species, of six geographical *proles* or forms. We continue to write Pyrus from old habit and custom, not doubting, however, that Pirus is the correct orthography.

Of Amelanchier, following Lindley, there are enumerated twelve species, six for the Old World and six for North America, and there are names for four more. Without being able to clear them up (and no wonder), Decaisne thinks that they may be distinguished into at least three groups, characterized by the distinct or united styles, and the glabrous or downy ovaries. We are continually impressed with the idea that there must be three or four American species, and the seeds may aid in their definition. But thus far the characters elude investigation. Peraphyllum, Nutt., referred by Bentham and Hooker to Amelanchier, has not been studied by Decaisne. When he examines the excellent specimens in flower and in fruit, which Mr. Siler has supplied from southern Utah, he will conclude that the genus must certainly be reinstated. The likeness is only in the peculiar structure of the fruit.

As respects the remaining genera, the difference between this monograph and the disposition in Bentham and Hooker's "Genera Plantarum" is mainly this: Eriobotrya, with its baccate fruit (what is termed endocarp reduced to a soft pellicle), large turgid seeds with thickened cotyledons, and undulate petals, is upheld as a good genus; Heteromeles is adopted from J. Rœmer for the Californian *Photinia arbutifolia*, and a second (probably not good) species, *H. Fremontiana*, is added. The characters appear to be the 10 instead of 20 stamens, in pairs opposite the calyx-lobes, their filaments dilated at base and somewhat monadelphous. In the tabular conspectus the petals are said to have "préfloraison tordue," but in the generic character it is "æstivatione imbricativa vel convolutiva," the latter term with the French botanists meaning the same as imbricated, only more enrolling. The diagram represents the whole five petals with one edge covered, *i. e.*, "tordue" or contorted (or, as we say, convolute), and so we find them in all the flower-buds now examined. But before adopting the genus it may be well to examine the *Photiniæ* generally. Photinia, of which *P. serrulata* is the type, is characterized as having imbricative æstivation, and Decaisne's diagram represents it as regularly (*i. e.*, quincuncially) so. But in *P. prunifolia* and in *P. Blumei* we find occasionally only one exterior petal, and the four others successively overlapping in the "contorted" way; and in one of Wallich's specimens of *P. integrifolia* the first flower-bud inspected showed the "contorted" æstivation complete. This is also the case in *P. dubia* (in one of Hooker's and Thompson's Khasyia specimens), and this Decaisne refers to Eriobotrya, which has imbricative æstivation. Next is Pourthiæa, a new genus of eleven Japanese and Indian species, the type being *Photinia arguta, villosa, lævis*, etc., and the character, among others, "æstivatione contorta." But we as commonly find one petal wholly exterior. So we think it evident that the æstivation of the corolla furnishes no characters for the division of the genus Photinia.

Finally, as to the proper stone-fruited genera, Pyracantha is adopted from J. Rœmer for *Cratægus Pyracantha* and an allied Indo-Chinese species, and placed near Cotoneaster; and a character not before used is introduced, namely, the position of the cotyledons, which in this genus are, as regards the rhaphe, accumbent.

There are eight plates, six of them filled with admirable dissections, neatly done upon stone by Riocreux from the author's sketches.

ENGELMANN'S NOTES ON THE GENUS YUCCA.[1]

THIS modest title comprises the principal results of Dr. Engelmann's long study of a difficult genus of plants. Pursuing his botanical investigations now for many years only

[1] *Notes on the Genus Yucca.* By George Engelmann, in Trans. St. Louis Acad. iii. St. Louis, 1873. (American Journal of Science and Arts, 3 ser., vi. 468.)

in the intervals and spare moments of a busy and exacting professional life, Dr. Engelmann has made them tell most effectively and advantageously upon the science which numbers him as a distinguished votary, by taking up one subject at a time and investigating it as thoroughly as possible. In this way he has mastered, in turn, our *Cuscutæ* (upon which his earliest essay was published in this Journal, thirty-one years ago, and his latest was a full monograph of all the known species throughout the world), our *Cactaceæ*, our Mistletoes, Euphorbias, Junci, Callitriches, etc., not to speak of several other genera or groups, or taking account of his sedulous and long-continued study of our Oaks, and, above all, of our *Coniferæ*. Nor need we look to this as the close of the series, but rather see before him "fresh fields and pastures new," and wish for him more time to expatiate in them. Upon the principle "to him that hath shall be given," he well deserves it, as having accomplished far more in these rescued moments than others who could mainly devote their days as well as nights to scientific work. Almost without exception these monographs relate to difficult subjects, and such as require long-protracted investigation. This is also true of the present essay upon the genus Yucca. It is not a large one, only a dozen species being clearly made out; but those of long cultivation in Europe have been much confused, and recent ones described without flowers, while fruit is rarely formed out of their native stations, and dried specimens of any completeness are difficult to make, so that means of comparison are much restricted.

The true anthesis, as is now shown, is nocturnal, the flowers remaining half closed during the day. The anthers, with comparatively large and few grains of glutinous pollen, open rather earlier than the flower. The tips of the style, which were naturally taken for stigmas, are now shown to be functionless, the stigmatic surface being the moist and glutinous lining of a stylar tube, which extends downward nearly to the cells of the ovary and even communicates directly with them. As soon as it became evident that fertilization must depend upon nocturnal insects, it was found that they were most frequently and regularly visited by "white moths, which, usually in pairs, disported in the open flowers at dusk, and were found quietly ensconced in them when closed in the day-time." Professor Riley of St. Louis, the distinguished entomologist, was at this point associated with Dr. Engelmann in the investigation. The result has been given to the scientific world in his interesting memoir on *Pronuba yuccasella*, first read at the Dubuque meeting of the American Association for the Advancement of Science, in August, 1872, and now also published, as a sequel to Dr. Engelmann's monograph, in the "Transactions of the St. Louis Academy."

"The rootstock of all the Yuccas is, under the name of 'Amole,' an important article in a Mexican household, being everywhere used as a substitute for soap, as it is replete with mucilaginous and saponaceous matter, probably a substance analogous to the saponine of the Saponaria root. It is curious to learn that the negroes of the coast of South Carolina repeatedly destroyed Dr. Mellichamp's carefully observed clumps of Yuccas, in order to obtain the saponaceous rootstock." In Colorado Territory we found that *Yucca angustifolia* is as generally called "Soap-plant" as is the Chlorogalum in California.

While the nature of the fruit, whether capsular or baccate, is a tribal character in *Liliaceæ* generally, Yucca has both kinds; and Dr. Engelmann turns this character, with accompanying differences in seeds, to good account in the systematic arrangement of the species. The common "Spanish bayonet," *Y. aloifolia* of the Southern States, and some related Texan and Mexican species, represent the pulpy-fruited section; *Y. brevifolia*, which ranges across the Arizonian border of the United States, has a spongy indehiscent pod, probably at first more or less drupaceous; while the Bear-Grass, *Y. filamentosa*, and its allies bear a dry capsule. It may here be recorded that the name *Y. canaliculata* of Hooker must replace that of *Y. Treculiana* of Carrière, the latter being a name published without characters, in 1858, the former described and figured in 1860, in the "Botanical Magazine." The prince of Yuccas must be *Yucca baccata*, which, in its variety *australis*, forms "trees twenty-five to thirty feet high, and two or three feet in diameter, with ten or a dozen branches," or sometimes reaches to even fifty feet of elevation according to the late Dr. Gregg, although the most northern form of it is almost stemless. Its pulpy fruits are "savory, like dates," are eaten fresh by both whites and Indians, and are cured by the latter for winter provisions. They also make a stew of the flower-buds and flowers, which Dr. Palmer found to be pleasant and nourishing. The seeds are said to be actively purgative. The fibers of the leaves are used for cordage, the trunks for palings, or are riven into slabs for the coverings of huts, and the tender top of the stem is roasted and eaten.

Professor Riley's curious paper upon the mutual relations of Yuccas with Pronuba, a Tineideous moth that does the work of pollination, will be read with interest.

RUSKIN'S PROSERPINA.

Mr. Ruskin, "having been privileged to found the School of Art in the University of Oxford," now proposes to found a new school of botany.[1] Of course, it will be a vagarious school. One crying evil to be remedied is "that there are generally from three or four up to two dozen Latin names current for every flower," and "the most current and authoritative names" are "of the devil's own contriving." This is not seemly. As Wesley would not allow the devil to have the singing of all the good tunes, so neither will Ruskin allow him to have the naming of all the sweet flowers. He proposes "to substitute boldly . . . other generic names for the plants thus faultfully hitherto titled." He "will not even name the cases in which they have been made," but will "mask those which there was real occasion to alter by sometimes giving new names in cases where there was no necessity of such kind."

[1] *Proserpina*, Studies of Wayside Flowers, etc. By John Ruskin. London and New York, 1875. (The Nation, No. 528, August 12, 1875.)

That is to say, the evil of a redundancy of botanical names is to be overcome by making more, some of them avowedly needless; and innocent names are to suffer, lest bad ones should become notorious by being discarded without company. For it appears that the diabolical names to be sent to their own place in this reform are not discarded because they are cacophonous, although that is the common charge, but because they are immoral. Of the two evils to be dealt with, the first is simply a superfluity of Latin names; the second, and the worse, a superfluity of naughtiness. As Mr. Ruskin forcibly puts it: —

"The second, and a much more serious one, is of the devil's own contriving (and, remember, I am always quite serious when I speak of the devil), namely, — that the most current and authoritative names are apt to be founded on some unclean or debasing association, so that to interpret them is to defile the reader's mind."

This reminds us of the fine lady who thanked Dr. Johnson for omitting indelicate words from his dictionary, to whom the blunt moralist rejoined: "I perceive, madam, that you have been looking for them." Now, if the case be really as it is represented, the sound practical axiom, "Quieta non movere," would seem to suggest the proper treatment; for his purposes, one would think the Latin and Greek names might be left untranslated; and the reform might begin, and end, with the popular English names, — almost the only ones the author may need to use, — some of which are coarse and vulgar enough. Indeed, as to botanical generic names, far from finding any confirmation of Mr. Ruskin's sweeping charges, we can recall barely one or two the translation or etymology of which would be embarrassing at the parlor-table. Moreover, if the following passages really refer to "nomenclature" (though terminology seems to be meant), our author, upon his own showing, need not waste his time in the endeavor to reform it: —

"The mass of useless nomenclature, now mistaken for science, will fall away, as the husk of a poppy falls from the bursting flower. . . . When the science becomes approximately perfect, all known plants will be properly figured, so that nobody need describe them, and unknown plants will be so rare that nobody will care to learn a new and difficult language in order to give an account of what in all probability he will never see."

Well, for that matter, the English Botany, in its various editions, furnishes fairly good figures of all British plants; and the "Botanical Magazine"—a page of an early number of which, eighty years old, is gibbeted by Mr. Ruskin—has gone on to figure more than 6000 cultivated exotics, and is continuing at the rate of nearly a hundred a year; so that our author's ideal is practically all but realized already. There are wellnigh pictures enough, if one knows how and where to find them. And it amusingly appears, from Mr. Ruskin's trouble with St. Bruno's Lily at the beginning, and from his investigation of moss further on, as well as from scattered statements, that his mode of proceeding in systematic botany is the simple one of searching high and low for a picture to match the specimen in hand. Accordingly, it is not surprising that his "botanical studies were, when [he] had attained the age of fifty, no further advanced than the reader will find them in the opening chapter of this book."

As to this, the conclusions which the reader will draw are all along anticipated by the author. Next to the pervading—well—bumptiousness, nothing is so prominent in the book as the profession, not to say the parade, of ignorance of the topics treated. As to "the elements of the science of botany," "I can scarcely say that I have yet any tenure of it myself." "And, meanwhile I don't know very clearly so much as what a root is or what a leaf is." "Some one said of me once, very shrewdly, When he wants to work out a subject, he writes a book on it. . . . This book will be nothing but processes. I don't mean to assert anything positively in it, from the first page to the last. Whatever I say is to be understood only as a conditional statement—liable to and inviting correction. And this the more because, as on the whole I am at war with the botanists, I can't ask them to help me, and then call them names afterwards." So "for many reasons, I am forced to print the imperfect statement, as I can independently shape it." To get at the facts, "I should have to write a dozen of letters before I could print a line, and the line at last would be only like a bit of any other botanical book — trustworthy, it might be perhaps, but certainly unreadable." The converse is preferred. Readable it certainly is, and in its way interesting, not so much for what it tells about botany as for what it tells about Mr. Ruskin; and the art student, out of the abundance of golden chaff, may pick some grains of knowledge that might not otherwise fall in his way. But the seeker of botanical information must glean warily, especially where the author grows positive. For almost the only instance in which he does pronounce decidedly happens to be a vexed question in vegetable physiology, and there is reason to fear that he decides it wrongly. At least the recent investigators who have had the matter in hand in the way of experimental inquiry, will not agree with him that the plant can get water from the atmosphere directly and "for the most part does so; though when it cannot get water from the air, it will gladly drink by its roots." Still "our natural and honest mistakes will often be suggestive of things we could not have discovered but by wandering." Very likely; but why invite learners to go forth with him upon his wanderings? In many a book the want of sufficient knowledge is pleaded as an excuse; in this, it is paraded as a recommendation. Ignorance, no doubt, has its uses; but it is questionable whether teaching is altogether the best use to put it to. As the member of Parliament who yawned desperately while delivering his speech was thought to trench upon the privilege of his hearers, so the students of "Proserpina" may complain that the playing of the rôle both of teacher and learner at once involves some incongruity and inconveniences.

The second part of "Proserpina" has just come to hand. It treats of the leaf and the flower, in a discursive and oracular way, leading into æsthetical questions, where we need not follow and do not greatly admire. Now and then a scientific topic is taken up, and the point missed, as usual. Treating of foliage and its office, we are bid "to think awhile of its dark clear green, and the good of it to you." We look for an exposition of the fact, in which the whole meaning of vegetation inheres, that leaves under the sun's influence create all the food of the world, and are therefore the basis of all animal existence. Instead of which we have: —

"Scientifically, you know, green in leaves is owing to 'chlorophyll,' or, in English, to 'green leaf.' It may be very fine to know that; but my advice to you, on the whole, is to rest content with the general fact that leaves are green when they do not grow in or near smoky towns, and not by any means to rest content with the fact that very soon there will not be a green leaf in England, but only greenish-black ones. . . . Well, this much the botanists really know and tell us" — that vegetation "is made chiefly of the breath of animals. . . . So that you may look upon the grass and forests of the earth as a kind of green hoar-frost, frozen upon it from our breath, as, on the window panes, the white arborescence of ice."

Mr. Ruskin evidently has no idea of the essential independence of the vegetable kingdom; that, as all the carbon of the breath of animals comes from plants, so they, in their decay, would furnish this material for succeeding vegetation perhaps as rapidly, on the whole, without the intervention of animals. At most, the latter somewhat expedite the decomposition.

"But how is it made into wood?" As to that and matters therewith connected, "under the impression that it had been ascertained, and that I could at any time know all about it, I have put off till to-day the knowing of anything about it at all. But I will really endeavor now to ascertain something, and take to my botanical books accordingly."

Behold the result of the cram, "the gist of the matter": —

"Hence generally, I think we may conclude thus much, that at every pore of its surface, under ground and above, the plant in the spring absorbs moisture, which instantly disperses itself through its whole system 'by means of some permeable quality of the membranes of the cellular tissue invisible to our eyes even by the most powerful glasses'; that in this way subjected to the vital power of the tree, it becomes sap, properly so called, which passes downwards through this cellular tissue, slowly and secretly; and then upwards, through the great vessels of the tree, violently, stretching out the supple twigs of it as you see a flaccid water-pipe swell and move when the cock is turned to fill it. And the tree becomes literally a fountain, of which the springing streamlets are clothed with new-woven garments of green tissue, and of which the silver spray stays in the sky — a spray, now, of leaves."

Then as to the blossom: "The flower exists for its own sake, not for the fruit's sake. . . . But the flower is the end of the seed, not the seed of the flower." "The corolla leads and is the object of final purpose. The stamens and the treasuries [Mr. Ruskin's new term for pistils] are only there in order to produce future corollas." Without criticising anybody's notion of final causes, we only notice how Mr. Ruskin fails to make his own point. He has seen "among the speculations of modern science, several, lately, not uningenious, and highly industrious, on the subject of the relation of color in flowers to insects, to selective development," etc. And he proceeds to intimate that even Mr. Darwin must be ranked among "the men of semi-faculty or semi-education who are more or less incapable of so much as seeing, much less thinking about color," etc., referring merely to the latter's speculations upon the ocelli of the Argus Pheasant, in blissful ignorance, it would seem, that he has to deal with Mr. Darwin upon this very subject of color and use in flowers, and that he is not prepared even to state his own side of the question.

EMERSON'S TREES AND SHRUBS OF MASSACHUSETTS.

TREE-LORE[1] is no longer confined to the few, and books like this address a large and various audience, or will do so when they become better known. Mr. Emerson's original volume

[1] *A Report on the Trees and Shrubs of Massachusetts*, growing naturally in the forests of Massachusetts. By George B. Emerson. 2d ed. Boston, 1875. (The Nation, No. 539, October 8, 1875.)

was published twenty-five years ago, "agreeably to an order of the Legislature, by the Commissioners on the Zoölogical and Botanical Survey of the State" of Massachusetts, being a supplement to the Geological Survey of that period, prosecuted under Edward Everett's governorship. It was the most popular report of the series, and the edition was ere long exhausted. When this came to pass — without waiting for the new survey which the State last year came near authorizing but failed to do so — Mr. Emerson, unassisted, set about the preparation of a new edition, devoted several years to it, and to the study of what had been done for the preservation and utilization of forests in the Old World, and for their waste and destruction here in the New; and he has at length brought out this second edition, in two goodly octavo volumes, illustrated and adorned by a large number of well-executed plates. These being interspersed through the pages, unnumbered, and nowhere enumerated, the only way of ascertaining their actual amount was to count them. We find 144 plates, of various kinds and merits. The least satisfying to us are those of portraitistic or scenic character, borrowed from the German "Der Wald" and the very French "Vegetable World" of Figuier; yet to others these may be the most attractive. Very good, though unpretending, are figures, mainly in outline, contributed by Mr. Isaac Sprague to the first edition, here reproduced. Best of all are those contributed by the same hand to the new edition, original figures of the foliage, flowers, and fruit of many of our trees and shrubs not before illustrated, transferred from Sprague's drawings to stone, and printed in colors. The plates representing our two northern Azaleas, the Roxbury Waxwork (as they name it around Boston), the Virginian Creeper in its autumn dress; the Red Maple, both in vernal and autumnal robes; and the Flowering Raspberry, from which seemingly one may almost shake the mountain dew, are good illustrations of what may be done in this way. Hand-coloring is too expensive, and chromo-lithography can really be turned to excellent account in its place for natural-history illustrations, whatever be its merits or demerits in other regions of art. The letter-press we find is a reprint of the first edition, as to the descriptive part. The popular descriptions seem to have hit the mark.

Mr. Emerson's instructions and appeals for the planting and care of trees, and for the renovation of our woods, wherever practicable and profitable, are worthy of all attention, as his efforts in this regard are worthy of all honor. They began long ago, and have been redoubled now in his later years, in this work and elsewhere. The memorial addressed to the President and Congress by the American Association for the Advancement of Science he took a large part in preparing; and his personal furtherance of it at Washington, of which he makes modest mention in the preface, may yet be fruitful of benefit. It may not be improper to add that the only permanently endowed arboretum in America — the Arnold Arboretum, entrusted to Harvard University — owes its existence to our author's thoughtfulness and sense of the importance of tree-culture.

DARWIN'S INSECTIVOROUS PLANTS.[1]

THIS long expected work appeared last autumn, was immediately reprinted by the American publishers, and before this time has been so widely read that no detailed account of it is at all necessary. Its main topic is Drosera or Sundew, upon which the vast number and diversity of the observations and experiments — at once simple, sagacious, and telling — which it records, are about as wonderful as the results. As to the latter, it is established beyond question that the common Sundews are efficient fly-catchers; that the stalked glands, or tentacles as Mr. Darwin terms them, are sensitive and turn inward or even in other required directions in response to irritation; that they equally respond and move in obedience to a stimulus propagated from a distance through other tentacles and across the whole width of the leaf; that the sensi-

[1] *Insectivorous Plants.* By Charles Darwin. London and New York, 1875. (American Journal of Science and Arts, 3 ser., xi. 69.)

tiveness belongs only to the glands and tips of the tentacles, but is propagated thence down their stalks and across the blade of the leaf through the cellular tissues; that they accurately and delicately discriminate animal or other nitrogenous matter from anything else; that the glands absorb such matter; that when excited by contact, or by the absorption of nitrogenous matter by the viscid enveloping liquid, an acid secretion is poured out and a ferment analogous to pepsin, the two together dissolving animal matter; so that the office and action of these glands are truly analogous to those of the glands of the stomach of animals. Finally, that animal or nitrogenous matter, thus absorbed and digested in the glands, is taken in, and conveyed from cell to cell through the tentacles into the body of the leaf, was made evident by ocular inspection of the singular changes in the protoplasm they contain. So particularly have the investigations been made and so conscientiously recorded, that the account of those relating to one species of Sundew, *Drosera rotundifolia*, fills 277 pages of the English edition, or more than half of the book. After all it ends with the remark: "and we see how little has been made out in comparison with what remains unexplained and unknown." The briefer examination of six other Sundews follows, some of them equally and others less efficiently fly-catchers and feeders.

Dionæa is next treated, but with less detail. Indeed, except as to the particular nature of the secreted digesting fluid, there is little in this chapter that had not been made out or already become familiar here. That the secretion has digestive powers, and that it is reabsorbed along with whatever has been digested, is now proved beyond reasonable doubt. That the motor impulse is conveyed through the cellular parenchyma, and not through the vascular bundles, or spiral vessels, and that the latter do not originate the secretion, as Rees and Wills in a recent paper seem to suppose they must, appears to be shown by the facts, and was antecedently probable. "The wonderful discovery made by Dr. Burdon Sanderson is now universally known: namely, that there exists a normal electrical current in the blade and footstalk, and that when the leaves are irritated the current is disturbed in the same manner as takes place during the contraction of the muscle of an animal." The conclusion here needs to be checked by parallel experiments, to see whether the same reversion of current does not take place whenever a part of any leaf or green shoot is forcibly bent upon itself.

Aldrovanda vesiculosa, of the Drosera family, "may be called a miniature aquatic Dionæa;" for, as discovered by Stein in 1873, "the bilobed leaves open under a sufficiently high temperature, and when touched suddenly close." Being submerged, their prey is confined to minute aquatic animals. For want of proper material and opportunity, Mr. Darwin was able to follow up only for a little way the observations of Stein and Cohn, — enough, however, to show that it also captures and consumes animals, but perhaps avails itself of the nitrogenous matter only when passing into decay.

Drosophyllum, a rare representative of the order, confined to Portugal and Morocco, grows on the sides of dry hills near Oporto; so that, as to station, it is the very counterpart of Aldrovanda. Its leaves are long and slender, in the manner of our *Drosera filiformis*, and are covered with much larger glands. To these, flies adhere in vast numbers. "The latter fact is well known to the villagers, who call the plant the 'fly-catcher,' and hang it up in their cottages for this purpose." Mr. Darwin found the glands incapable of movement, and their behavior in some other respects differs from that of Drosera; but they equally secrete a digestive juice. Insects usually drag off this secretion instead of being fixed on the glands by it; but their fate is no better; for as the poor animal crawls on and these viscid drops bedaub it on all sides, it sinks down at length exhausted or dead, and rests on a still more numerous set of small sessile glands which thickly cover the whole surface of the leaf. These were till then dry and inert, but as soon as animal matter thus comes in contact with them, they also secrete a digestive juice, which, as Mr. Darwin demonstrated, has the power of dissolving bits of coagulated albumen, cartilage, or meat, with even greater readiness than that of Drosera.

Mr. Darwin next records various observations and experiments upon more ordinary glandular hairs of several plants. To certain Saxifrages his attention was naturally called, on account of the presumed relationship of *Droseraceæ* to this genus. He declares that "their glands absorb matter from an infusion of raw meat, from solutions of nitrate and carbonate of ammonia, and apparently from decayed insects." To such plants the vast number of little insects caught may not be useless, as they may be to many other plants (Tobacco, for instance) with sticky glands, in which Mr. Darwin could detect no power of absorption. The prevalent idea, that glandular hairs in general serve merely as secreting or excreting organs, and are of small or no account to the plant, must now be reconsidered. Those of the common Chinese Primrose (*Primula Sinensis*), although indifferent to animal infusions, were found to absorb quickly both the solution and vapor of carbonate of ammonia. Now, as rain-water contains a small percentage of ammonia, and the atmosphere a minute quantity of the carbonate or nitrate, and as a moderate-sized plant of this Primrose was ascertained (by estimate from a count on small measured surfaces by Mr. Francis Darwin) to bear between 2½ to 3 millions of these glands, it begins to dawn upon us that these multitudinous organs are neither mere excrescences nor outlets, nor in any just sense insignificant.

Mr. Darwin next investigates the densely crowded short glandular hairs, with their secretions, which form the buttery surface of the face of the leaves of Pinguicula, the Butterwort. He finds that the leaves of the common Butterwort have great numbers of small insects adhering to them, as also grains of pollen, small seeds, etc.; that most substances so lodged or placed, if yielding soluble matter to the glands, excite them to increased secretion; but that if non-nitrogenous the viscid fluid poured out is not at all acid, while if nitrogenous it invariably has an acid reaction and is more copious; that in this state it will quickly dissolve the muscles of insects, meat, cartilage, fibrin, curds of milk, etc.; that when the surface of a plane leaf is fed, by placing upon it a row of flies along one margin, this margin, but not the other, folds over within twenty hours to envelop them; and when placed on a medial line, a little below the apex, both margins incurve. He concludes "that *Pinguicula vulgaris*, with its small roots, is not only supported to a large extent by the extraordinary number of insects which it habitually captures, but likewise draws some nourishment from the pollen, leaves, and seeds of other plants, which often adhere to its leaves. It is therefore partly a vegetable as well as an animal feeder." The leaves in one or two other species were found capable of greater and more enduring inflection, and the glands excitable to increased secretion even by bodies not yielding soluble nitrogenous matter.

The aquatic type of this family is Utricularia; and the bladder-bearing species of this genus are to Pinguicula nearly what Aldrovanda is to Dionæa and Drosera — the bladders imprisoning minute aquatic animals by a mechanism almost as ingenious as that of Dionæa itself. Observations of the same kind were made in this country by Mrs. Treat, of Vineland, New Jersey, before Mr. Darwin's investigations were made known. These submerged aquatic stomachs, ever deluged with water, apparently do not really digest their captures, but merely absorb the products of their decay.

The same must in all probability be said of such Pitcher-plants as Sarracenia and Darlingtonia, which Mr. Darwin merely alludes to at the close of his volume but does not treat of. Nepenthes, however, according to Dr. Hooker's investigations, has attained a higher dignity, and converted its pitcher into a stomach. This parallelism, and this higher and lower mode of appropriating organic products by each of the three well-marked carnivorous families of plants, are highly suggestive.

In concluding this notice of a book for which we have no room to do justice, — but which is sure to be in the hands of many interested readers, — there is something to be said in regard to the discovery of the lure in some of our Sarracenias. We have by degrees to discover our discoverers. In this Journal, only so far back as the number for August, 1873, is a notice of the discovery of a sweet secretion at the orifice of the pitcher of *Sarracenia flava*, by Mr. B. F. Grady, of Clinton, North Carolina (in the article by an oversight called "Mr. Hill"), which effectively lures flies to their destruction. This statement, made in a letter, had been for several months in our hands, awaiting the opportunity of confirmation, when an allusion to the same thing appeared in the English edition of LeMaout and Decaisne's System of Botany, without reference to any source, and on inquiry we learned that the authority for the statement was forgotten. But early in the following year, when the monograph of the order appeared in the last volume of De Candolle's "Prodromus," a reference was found to a paper by Dr. Macbride in the Transactions of the Linnæan Society. His observations (made upon *S. variolaris*), it appears, were communicated to Sir J. E. Smith, read before the Linnæan Society in 1815, and published soon after. They are referred to by his surviving friend and associate, Eliott, in his well-known work, and therefore need not have gone to oblivion, or needed rediscovery here in our days by Mr. Grady and Dr. Mellichamp, the latter greatly extending our knowledge of the subject. Probably the main facts were all along popularly known in the regions these species affect, and where their use as fly-traps is almost immemorial. But the gist of these remarks is, that a colleague has just called our attention to an earlier publication than that of Dr. Macbride, namely, an article on "Certain Vegetable Muscicapæ," by Benjamin Smith Barton (one of our botanical fathers), published in "Tilloch's Philosophical Magazine" for June, 1812. Among other matters not bearing directly upon this point, he says of Sarracenia, without reference to any particular species: "A honeyed fluid is secreted or deposited on the inner surface of the hollow leaves, near their *faux* or opening; and this fluid allures great numbers of the insects which they are found to contain into the ascidia."

Here is earlier publication by three years. Yet we suspect that Dr. Barton knew little about it at first hand, and we find clear evidence that he had not anticipated Dr. Macbride. All his references have an indefiniteness quite in contrast with

Dr. Macbride's narrative; he says that "some if not all the species of the genus appear to possess a kind of glandular function," without mentioning those that have it, or the absence of it in the only species growing around him at the north; and he adds that he "was entirely unacquainted with this curious economy . . . when I published the first edition of my 'Elements of Botany,' and even when I printed the appendix (in vol. i.) to the second edition of this work." Now his paper is dated September 11, 1811; and the volume referred to, as just printed, is dated 1812. But Macbride states that his observations were chiefly made 1810 and 1811; he corresponded intimately with Eliott, through whom, if not directly, his observations would probably find their way at once to the Philadelphia naturalists.

NAUDIN ON THE NATURE OF HEREDITY AND VARIABILITY IN PLANTS.[1]

Why is it the nature and essence of species to breed true, and why do species sometimes vary? In other words, why is offspring like parent, and when unlike in certain particulars, what is the cause and origin of the difference? We commonly and properly enough take these two associated yet opposed facts as first principles. But it is equally proper and legitimate to enquire after the cause of them.

M. Naudin, a good many years ago, took up the study of hybrid plants, and followed up for a series of generations, the course of life of certain self-fertile ones, notably of Datura. We gave at the time an abstract of his observations of the manner in which the characters of two closely related common species, *D. Stramonium* and *D. Tatula*, were mixed, and in which the characters of the two began to separate in the close-bred progeny of the next generation, ending in a complete division of the amalgamated forms into those of the two constituent species after a few generations.

[1] American Journal of Science and Arts, 3 ser., xi. 153.

The "Comptes Rendus" of September 27 and October 4, 1875, contain an abstract of a paper communicated by M. Naudin to the French Académie des Sciences, of which the text was suggested by a hybrid between the wild *Lactuca virosa* and a variety of *L. sativa*, the common Lettuce. The hybrid was an accidental one: its seeds were fully fertile; a great number of young plants were raised from them, of which twenty were preserved for full development and study. Like other hybrids the original showed no character which was not evidently derived from the two parents; and, fertilized by its own pollen, the offspring all agreed in this respect, although they varied exceedingly among themselves in the division of the parental heritage, no two being quite alike. This exceeding vacillation between the two parental forms, but not overpassing the limits on either hand, — which Naudin finds to be the common characteristic of fertile hybrids, close-bred, — he names disordered variation (*variation désordonnée*). His explanation is that the hybrid is a piece of living mosaic, that two specific natures are at strife in it; in the progeny each endeavors to reclaim its own, like seeks like; whence in the course of a very few generations (as he first showed in Datura) a segregation takes place, part of the progeny reverting completely to one ancestral type, part to the other. What Naudin now insists upon is that out of all this disturbance comes nothing new; that there is here no variation beyond the line of inheritance; and therefore from crossing no possible development of species.

To this proposition we accede, so far as respects the direct consequence of crossing. To fill up the interval more or less between two forms or species with intermediate patterns may tend to the fusion or confusion of the two, but not to the origination of new forms or species. Although Naudin's own experiments lead him to deny all tendency to variation overpassing these limits, we do not forget that his countryman, the late M. Vilmorin, — working in a different way and with another object, — arrived at a different conclusion. He succeeded, as we understand, in originating floricultural novelties from species which refused to vary *per se*, by making a cross,

— not to infuse the character of the male parent, for he fertilized the progeny with the pollen of the female parent, and thus early bred out the other blood, but to induce variation, which, once initiated in the internal disorder consequent upon the crossing, was apt to proceed, or might be led on by selection, to great lengths, according to Vilmorin. The variations in question, being mainly such as are sought in floriculture, may not have passed the line laid down by Naudin, or actually have introduced new features. But such plants would surely have no exemption from the ordinary liability to variation. If other plants vary, in the sense of producing something new, so may these.

This brings us to another inference which Naudin draws. Having observed that his hybrids in their manifold variation exhibited nothing which was not derivable from their immediate ancestry, he directly (and in our opinion too confidently) concludes that all variation is atavism, — that when real variations are set up in ordinary species, this is not an origination but a reversion, a breaking out of some old ancestral character, a particular and long deferred instance of this *variation désordonnée*, which would thus appear to be the only kind of variation. This view has been presented before, but not, perhaps, so broadly. Adducing some theoretical considerations in its favor — to which we may revert — and some sound reasons against the view that variation is caused by external influences, he declares it "infinitely more probable that variation of species properly so called is due to ancestral influences rather than to accidental actions." We might think so if these two categories were exhaustive, and external conditions must be supposed to act immediately, as the cause rather than the occasion of variation. But the supposition that "accidental actions," whatever they may be, and external influences of every sort do not produce but educe and conduct variation — which is our idea of what natural selection means — avoids the force of Naudin's arguments.

Moreover, Naudin's view, regarded as an hypothesis for explaining variation, leaves the problem just where it finds it. To explain the occurrence of present and actual variations, hypothetical ones like those of a former time are assumed; the present diversity implies not only equal but the very same anterior diversity, and so on backwards. Or rather it demands a much greater diversity at the outset than now; for these aberrant forms are the rare exception, and if due to atavism they imply the loss of the many and the incidental reappearance of the few. Else they would be the rule instead of the exception, and atavism would be simply heredity. This comes to the view which Mr. Agassiz strongly maintained, that really there are no varieties, — meaning, we understand, that all the forms are aboriginal, except the transient ones evidently due to circumstances.

That some variation is atavism is clear enough. This is the natural explanation of the appearance of characters wanting in the immediate parents but known in their ancestors or presumed ancestors. But the assumption of hypothetical ancestors to account for variation generally is quite another thing. Besides its inutility as an explanation, to which we have adverted, its improbability as an hypothesis is set in a strong light by Naudin's own forcible conception of the nature of heredity. What is heredity? he asks. In other words, what keeps species so true, offspring like parent, through the long line of generations? He illustrates hereditary force by comparing its action with that of physical force, in which the movement from one state of equilibrium to another is always that in which there is least resistance. From which it follows that when it has once begun to proceed in a certain course, its tendency to continue in that direction increases, because it facilitates its way as it overcomes obstacles. In other words, this line becomes fixed by habit; *vires acquirit eundo;* the stream deepens its bed by flowing; and the more remote the commencement of a certain course, the more fixed its direction, and the greater its power of overcoming opposition. The species is kept true in its course by the sum of the heredities which press each individual forward in its actual direction. So that, as Naudin remarks, if we could calculate the energy with which millions of ancestors tend to impel the living representative of the line onward in the same direction, we should better apprehend the persistence of species, and feel the great improbability that the stream will ever escape from its ancient and well-worn bed, and strike into new courses.

Now, in the first place, the more lively the conception we thus form of the invariability of species, through a happy metaphorical illustration of it, the more unlikely does it appear that early characters, long lost in the flow, should reappear through atavism as varieties. To continue the simile, the more impetuous the stream, the less the possibility of its turning back upon itself, and resuming old characteristics. The eddies of atavism (the resumption of dropped characters) are not likely to extend back very far; and it seems gratuitous to have recourse to them in explanation of new forms. Moreover, although the stream has made its bed and lies in it, not escaping from its own valley, it is flexible enough to obstacles, is ever changing its particular course as it flows, and may by its own action send off here and there a bayou (variety) or branch into a delta of channels (derivative species).

Like Agassiz, Naudin conceives of species as originating with a large number of individuals of the same structure, and of which numerous reciprocal crosses have determined the direction of the line in which their posterity have evolved. But he maintains that these individuals, and all existing species, had a common origin in a "proto-organism"; and that the various lines of descent acquired fixity into species only as they acquired sexuality. If we rightly apprehend it, Naudin's idea of the purport of sexual reproduction (as contrasted with that by buds) is to give fixity to species. Our idea is a different one, both as to the essential meaning of sexuality and as to its operation in respect to fixity. His conception may be tested by inquiring which are the more variable, or sportive, seedlings or plants propagated from buds. This we suppose can be answered only one way.

M. Naudin is a veteran and excellent investigator, and nothing which he writes is to be slighted. We have frankly set down our impressions upon a first perusal of his important communication; but are ready to revise them, if need be, upon more deliberate consideration.

CROSS AND SELF-FERTILIZATION IN THE VEGETABLE KINGDOM.

MR. DARWIN, in the title of his new work,[1] refers only incidentally to adaptations for cross-fertilization, — a subject which has given origin to a copious literature since he opened it anew in his book on the Fertilization of Orchids, in 1862. A new edition of this latter book is on the eve of publication in England, and we believe that this author's scattered papers on cross-fertilization, as secured by various contrivances, are about to be collected, revised, and published in book form. In the volume now before us, Mr. Darwin deals with the effects of cross and self-fertilization, recounts at length the experiments he has devised and carried on, collects and criticises the results, glances at the means of fertilization, and the habits of insects in relation to it, and ends with some theoretical considerations of inferences suggested by or deduced from the facts which have been brought to light.

If writing for the popular press, we should be bound to say that the book is not light reading. Three fourths of its pages and of the chapters are devoted to the details of the experiments and the sifting and the various presentation of the results; and the remainder, although abounding in curious facts and acute suggestions, is yet of a solid character. The bearings of various points upon what is called "Darwinism" are merely touched or suggested, here and there, in a manner more likely to engage the attention of the thoughtful scientific than of the general reader.

That cross-fertilization is largely but not exclusively aimed at in the vegetable kingdom, is abundantly evident. As Mr. Darwin declares, "it is unmistakably plain that innumerable flowers are adapted for cross-fertilization, as that the teeth and talons of a carnivorous animal are adapted for catching prey, or that the plumes, hooks, and wings of a seed are adapted

[1] *The Effects of Cross and Self-Fertilization in the Vegetable Kingdom.* By Charles Darwin. London and New York, 1876. (American Journal of Science and Arts, 3 ser., xiii. 125.)

for its dissemination." That the crossing is beneficial, and consequently the want of it injurious, is a teleological inference from the prevalence of the arrangements which promote or secure it, — an inference the value of which increases with the number, the variety, and the effectiveness of the arrangements for which no other explanation is forthcoming. That the good consisted in a re-invigoration of progeny, or the evil of close-breeding in a deterioration of vigor, was the suggestion first made (so far as we know), or first made prominent, by Knight, from whom Darwin adopted it. However it be as to animals, there was until now no clear and direct evidence that cross-fertilization in the vegetable kingdom did re-invigorate. Indeed, the contrary might be inferred from the long and seemingly indefinite perpetuation of bud-propagating varieties, which have no fertilization at all. But the inference from this is not as cogent as would at first appear. For, although bud-propagation is, we think, to be considered as the extreme of close-breeding, yet in it the amount of material contributed by parent to offspring is usually vastly more than in sexual reproduction; and, accordingly, the diminution to an injurious degree of any inherited quality or essence might be correspondingly remote. Yet, as sexual reproduction may be and often must be much closer in plants than it can be in most animals, the ill effects of self-fertilization, or the good of cross-fertilization, might the sooner be noticeable. Mr. Darwin arranged a course of experiments to test this question, prosecuted it as to some species for eleven years; and the main object of this volume is to set forth the results.

Ipomœa purpurea, the common Morning Glory of our gardens, was the leading subject. The flowers of this species self-fertilize, but must also be habitually cross-fertilized, as they are visited freely by bumble-bees and other insects. Ten flowers of a plant in a green-house were fertilized with their own pollen; ten others were crossed with pollen from a different plant. The seeds from both were gathered, allowed to germinate on damp sand, and as often as pairs germinated at the same time the two were planted on opposite sides of the same pot, the soil in which was well mixed, so as to be uniform in composition. "The plants on the two sides were always watered at the same time and as equally as possible, and even if this had not been done the water would have spread almost equally to both sides, as the pots were not large. The crossed and self-fertilized plants were separated by a superficial partition, which was always kept directed towards the chief source of light, so that the plants on both sides were equally illuminated." Five pairs were thus planted in two pots, and all the remaining seeds, whether or not in a state of germination, were planted on the opposite sides of a third pot, so that the plants were crowded and exposed to a very severe competition. Rods of equal diameter were given to all the plants to twine up, and as soon as one of each pair had reached the summit, both were measured. But a single rod was furnished to each side of the crowded pot, and only the tallest plant on each side was measured. This was followed up for ten generations; the close-fertilization being always self-fertilization, *i. e.*, by pollen to stigma of the same flower; the crossing, between individuals in successive generations of this same stock, except in special instances, when an extraneous stock was used as one parent, — to eminent advantage, as will be seen.

The difference in vigor between the cross-bred and the close-bred progeny, as measured by early growth, was well marked throughout. In the mean of the ten generations it was as 100 to 77. In the tenth generation it was 100 to 54, that is, five cross-bred plants grew to the average height of 93.7 inches while the close-bred were reaching the average of 50.4 inches. This was a notably greater difference than in any previous generation. But this was probably accidental or anomalous, for it was not led up to by successive steps. Indeed, the difference in the first generation was a trifle greater than the average of all ten, being as 100 to 76. The second generation was as 100 to 79; the third as 100 to 68; the fourth as 100 to 86; the fifth as 100 to 75; the sixth as 100 to 72; the seventh as 100 to 81; the eighth as 100 to 85; the ninth as 100 to 79; the tenth as already stated, 100 to 54. The general result is made striking in the following illustration.

"If all the men in the country were on an average six feet high, and there were some families which had been long and closely interbred, these would be almost dwarfs, their average height during ten generations being only four feet eight and one-quarter inches." (p. 53.)

It is remarkable that the difference between the close-bred and the cross-bred individuals should have been as great as it was in the first generation; and, this being the case, it might have been expected that the difference would have gone on increasing in the succeeding generations. If self-fertilization is injurious, the ill effects would be expected to be cumulative. "But," instead of this, "the difference between the two sets of plants in the seventh, eighth, and ninth generations taken together is less than in the first and second generations together." Upon this Mr. Darwin remarks: "When, however, we remember that the self-fertilized and crossed plants are all descended from the same mother plant, that many of the crossed plants in each generation were related, often closely related, and that all were exposed to the same conditions, which, as we shall hereafter find, is a very important circumstance, it is not at all surprising that the difference between them should have somewhat decreased in the later generations." (p. 56.)

Further light was thrown upon these points by two kinds of subsidiary experiments. In one case, the cross was made between two flowers of the same plant of Ipomæa, while other flowers were self-fertilized as before. On raising seedlings from the two lots, it was found that such crossing gave no superiority; indeed, the offspring of the self-fertilized flowers appeared to be rather more vigorous than the close-crossed. And other experiments led to the same conclusion, namely, that there was no particular benefit from cross-fertilization on the same plant. In the other case, the cross was made not only between the flowers of distinct plants, but between those from different sources, and which had presumably grown under somewhat different conditions. For instance, several flowers of the ninth generation of crossed plants of Ipomæa were crossed with pollen taken from the same variety but from a distant garden. The resulting seedlings showed the benefit of the fresh stock remarkably, being as much superior in vigor to those of the tenth intercrossed generation as the latter were to the self-fertilized plants of a corresponding generation. In height they were as 100 to 78, over the ordinary intercrossed; and in fertility, as 100 to 51. Indeed, Mr. Darwin's main conclusion from all his observations is, "that the mere act of crossing by itself does no good. The good depends on the individuals which are crossed differing slightly in constitution, owing to their progenitors having been subjected during several generations to slightly different conditions, or to what we call in our ignorance spontaneous variation."

The greater constitutional vigor of the crossed plants of Ipomæa was manifested in other ways than their rate or amount of growth: they better endured exposure to a low temperature or sudden changes of temperature; they blossomed earlier; and they were more fertile. The difference in fertility varied greatly in degree (the extremes in different experiments and in different generations being 100 to 99 and 100 to 26), but was always sustained. Also, "the impaired fertility of the self-fertilized plants was shown in another way, namely, by their anthers being smaller than those in the flowers on the crossed plants. This was first observed in the seventh generation, but may have occurred earlier. . . . The quantity of pollen contained in one of the self-fertilized was, as far as could be judged by the eye, about half of that contained in one from a crossed plant. The impaired fertility of the self-fertilized plants of the eighth generation was also shown in another manner, which may often be observed in hybrids — namely, by the first-formed flowers being sterile."

Similar experiments were made, but not carried to the same extent, upon fifty-seven other species of plants, belonging to fifty-two genera, and to thirty great natural families, the species being natives of all parts of the world. The results — the details and discussion of which occupy the bulk of this volume — vary greatly, some plants making a better and others a less good showing for the advantage of cross-fertilization, and this advantage manifesting itself in different ways, some in vigor or amount of growth, some in hardiness, most in fertility; but with twelve cases in which the crossed plants showed no marked advantage over the self-fertilized. There were, however, fifty-seven cases in which the crossed exceeded the self-fertilized by at least five per cent., generally by much more.

Increase of vigor, as evinced in growth, appears generally to be accompanied by increased fertility; but sometimes the good of crossing was manifested only in productiveness, *i. e.*, in a larger amount of seed. This proved to be the case in Eschscholtzia, in which — strange to say — self-fertilized plants of several generations were superior in size and weight to intercrossed plants, even when the crossing was between flowers derived on one side from American, on the other from English seed, from which, upon Mr. Darwin's view, the maximum benefit should be gained. This instance, however, stands alone. Yet it is approached by several others, in a manner which might have negatived the general conclusions of the research, if they had been hastily gathered from a small number of trials.

For example, in the sixth self-fertilized generation of *Ipomœa purpurea*, one of these plants took the lead of its competitor, kept it almost to the end, and was ultimately overtopped only by half an inch on a total height of several feet. To ascertain whether this exceptionally vigorous plant would transmit its power to its seedlings, several of its flowers were fertilized with their own pollen, and the seedlings thus raised were put into competition with ordinary self-fertilized and with intercrossed plants of the corresponding generation. The six children of Hero (the name by which this individual was designated) beat the ordinary self-fertilized competitors at the rate of 100 to 84, and the intercrossed competitors at the rate of 100 to 95; and in the next generation the self-fertilized grandchildren beat those from a cross between two of the children at the rate of 100 to 94. In the next generations the seedlings were raised in winter in a hot-house, became unhealthy, and the experiment terminated without marked result. Moreover, the remarkable vigor of growth in Hero and its progeny was attended by somewhat increased fertility. Here, then, an idiosyncrasy arose from some utterly unknown cause, — a spontaneous variation of constitution, which was transmitted to posterity, and which gave all the benefit of cross-fertilization, and somewhat more, both as to vigor and fertility. A similar idiosyncrasy made its appearance in the third generation of seedlings of *Mimulus luteus.*

Discordant or anomalous facts like these seem confusing, even though too few to affect seriously the grand result of the numerous experiments; but upon Darwinian principles, in which adaptations are ultimate results, they are to be expected, as a consequence of the general and apparently vague proclivity to vary.

In Foxglove, — the flowers of which are naturally self-sterile or nearly so, and in which crossing gave a marked advantage over self-fertilizing, both as to growth and productiveness, — a decided, though small advantage appeared to come from the crossing of flowers on the same plant.

In *Origanum vulgare*, crosses were made between different plants of a large clump, long cultivated in a kitchen-garden, which had evidently spread from a single root by stolons, and which had become in a good degree sterile, as is usual under such conditions. The crossing caused rather more seed to form; but the seedlings from the crossed did not surpass in growth those of the self-fertilized; "a cross of this kind did no more good than crossing two flowers on the same plant of Ipomæa or Mimulus. Turned into the open ground, and both self and cross-fertilized the following summer, and equal pairs of the resulting seeds planted on opposite sides of two very large pots, the crossed plants from seed showed a clear superiority over their self-fertilized brethren, at the rate of 100 to 86. But this excess of height by no means gives a fair idea of the vast superiority in vigor of the crossed over the self-fertilized plants. The crossed flowered first and produced thirty flower-stems, while the self-fertilized produced only fifteen, or half the number. The pots were then bedded out, and the roots probably came out of the holes at the bottom, and thus aided their growth. Early in the following summer, the superiority of the crossed plants, owing to their increase by stolons, over the self-fertilized plants, was truly wonderful. . . . Both the crossed and the self-fertilized plants being left freely exposed to the visits of bees, manifestly produced much more seed than their grandparents, — the plants of original clumps still growing close by in the same garden, and equally left to the action of bees."

These few cases must here suffice, and they give a fair general idea of the main results reached, — somewhat qualified, however, by certain instances in which little or no benefit was observed. Let it be remarked that while most of the cases show decided and unequivocal good from the crossing, none of them unequivocally tell to the contrary, as the advantage appears sometimes in one direction, sometimes in another. "Thus, the crossed and self-fertilized plants of Ipomæa, Papaver, *Reseda odorata*, and Limnanthes were almost equally fertile, yet the former exceeded considerably in height the self-fertilized plants. On the other hand, the crossed and self-fertilized plants of Mimulus and Primula differed in an extreme degree in fertility, but by no means to a corresponding degree in height or vigor."

We must wholly omit — among many other things — the interesting account of self-sterile plants, meaning here not those in which the pollen does not reach the stigma unaided, but those in which it is impotent, or nearly so, when applied, although efficient upon the stigma of another individual. Verbascum, Passiflora, Corydalis, and many Orchids afford instances of this sort. In these, the advantage of cross-fertilization arises to a necessity. A noteworthy fact respecting them (of which Mr. Darwin makes much) is, that such self-sterility, or the reverse, is influenced by slight changes in the conditions, such as difference in temperature, grafting on another stock, and the like. In South Brazil, Fritz Müller found that for six generations all his plants of *Eschscholtzia Californica* were completely sterile, unless supplied with pollen from a distinct plant, when they were completely fertile. This was not the case in English plants, which, when covered by a net, set a considerable number of capsules, the seeds of which, by weight, were as 71 to 100 of those on plants intercrossed by bees. These Brazilian seeds, sent to England, yielded plants with moderately self-fertile flowers, and this limited self-fertility was increased in two generations of English growth. Conversely, seeds from English plants grown in Brazil were more self-fertile than those reared in Brazil for several generations; yet "one which did not flower the first year, and was thus exposed for two seasons to the climate of Brazil, proved quite self-sterile, like a Brazilian plant, showing how quickly the climate had acted on its sexual constitution." Having observed that certain individuals of Mignonette were self-sterile, Mr. Darwin secured several such plants under separate nets, and by intercrossing these for a few generations, obtained plants which inherited this peculiarity, so that "without doubt a self-sterile race of Mignonette could easily have been established."

Nine of the twelve chapters are devoted strictly to the effects of cross and self-fertilization. The tenth considers the "means of fertilization." Cross-fertilization is favored or ensured by: 1, the separation of the sexes; 2, the maturity of the male and female sexual elements at different periods; 3, dimorphism, or even trimorphism; 4, various mechanical contrivances; 5, the more or less complete inefficiency of the flower's own pollen on its stigma, and the prepotency of pollen from any other individual over that from the same plant. We understand that Mr. Darwin is just now occupied in revising and extending his various papers upon these topics, with a view to their publication in a volume. Here he gives a list of plants which, when insects are excluded, are either quite sterile or produce less than half the number of seeds yielded by unprotected plants. This is followed by a list of plants which, when protected from insects, are either quite fertile or yield more than half the number of seeds produced by unprotected plants.

"Each of these lists contains by a mere accident the same number of genera, namely, forty-nine. The genera in the first list include sixty-five species, and those in the second sixty species; the *Orchideæ* in both being excluded. If the genera in this latter order, as well as in the *Asclepiadeæ* and *Apocynaceæ*, had been included, the number of species which are sterile if insects are excluded would have been greatly increased; but the lists are confined to species which were actually experimented upon. The results can be considered as only approximately accurate, for fertility is so variable a character, that each species ought to have been tried many times. The above number of species, namely, 125, is as nothing to the hosts of living plants; but the mere fact of more than half of them being sterile within the specified degree, when insects are excluded, is a striking one; for whenever pollen has to be carried from the anthers to the stigma in order to insure full fertility, there is at least a good chance of cross-fertilization. I do not, however, believe that if all known plants were tried in the same manner, half would be found to be sterile within the specified limits; for many flowers were selected for experiment which presented some remarkable structure; and such flowers often require insect-aid." — (p. 370.)

It is worth noticing that *Trifolium repens* and *T. pratense* (the common White and Red Clovers) have a place in the first list; *T. arvense* and *T. procumbens* in the second. Darwin refers to Mr. Miner's statement that "in the United States hive-bees never suck the Red Clover," and says it is the same in England, except from the outside through holes bitten by humble-bees; yet that H. Müller has seen them visiting this plant in Germany for the sake both of pollen and nectar, which latter they obtained by breaking apart the petals. Darwin has not qualified his statement, long ago made, of the complete sterility of Red Clover protected from insects; but Mr. Meehan asserts that protected plants are fertile in this country, without, however, giving details or the rate of fertility. In *T. arvense*, "the excessively small flowers are incessantly visited by hive and humble-bees; when insects were excluded the flower-heads seem to produce as many and as fine seeds as the exposed heads."

As to cross-fertilization, "the most important of all the means by which the pollen is carried from the anthers to the stigma of the same flower, or from flower to flower, are insects, belonging to the orders of Hymenoptera, Lepidoptera, and Diptera; and in some parts of the world, birds." In a note the author cites all the cases known to him of birds fertilizing flowers. These are chiefly humming-birds. "In North America they are said to frequent the flowers of Impatiens" (for which Gould, "Trochilidæ," is referred to as authority, and a reference is given to the "Gardener's Chronicle," which we find relates to something else in South America); and this is all concerning the United States. Can it be that there are no references in print to the most familiar fact that our humming-bird is very fond of sucking the blossoms of Trumpet Creeper (*Tecoma radicans*) and of Honeysuckles? Both these are, in size and arrangement of parts, well adapted to be thus cross-fertilized.

Flowers are rendered conspicuous to birds and still more to insects, by bright colors. And as "almost every fruit which is devoured by birds presents a strong contrast in color with the green foliage, in order that it may be seen and its seeds disseminated," so the proportionally large size and the bright colors of the corolla, or, in some cases, the equally bright hues of the adjoining parts of the flower, or of the inflorescence, are correlated to visiting insects, — have come to pass, as Darwin would say, in consequence of the visits of insects, through the advantages in vigor and productiveness gained by cross-fertilization. He is ready to adopt even the idea of Conrad Sprengel, which seemed to be so fanciful, that marks and streaks on the corolla serve as guides to the nectary; for, although insects are well able to discover the nectar without the aid of guiding marks, yet they are of service by facilitating the search and enabling insects to suck a greater number of blossoms within a given time, which is tantamount to greater opportunity for cross-fertilization.

That odors attract insects is certain, and many flowers are both conspicuous and odoriferous, while others make up in fragrance what they lack in show. "Nägeli affixed artificial flowers to branches, scenting some with essential oils, and leaving others unscented; and the insects were attracted to the former in an unmistakable manner."

"Of all colors white is the prevailing one; and of white flowers a considerably larger proportion smell sweetly than of any other color, namely, 14.6 per cent.; of red, only 8.2 per cent. are odoriferous. The fact of a larger proportion of white flowers smelling sweetly may depend in part on those which are fertilized by moths requiring the double aid of conspicuousness in the dusk and of odor. So great is the economy of nature, that most flowers which are fertilized by crepuscular or nocturnal insects emit their odor chiefly or exclusively in the evening. Some flowers, however, which are highly odoriferous depend solely on this quality for their fertilization, such as the night flowering-stock (Hesperis) and some species of Daphne; and these present the rare case of flowers which are fertilized by insects being obscurely colored."

"The shape of the nectary and of the adjoining parts are likewise related to the particular kinds of insects which habitually visit the flowers: this has been well shown by H. Müller, by his comparison of lowland species, which are chiefly visited by bees, with alpine species belonging to the same genera, which are visited by butterflies."

"Pollen contains much nitrogen and phosphorus,—the two most precious of all the elements for the growth of plants,—but in the case of most open flowers, a large quantity of pollen is consumed by pollen-devouring insects, and a large quantity is destroyed during long-continued rain. With many plants this latter evil is guarded against, as far as possible, by the anthers opening only during dry weather, by the position and form of some or all of the petals, by the presence of hairs, etc.; also, as Kerner has shown in his interesting essay, by the movements of the petals or of the whole flower during cold and wet weather. In order to compensate the loss of pollen in so many ways, the anthers produce a far larger amount than is necessary for the fertilization of the same flower. I know this from my own experiments on Ipomæa, given in the Introduction; and it is still more plainly shown by the astonishingly small quantity produced by cleistogene flowers, which lose none of their pollen, in comparison with that produced by the open flowers borne by the same plants; and yet this small quantity suffices for the fertilization of all their numerous seeds. Mr. Hussall took pains in estimating the number of pollen-grains produced by a flower of the Dandelion, and found the number to be 243,600, and in a Peony 3,654,000 grains. The editor of the 'Botanical Register' counted the ovules in the flowers of *Wistaria sinensis*, and carefully estimated the number of pollen-grains, and he found that for each ovule there were 7,000 grains."—(pp. 376, 377.)

These are probably fair averages of the numerical ratio of pollen to ovules in flowers which are adapted to be fertilized by insect agency. Their meaning in the "economy of nature" is seen by a comparison on the one hand with *anemophilous, i. e.* wind-fertilized, flowers, in most of which there is a vastly greater disproportion between the numbers, — compensating for inevitable waste, — and on the other hand with *cleistogenous* flowers, namely, those small and less developed blossoms which some plants produce in addition to the ordinary sort, and which fertilize as it were in the bud, necessarily by their own pollen. Here is no waste, and accordingly the anthers are very small, and the pollen-grains are not many times more than the ovules: also such flowers are never brightly colored, never odoriferous, and they never secrete nectar.

The only advantages of this close-fertilization which we can think of are sureness and strict likeness; both of which are quite as well secured by budding-reproduction. Now, as cleistogene flowers are borne, we believe, chiefly and perhaps only, by species whose normal blossoms are adapted for insect-fertilization, they must be regarded as a subsidiary arrangement, a safeguard against failure of proper insect-visitation. As the volume before us amply shows, this failure is in general provided for by a more or less wide margin of self-fertilization in the very flowers which are adapted for crossing. In Impatiens, Viola, and the like, it is provided for by separate flowers, the special adaptations of which are unmistakable.

H. Müller appears to have shown "that large and conspicuous flowers are visited much more frequently and by many more kinds of insects than are small inconspicuous flowers. He further remarks that the flowers which are rarely visited must be capable of self-fertilization, otherwise they would quickly become extinct." Mr. Darwin's list seems to show that, as a rule, they are so; yet many very small flowers, like those of *Trifolium arvense*, and small and dingy ones, like those of Asparagus, are freely visited by bees; and, conversely, many large and conspicuous flowers which are frequented by insects are none the less self-fertilizable. Throughout we find that such things do not conform to arbitrary or fixed rules; and this favors the idea that the differences have been acquired. Mr. Darwin conjectures that the self-fertilizing capabilities of many small and inconspicuous flowers may be comparatively recent acquisitions, on the ground that, if they were not occasionally intercrossed, and did not profit by the process, all their flowers would have become cleistogenous, "as they would thus have been largely benefited by having to produce only a small quantity of safely protected pollen."

Mr. Darwin's experiments tending to prove that cross-fertilization between flowers on the same plant is of little or no use, he is naturally led to consider the means which favor or ensure their fertilization with pollen from a distinct plant. This must needs take place with diœcious plants, and is likely to occur with the monœcious, and is in some cases secured (as in Walnut and Hazelnut) by some trees being proterandrous and others proterogynous, so that they will reciprocally fertilize each other. In ordinary hermaphrodite species the expansion of only a few blossoms at a time greatly favors the intercrossing of distinct individuals, although in the case of small flowers it is attended with the disadvantage of rendering the plants less conspicuous to insects. Our common Sundews furnish a good illustration of this. They abound wherever they occur, and are for a long while in blossom, but each plant or spike opens but one flower at a time. The fact of bees visiting the flowers of the same species as long as they can, instead of promiscuously feeding from the various blossoms nearest within reach, greatly favors such intercrossing. So does the remarkable number of flowers which bees are able to visit in a short time (of which mention will be made), and the fact that they are unable to perceive without entering a flower whether other bees have exhausted the nectar. Then dichogamy (the maturation of one sex in a hermaphrodite flower earlier than the other) is so prevalent that it may almost be regarded as the rule; and this ensures such crossing between few-flowered plants, and greatly favors it in the case of spikes, racemes, and the like. For, proterandry being the commonest arrangement, so that the younger flowers act as male, and the older as female, the bees habitually alighting at the bottom and proceeding upward, they carry the pollen from the upper and younger flowers to stigmas of the lower and older flowers of the next spike, and so on. Heterogonism, which is less common, operates precisely like complete diœcious separation of the sexes in this respect, and with the advantage that all the individuals are seed-bearing. Most of the special arrangements peculiar to certain families, such as Orchids, — or to plants, such as Posoqueria, with its wondrous mechanism for quickly stopping out access to the stigma when the pollen is violently discharged upon some insect, but opening the orifice the next day, — are of a kind to favor the crossing of distinct plants. Prepotency of other pollen, which may accompany the other arrangements or exist independently, acts largely and powerfully toward the same end. Our author investigates this at some length: we cite for illustration a single but strong case. The stigmas of a long-styled Cowslip were supplied with pollen from the same plant, and again after twenty-four hours, with pollen of a short-styled, dark-red Polyanthus, a variety of the same species: from the resulting seeds twenty seedlings were raised, and all of them bore reddish flowers; so that the effect of the plant's own pollen, though placed on the stigmas twenty-four hours previously, was destroyed by that of the red variety. The same thing is shown by the impossibility in many cases of raising two varieties of the same species pure if they grow near each other. "No one who has had any experience would expect to obtain pure cabbage-seed, for instance, if a plant of another variety grew within 200 or 300 yards." And a veteran cultivator once had his whole stock of seeds seriously bastardized by some plants of purple Kale which flowered in a cottager's garden half a mile away. Mr. Gordon records a case of the crossing between Primroses and Cowslips through pollen carried by bees over more than two kilometers, or an English mile and a quarter.

We must copy the close of this section — long though it be — because of its capital illustration of the topic in hand, and for the teleological lesson which it teaches.

"The case of a great tree covered with innumerable hermaphrodite flowers, seems at first sight strongly opposed to the belief in the frequency of intercrosses between distinct individuals. The flowers which grow on the opposite sides of such a tree will have been exposed to somewhat different conditions, and a cross between them may perhaps be in some degree beneficial; but it is not probable that it would be nearly so beneficial as a cross between flowers on distinct trees, as we may infer from the inefficiency of the pollen taken from plants which have been propagated from the same stock though growing on different roots. The number of bees which frequent certain kinds of trees when in full flower is very great, and they may be seen flying from tree to tree more frequently than might have been expected. Nevertheless, if we consider how numerous are the flowers, for instance, on a Horse-Chestnut or Lime-tree, an incomparably larger number of flowers must be fertilized by pollen brought from other flowers on the same tree, than from flowers on a distinct tree. But we should bear in mind that with the Horse-Chestnut, for instance, only one or two of the several flowers on the same peduncle produce a seed; and that this seed is the product of only one out of the several ovules within the same ovarium. Now we know from the experiments of Herbert and others that if one flower is fertilized with pollen which is more efficient than that applied to the other flowers on the same peduncle, the latter often drop off; and it is probable that this would occur with many of the self-fertilized flowers on a large tree, if other and adjoining flowers were cross-fertilized. Of the flowers annually produced by a great tree, it is almost certain that a large number would be self-fertilized; and if we assume that the tree produced only 500 flowers, and that this number of seeds were requisite to keep up the stock, so that at least one seedling should hereafter struggle to maturity, then a large proportion of the seedlings would necessarily be derived from self-fertilized seeds. But if the tree annually produced 50,000 flowers, of which the self-fertilized dropped off without yielding seeds, then the cross-fertilized flowers might yield seeds in sufficient number to keep up the stock, and most of the seedlings would be vigorous from being the product of a cross between distinct individuals. In this manner the production of a vast number of flowers, besides serving to entice numerous insects and to compensate for the accidental destruction of many flowers by spring-frosts or otherwise, would be a very great advantage to the species; and when we behold our orchard-trees covered with a white sheet of bloom in the spring, we should not falsely accuse Nature of wasteful expenditure, though comparatively little fruit is produced in the autumn."

The Horse-Chestnut is not altogether a well-chosen example, for in it, as in our Buckeyes, a very large proportion of the flowers in the thyrsus are usually male, with barely a vestige of pistil. These serve, however, to increase the show, in the manner here illustrated, as well as to furnish abundance of pollen.

The section on anemophilous (wind-fertilized) plants,—their interest as survivals of the earlier phænogamic vegetation, the speculation as to how, when flying insects came to prevail, an anemophilous plant may have been rendered entomophilous,—how pollen, being a most nutritious substance, would soon have been discovered and devoured by insects, and by adhering to their bodies be carried from anthers to stigma and from one flower to another,—how a waste secretion, such as honey-dew or glandular exudations, may have been developed into nectar and utilized as a lure,—the interesting illustrations of the vast amount of pollen produced by anemophilous plants, and the great distances to which their light pollen is often carried by the wind,—all these inviting topics we must now pass by.

In passing we note the remark that "the excretion of a sweet liquid by glands seated outside of a flower is rarely utilized as a means of cross-fertilization by the aid of insects;" and the sole exception alluded to is that of the bracts of *Marcgraviaceæ*. But a parallel case is afforded by many species of Euphorbia, and notably in a striking species cultivated in conservatories, under the name of Poinsettia. Here the attraction to the eye is supplied by the intense red coloration of ordinary leaves placed next to the inflorescence, and that to the palate or tongue (if either term be allowed), by a large cup-shaped gland on the side of the involucre, which contains or surrounds the naked and greatly simplified flowers of both sexes.

That anemophilous plants are prevailingly declinous (either monœcious or diœcious) is speculatively connected with their antiquity; that they are very largely trees or shrubs is because "the long life of a tree or bush permits of the separation of the sexes with much less risk of evil from impregnation occasionally failing, and seeds not being produced, than in the case of short-lived plants. Hence it is probably, as Lecoq has remarked, that annual plants are rarely diœcious." The number of anemophilous species is comparatively small, but that of individuals of the species strikingly large, so that they form of themselves, in cold and temperate regions, where plant-fertilizing insects are fewer, either vast forests, as of *Coniferæ*, Birches, Beeches, etc., or meadows and glades, as of Grasses, Sedges, and Rushes. Being thus either necessarily or prevailingly cross-fertilizable and gregarious, it is not wonderful that they should hold their own unchanged in various parts of the world. Still their advantage is gained at the expense of the production of an enormous superfluity of pollen, a costly product; and, when diœcious, half the individuals produce no seed. Hermaphroditism with dichogamy, or some equivalent, and transportation by an appeal to the senses and appetites of insects, secures all the advantages with least expenditure. The earliest fertilization in plants took place by the locomotion of the fertilizing or even of the fertilized material, in manner of most of the *Algæ:* mainly losing this as vegetation became terrestrial, the transportation was committed to the winds, and finally in the higher plants more economically consigned to insects.

The eleventh chapter, on the habits of insects in relation to the fertilization of flowers, is one of the interesting and readable although one of the shortest. It appears that the prince of naturalists, Aristotle, had observed more than two thousand years ago that the hive-bees visited the flowers of the same species as long as possible before going to a different species. This holds true of all kinds of bees and certain other insects, generally, but not absolutely; although, as Lubbock has recently proved, bees are much guided by color, yet they hold to the practice just mentioned in spite of difference in this respect, being botanists enough to know that color is not a good specific character. Mr. Darwin has repeatedly seen humble-bees flying straight from a red Fraxinella to a white variety, from one Larkspur to a different-colored variety; and the same as to Primroses and Pansies. But two species of Poppy were by some bees treated as one; and H. Müller traced hive-bees from blue Hyacinths to blue Violets. On the other hand, Darwin's bees fly straight from clump to clump of yellow Œnothera without turning an inch in their course to Eschscholtzias with yellow flowers which abound on either side. This constancy to species, however, is manifested only when their flowers abound; a fact which may have led Mr. Darwin to his explanation of the reason of it.

"The cause probably lies in insects being thus enabled to work quicker; they have just learned how to stand in the best position on the flower, and how far and in what direction to insert their proboscides. They act on the same principle as does an artificer who has to make a half a dozen engines, and who saves time by making consecutively each wheel and part for all of them. Insects, or at least bees, seem much influenced by habit in all their manifold operations;

and we shall presently see that this holds good in their felonious practice of biting holes through the corolla." — (p. 420.) As to this latter practice —

"The motive which impels bees to gnaw holes through the corolla seems to be the saving of time, for they lose much time in climbing into and out of large flowers, and in forcing their heads into closed ones. They were able to visit nearly twice as many flowers, as far as I could judge, of a Stachys and Pentstemon by alighting on the upper surface of the corolla and sucking through the cut holes, than by entering in the proper way. Nevertheless each bee before it has had much practice, must lose some time in making each new perforation, especially when the perforation must be made through both calyx and corolla. This action therefore implies foresight, of which faculty we have abundant evidence in their building operations; and may we not further believe that some trace of their social instinct, that is, of working for the good of other members of the community, may here likewise play a part? Many years ago I was struck with the fact that humble-bees as a general rule perforate flowers only when these grow in large numbers near together," etc., etc. (p. 433.)

It appears that the cutting of these holes is done only by humble-bees, never by hive-bees. Yet the latter are quick to take advantage of them.

"In the early part of the summer of 1857 I was led to observe during some weeks several rows of the scarlet Kidney-bean (*Phaseolus multiflorus*), whilst attending the fertilization of this plant, and daily saw humble and hive-bees sucking at the mouths of the flowers. But one day I found several humble-bees employed in cutting holes in flower after flower; and on the next day every single hive-bee, without exception, instead of alighting on the left wing petal and sucking the flower in the proper manner, flew straight without the least hesitation to the calyx, and sucked through the holes which had been made only the day before by the humble-bees, and they continued this habit for many following days. Mr. Belt has communicated to me (July 28, 1874) a similar case, with the sole difference that less than half of the flowers had been perforated by the humble-bees; nevertheless all the hive-bees gave up sucking at the mouths of the flowers and visited exclusively the bitten ones. Now how did the hive-bees find out so quickly that holes had been made? Instinct seems to be out of the question as the plant is an exotic. The holes cannot be seen by bees whilst standing on the wing-petals, where they had always previously alighted. From the ease with which bees were deceived when the petals of *Lobelia Erinus* were cut off, it was clear that in this case they were not guided to the nectar by its smell; and it may be doubted whether they were attracted to the holes in the flowers of the Phaseolus by the odor emitted from them. Did they perceive the holes by the sense of touch in their proboscides, whilst sucking the flowers in the proper manner, and then reason that it would save them time to alight on the outside of the flower and use the holes? This seems almost too abstruse an act of reason for bees; and it is more probable that they saw the humble-bees at work, and understanding what they were about, imitated them and took advantage of the shorter path to the nectar. Even with animals high in the scale, such as monkeys, we should be surprised at hearing that all the individuals of one species within the space of twenty-four hours understood an act performed by a distinct species and profited by it." — (pp. 430, 431.)

But we must cut short our citations and remarks; passing by one of the most important points, relative to the amount of fertilizing work done by insects, namely, the evidence of the extraordinary industry of bees and the number of flowers visited within a short time; which, as well as the distance to which pollen is sometimes transported, is far greater than one would have supposed. But the volume is reprinting by the Appletons, and will soon be within the reach of all, — along with a new edition of the Orchid-fertilization book, the proper supplement to the present work, relating as it does to the class of plants in which the adaptation for fertilization by insects is carried to the highest degree of specialization and perfection.

THE HYBRIDIZATION OF LILIES.

MR. PARKMAN, under the above title, gives a summary[1] of his experiments, during ten or twelve years, in crossing Lilies. One of the earlier results, and that which the horticulturists count as the eminent one, was the production of that magnificent hybrid between *L. auratum* and *L. speciosum*, with flower resembling the former in fragrance and form and the most brilliant varieties of the latter in color, which was brought out in England under the name of *Lilium Parkmanni*. The interesting physiological point which Mr. Parkman here records is, that this striking novelty was wholly unique; that all the other seeds of the same parentage which germinated, over fifty in number, gave rise to plants which in the blossom showed no trace of the male parent, *L. auratum*, but were exactly like the female parent, *L. speciosum*. That these plants were truly hybrids, notwithstanding, is well made out: 1, by the precautions taken against any possible access of own pollen; 2, by the scantiness of seed, most of which was abortive; 3. "such good seed as there was differed in appearance from the seed of the same Lily fertilized by the pollen of its own species," which is smooth, while this was rough and wrinkled; and 4, the stems were mottled after the manner of the male parent.

It would naturally be thought that this slight but evident impression of the character of the male parent might be deepened by iteration. That was tried next year, when the flowers of several of these plants were fertilized with the pollen of *L. auratum* precisely as their female parent had been fertilized. The result was an extremely scanty crop of seed, "but there was enough to produce 8 or 10 young bulbs. Of these, when they bloomed, one bore a flower combining the features of both parents, but though large, it was far inferior to *L. Parkmanni* in form and color; the remaining flowers

[1] *The Hibridization of Lilies.* By Francis Parkman, in Bull. Bussey Institution, ii., 1877. (American Journal of Science and Arts, 3 ser., xv. 144.)

were not distinguishable from those of the pure *L. speciosum.*" The article records the results of various similar attempts to hybridize other Lilies. For instance, our *L. superbum* was pollenized with eight different old-world species. The result was, that capsules, apparently perfect, were abundantly produced; some of them contained nothing but chaff, others had a few imperfect seeds, still others gave a fair supply of good seed. From this seed several hundred young bulbs were produced. "But when these came into bloom, not a single flower of them all was in the least distinguishable from the pure *L. superbum.*" Moreover, in this case (different from the other) "not one of the eight different male parents had imposed his features on his hybrid offspring. Not only in their flowers, but in their leaves, stems, and bulbs, the young plants showed no variation from their maternal parent." The experiment proceeded one generation farther. "In the following year I set some of them apart from the rest, and applied to them, as to their mother before them, the pollen of several species of Lilies. This time the seeds were extremely scanty. A few, however, were produced; but the plants and flowers that resulted from them were, to all appearance, *L. superbum* pure and simple."

In trials of other species results intermediate between these two cases were obtained. For instance, the pure white of the perianth of *L. longiflorum* came out unstained in the progeny raised by a crossing with *L. speciosum*, and the herbage was equally unaffected; but in that or the next generation "distinct evidence could be seen of the action of alien pollen" in the changed color of many of the anthers, and in the abortion of others. They also showed differences of habit among themselves, some being very tall and vigorous, and others compact and bushy, with a tendency to bloom in clusters; but these may have been mere seedling variations, with which the hybridization had nothing to do. Yet some of these marks correspond with known results of hybridization.

That offspring should partake unequally of the characters of the two parents is a matter of common observation. That in the genus Lilium the hybrid offspring should in forty instances out of fifty take almost all its traits from the female parent, as Mr. Parkman has shown, is very remarkable. That, in not a few instances, it should take them all, so far as can be seen, — that the paternal influence should be represented by zero, — is most extraordinary. If parthenogenesis in plants were more unequivocally demonstrated, so as to be placed in certain instances quite beyond doubt (which is hardly the case), then we should regard the supposition which Mr. Parkman mentions as having been suggested to him, namely, that in the case of *L. superbum* the embryo was developed without male influence, to be quite as likely as the alternative of the progeny's inheriting everything from the female and nothing from the male parent; in fact the two suppositions approximate to the same thing. We are supposing the total absence of male parent's characters, and also that the alternative of fertilization by chance pollen of the species is absolutely excluded. Of this there is very high probability, yet not entire certainty. One of Mr. Parkman's "reasons for believing that parthenogenesis had nothing to do with the cases in question," namely, that some of the Lilies were young plants that never had bloomed before, has no application, but comes from a slight confusion of the idea of parthenogenesis with the effect in some animals of a previous male influence upon next succeeding progeny, which is quite a different thing.

The fact that more than one sort of hybrid may be generated between the same two species, copulated in the same way, must do away with the old mode of naming hybrids by a combination of the name of the two parents, that of the male preceding. The plan had the double advantage of indicating the origin of the cross, and of distinguishing hybrids from species in nomenclature; but in practice it proves insufficient.

PHYTOGAMY.

If this name has not been coined already it ought to be. For "the loves of the plants," so mellifluously sung by Dr. Erasmus Darwin in the days of our grandfathers, have been in our time, through a felicitous atavism, more scientifically, if prosaically, expounded by his grandson, in a series of articles and volumes, of which the subjoined are the principal titles.[1] If we have too long delayed our notice of these books, we make amends by calling attention to them at the season which invites and amply rewards the observations in field and garden which they suggest. Mainly in consequence of these writings, the subject which our new word connotes, namely, the connubial relations of plants, has become a popular and fruitful branch of biological science, which has its own laws and rules and technical terms, its distinction of legitimate and illegitimate unions, and tables of forbidden degrees. For example, it is not lawful, at least it is not *en règle* nor beneficial, for "thrum-eyed" Primroses to interbreed, nor for "pin-eyed" Primroses to interbreed. Such are illegitimate unions, seldom blessed with progeny. To the uncurious observer in Wordsworth's poem, —

> "A primrose by a river's brim
> A yellow primrose was to him,
> And it was nothing more."

But as concerns the Primrose, where seed-bearing is in question, if it be one of the thrum-eyed stock, the pollen brought to it must come from the pin-eyed, and vice versa, in order to secure full fertility. Tiny blue-eyed Houstonias, enamelling our meadows in early spring, and fragrant Mitchellas, carpeting Pine-woods in midsummer, are in a similar case. It is this kind of arrangement for cross-breeding to which

[1] *The Different Forms of Flowers on Plants of the Same Species.* — *The Various Contrivances by which Orchids are Fertilized by Insects.* Second Edition, revised. — *The Effects of Cross and Self-Fertilization in the Vegetable Kingdom.* By Charles Darwin. London and New York, 1876–77. (The Nation, No. 667, April 11, 1878.)

the larger part of Darwin's latest volume on "The Different Forms of Flowers on Plants of the Same Species" is devoted. In such flowers — and they are rather numerous and of many families — the advantage of cross-breeding between different individuals of the same species is unquestionable, for it is essential to full fertility. The differences in structure, which consist of relative and reciprocal length of stamens and style in blossoms otherwise alike, have long been known; the meaning of it was one of Darwin's happy thoughts, and the confirmation is due to his labors. He demonstrated that the structure was correlated to the transport by insects of the pollen of the one sort to the stigma of the other, and that each pollen was inert, or nearly so, upon the stigma of the flower it belonged to, but potent upon the stigma of the other sort, upon which, in passing from blossom to blossom among the plants (of about equal number as to sort), the visiting insects are pretty sure to deposit it.

It is noteworthy that this significant dimorphism belongs to certain species of a considerable number of natural families, while others, sometimes even of the same genus, and in most of their species, show no trace of it; as if certain favored species had acquired a peculiarity in which their brethren have not shared. We ourselves call to mind some species in which this acquisition is either incipient or the correlations imperfect. But in his earliest work of the present series, on "The Various Contrivances by which Orchids are Fertilized by Insects," — a fascinating volume, which has recently been brought out in a second edition, — the "contrivances," as they may well be termed, are the common property of the whole order, although each genus seems to have patented a modification of its own. Here there is no dimorphism, but (with rare exceptions) all the flowers are alike, and all agree in having the pollen placed tantalizingly near the stigma, but prevented from reaching it, as well as in having some arrangement for the pollen's being transported by insects from one flower to another, ultimately from one plant to another. Wonderful arrangements, indeed, they are, which it requires a volume to describe, and of which we can here offer no details. Suffice it to say that, in this great order, cross-fertilization must be all but universal as between different flowers of the same plant, and commonly between different individual plants.

In both these kinds of hermaphrodite flowers the practical separation of the sexes is hardly less than in Oaks, Willows, and other trees and herbs, in which the stamens and pistils occupy distinct plants or different blossoms. To these three classes, then, Mr. Charles Darwin's aphorism, "Nature abhors perpetual self-fertilization," undoubtedly applies. But there remains an equal number of plants with hermaphrodite blossoms, all alike, with no obvious obstacle to fertilization with their own pollen, while in many the adaptations are such as must apparently insure it, and indeed does very commonly insure it. Wherefore it is nowise surprising that self-fertilization was the orthodox doctrine — that there was thought to be a general adaptation for the falling of the pollen upon the stigmas of the same blossom. It is true that Christian Conrad Sprengel taught the contrary, in his work entitled "The Secret of Nature Discovered," published eighty-five years ago, and that he — mainly upon good observations — in a measure anticipated Mr. Darwin's aphorism; but he was accounted whimsical and untrustworthy by his own generation, and was forgotten by the next. Not so the contemporary "Loves of the Plants" — the hymnal of the old orthodox cult — which sings the —

> "Gay hopes and amorous sorrows of the mead,"

in verse which our fathers were fond of, but from which we will not further quote. Had Dr. Erasmus Darwin known Sprengel's book, and brought to it the insight of the grandson, how different and how much richer the poem might have been. What curious facts and teeming fancies have been left unsung!

To H. Müller and to Hildebrand, two of Sprengel's countrymen, in our own day, may be credited the confirmation of the latter's thesis as respects the general run of hermaphrodite flowers; and this by showing what a large proportion even of these are functionally unisexual, either by the shedding of their pollen before the stigma of that blossom is ready to receive it, or by the development and subsequent shrivelling of the stigma before the pollen matures, or by various other arrangements of like effect. And here, too, comes in the significant fact for the evolutionist, that these arrangements belong to widely different families, but only to certain of their species or groups of species, and not to their near relatives; also that they are more pronounced in some species than in others.

Yet, withal, there is much close-fertilization, and no one has demonstrated this better than Mr. Darwin, nor so well illustrated its meaning. The more particular and special the adaptations for cross-fertilization — depending, as they mainly do, upon insect-transportation, consequent upon visits for nectar or other floral products — the greater the chances of no fertilization through the failure of the proper insect visitation. So nature, not scorning a succedaneum, arranges for self-fertilization also as the next best thing, indicating her preference, however, by endowing the pollen with greater potency upon other stigma than its own; the principle throughout being to place the pollen where it will do the most good, all things considered. But Mr. Darwin insists, apparently with reason, that cross-breeding is the general plan, and close-breeding the subsidiary proceeding, or at least that no species of flowering plants is deprived of its chance of wide-breeding, or fails to receive the benefit of it for any long number of generations.

This assumes that wide-breeding is beneficial. The assumption is one which a teleologist like Darwin is bound to make, and which an investigator like Darwin is bound to verify, if possible. The assumption is that ends elaborately brought to pass in a large number of species, in a variety of ways, and by great nicety and exactness of adaptation, cannot be meaningless or useless — must somehow conduce to the well-being of the species. Happily, this inference holds equally good whether, with the old-fashioned teleologist, the word "end" denotes a result aimed at, or, as in Darwinian teleology, a result attained. The two senses are not contradictory, and, as concerns the validity of the inference, it matters not which sense is adopted, or whether the two are combined. Darwin's investigation, undertaken to determine by experiments whether such crossing is beneficial, is published in the remaining volume of the series under consideration — that on "The Effects of Cross and Self-Fertilization in the Vegetable Kingdom." It does not fall within the scope and limits of this notice to set forth the nature and the extent of these experiments. Readers interested will go to the book, and probably have done so already. As to the results we may only say that, on the whole, they corroborate the inference — in some cases unequivocally and strongly, in others feebly, while in a very few the result was simply negative. While the crossing in many cases showed astonishing reinvigoration, and self-fertilization evident injury, the maximum good was obtained at the first or second crossing; and some close-fertilized plants soon became tolerant of that condition, and retained their fertility for several close-bred generations. If the Darwinian thesis was on the whole maintained, yet it was also shown that plants have many inexplicable idiosyncrasies, and that many unknown or obscure factors enter into the results of the experiment. On looking over the series we are reminded of the late Jeffries Wyman's aphorism: "No single experiment in physiology is worth anything."

It seems reasonably made out that the benefit of cross is, *cœteris paribus*, in direct relation to a certain difference in constitution between the two parents, or to some difference in their surroundings or antecedents, from which diversity of constitution may be inferred. The benefit is more decided when the parents come together from a distance than when grown side by side for several generations, and "a cross between two flowers on the same plant does no good, or very little good." The qualification is a proper one. It would be hasty to infer that it does absolutely no good, even though the advantage be inappreciable in any single instance. Still, however just and fairly well sustained the principle of Darwin's aphorism may be, it is confronted by the immense and seemingly endless vitality of long-propagated varieties which do not seed at all.

If we were writing a popular review of this volume on cross and self-fertilization, we should make much of the tenth and eleventh chapters, on the means of fertilization, and especially of cross-fertilization; on the plants which are sterile, or more than half-sterile, without insect aid; and, above all, on the habits of insects in relation to the fertilization of flowers. A closing chapter in the volume, on the Forms of Flowers, should also receive attention — that in which cleistogamous blossoms are discussed, namely, small and inconspicuous ones which never open, but are far more fertile than the showy ordinary blossoms of the same plants; for capital converse testimony, to the effect that all ordinary flowers are in primary reference to cross-fertilization, may be derived from the structure and behavior of these blossoms, in which the contrary intent is unmistakable. When nature means close-fertilization she makes her purpose manifest. Also, we should note that this cleistogamy is sporadic, affects certain families only, and certain members only of families not otherwise particularly related; so that this peculiarity also seems to be of special and apparently late acquisition. When we gather into one line the several threads of evidence of this sort, to which we have barely alluded, we find that they lead in the same direction with the clews furnished by the study of abortive organs: slender, indeed, each thread may be, but they are manifold, and together they bind us firmly to the doctrine of the derivation of species.

BENTHAM'S FLORA OF AUSTRALIA.

THIS volume [1] brings a great undertaking to a happy completion. The first volume was issued in the year 1863, and the work has made steady progress to the end. It is the complete phænogamous Flora of a continent, and the only one;

[1] *Flora Australiensis: a Description of the Plants of the Australian Territory.* By George Bentham, assisted by Baron Ferdinand von Mueller. *Roxburghiaceæ* to *Filices.* London, 1878. (American Journal of Science and Arts, 3 ser., xvi. 237.)

is worked up by one mind and hand, within a time and at an age which allows no sensible change of ideas or points of view, so that it is throughout comparable with itself. It is the work of the most experienced and wise systematic botanist of the day, and when we know that fully as much other work, of equal character, has been done within these fifteen years, it will not be denied that the author's industry and powers of accomplishment are unrivalled. No one else has done such good botanical work at such a rate. If, as some fear, the race of first-class systematic (phænogamous) botanists is destined to die out or dwindle, it will not be for the lack in our day of a worthy model.

In the concluding Preface, Mr. Bentham turns over to his able and equally indefatigable coadjutor, Von Mueller, the duty of incorporating addenda and corrections, and suggests the preparation of a methodical synopsis, for convenient use, especially in Australia, where such a handbook will be most helpful and needful. This trust, we doubt not, Von Mueller will duly undertake, and may be expected worthily to accomplish. His fellow-workers over the world are not unmindful of their great obligations to him in the development of Australian botany, and in rendering practicable the production of this "Flora Australiensis" which has been equally enriched by his vast collections and facilitated by his preliminary study of them.

Mr. Bentham now declines to undertake "a detailed examination of the relations, as well of the whole flora to that of other countries, as of its component parts to each other," referring instead to "the principles laid down by J. D. Hooker in the admirable essay prefixed to his 'Flora Tasmaniæ,'" but recapitulating shortly the general characteristics of the chief component parts of the present flora of Australia, the most peculiar one of any large part of the globe. Let us still hope that he may some day reconsider this determination, so far as to discuss in a general way the relations of Australian botany to the history of vegetation on the globe.

Peculiar as the Australian vegetation is, its treatment not rarely touches points which concern the student of the American flora. Especially interesting to us is the elaboration, in the present volume, of the *Gramineæ*, in which General Munro's matured views — as yet little known by publication — have passed under the independent consideration of a veteran general botanist, and in which the author's own conclusions regarding the morphology and terminology of the floral parts and their accessories are practically applied. We duly noticed Mr. Bentham's essay on this subject, and had to acknowledge that its conclusions are apparently incontrovertible.

Next to this order in importance is the order *Cyperaceæ*, upon the arrangement of which sound judgment is brought to bear. The great order *Liliaceæ* is made to include the *Smilaceæ*, and not the *Roxburghiaceæ*. We should have excluded both, but Smilax in preference. Contrary to Mr. Bentham's opinion, we should insist that the anthers in Smilax are unilecular but bilocellate. The diagnosis of *Roxburghiaceæ* in the conspectus distinguishes the order from Australian Liliaceæ only, and by an oversight the second genus of the order is said to be restricted to Japan, whereas it was founded on a North American plant.

DE CANDOLLE'S NEW MONOGRAPHS.

In this form[1] and way we may hope to see the Monocotyledonous orders elaborated, and some of the earlier Dicotyledonous ones re-elaborated. The middle of this volume is filled by the monograph of *Restiaceæ*, by Dr. Masters. This is an order allied on the one hand to *Juncaceæ*, on the other to *Cyperaceæ*, of twenty genera and two hundred and thirty-four species, wholly of the southern hemisphere, divided between South Africa (which has much the larger share),

[1] *Monographiæ Phanerogamarum Prodromi nunc continuatio, nunc revisio*, auctoribus Alphonso et Casimir De Candolle, aliisque Botanicis ultra memoratis. Vol. I. *Smilaceæ, Restiaceæ, Meliaceæ*, cum tabulis ix. Paris, June, 1878. (American Journal of Science and Arts, 3 ser., xvi. 325; xxxiv. 490.)

and Australia with New Zealand, and a single species in Chili. It is not a prepossessing family, and presents peculiar difficulties to the systematist, on account of the diœcious character of most of them, and a striking difference between the plants of the two sexes, which in collections are hard to match. Much praise is due to Dr. Masters for his great labor, patience, and skill. The latter half of the volume is occupied by Casimir De Candolle with his neat revision of the *Meliaceæ*, chiefly a tropical order. The stamineal tube in the monadelphous *Meliaceæ* is concluded to be a staminiferous disk. The *Smilaceæ* by Alphonse De Candolle form the smaller but to us the most interesting part of the volume.

This order is restricted to three genera: two of them diœcious, Heterosmilax with united sepals, no petals, and three monadelphous stamens (east Asiatic), Smilax with separate sepals, petals, and (6–15) stamens; the third, Rhipogonum (of New Zealand and Australia), with hermaphrodite flowers. Of Smilax one hundred and eighty-six species are characterized, and a dozen or two more are obscure or doubtful. There are thirty-eight pages of prefatory *generalia*, in De Candolle's best manner. We are pleased to find that he keeps up the *specific phrase*, and with true Linnæan curtness, relegating all particulars, not truly diagnostic under the sections and other divisions, to the description. In discussing the nature and characters of the leaf (which in its general sense is called "récentement et assez inutilement phyllome") the morphology of the petiolar tendrils has to be considered; the conclusion is that these answer rather to leaflets than to stipules, and the articulation, in some species well marked, between the blade and the petiole, or in the petiole, is noted as supplying good specific characters, which have been overlooked. The umbels are centrifugal or cymose. To distinguish, as is here done, the perianth into sepals and petals and to use these names when practicable, is most proper; but it hardly follows that the term perianth or perigone will then have no *raison d'être*. Whatever the number and position of the stamens, the carpels are superposed to the sepals, as indeed is the case in most Monocotyledons. It is pertinently noted that in Smilax, always diœcious, and with dull-colored perianth, the pollen is papillose as in most entomophilous flowers; but that Rhipogonum, the only hermaphrodite genus, has a smoothish pollen, more like that transportable by the winds. Most have odorous blossoms, some pleasantly, some the reverse. De Candolle asks whether in our Coprosmanthus (the name of which indicates the ill odor) this is common to both sexes and the same in both. Can any of our readers speak to this? An exposition of the geographical distribution of the order, and of what is known of it in a fossil state, is followed by a statement that all the four natural sections of Smilax and the two other genera—*i. e.*, all the types of the order—coexist in the comparatively small area comprised between the north of New Holland, the Fiji Islands, the Sandwich Islands, and Japan; that India has four of these six types, New Holland three, North America two, all Europe and Africa one; South America only one, but is rich in species. The speculative inference is, that, anterior to the eocene formations of Europe, the ancestors of the family occupied a continent situated in the region above indicated, of which the most ancient form was probably monœcious, gamosepalous, apetalous, monadelphous, and with more or less volatile pollen,—in short was like Heterosmilax; that this ancestor was in that region diversified, giving origin to the five other groups, beginning with Eusmilax, the widest diffused and most numerous in species, and finishing with Rhipogonum, which with Heterosmilax has clung to its birthplace. The sole Californian Smilax is referred, as a variety, to *S. rotundifolia*, but is nearer *S. hispida*, although distinct from both.

Vol. V. *Pars secunda: Ampelideæ*; by J. E. Planchon, has at length appeared. It occupies 350 pages; and it represents a great amount of labor, the permanent value and complete acceptance of which cannot be adjudicated off-hand. The plan of merging all the forms into one genus, Vitis, has been abundantly tried, not with very satisfactory results,—partly, it may be, because the groups have not been well worked out. Professor Planchon, a most experienced and keen botanist, who has especially investigated the Vines for a good many years, has very naturally tried the other tack, and has developed the Linnæan Vitis and Cissus into ten genera. The principles upon which he has proceeded, as explained in the preface, are wholly legitimate; and one could wish that they have been successfully applied. This, only use can determine. We may be confident, however, that if this monograph had been in the hands of the authors of the latest Genera Plantarum, they would not have bodily adopted its conclusions, although they would have been much helped by the elaborate investigations, and might have seen their way to admit three or four genera. They would not have trusted over-much to the difference between polygamo-diœcious, polygamo-monœcious and partly pseudo-hermaphrodite, hermaphrodite and probably some pseudo-hermaphrodite, and hermaphrodite or rather physiologically polygamo-monœcious and with some blossoms pseudo-hermaphrodite,—differences which must be shadowy,— nor to variations in the mere shape of style and stigma. And as to the disk, which should be more tangible and hopeful, we gather from Planchon's synopsis and from our own observations that there are only three types. In the true Grapevines the disk is represented by nearly distinct and free nectariferous glands, alternate with the stamens. In most other *Ampelideæ*, it is cupular or annular (entire or crenate or lobed), with base or lower half more adnate to the base of the ovary, but at least the margin or lobes free. In the Virginia Creeper there is really no disk at all, as was first noted by Dr. Torrey in his "Flora of the Northern States," in 1824, and insisted on in the "Flora of North America," in 1838, and again in the "Genera Illustrata," where there are correct figures. Dr. Planchon expresses the same opinion in essence but in different language, *i. e.*, "Discus obsoletus ovarii basi plane adnatus et tantum colore proprio subdistinctus." We could not make much of the color; but the tissue does thicken more or less, and possibly may become obscurely nectariferous; but the flowers are not attractive to bees, as the allied species from Japan is. In the latter, while there is equally no hypogynous disk, there is much thickening of nectariferous tissue over all the lower part of the ovary more or less in longitudinal ridges, the whole "plane adnatus" throughout. Now we should make more of these three types than Dr. Planchon does. For the first goes with the calyptrately caducous corolla and polygamo-diœcious flowers of true Vitis. The third with disk, if so called, wholly confluent with the ovary itself, belongs to and includes all of the few known species (including Planchon's Landukia), which have the striking biological character of climbing by the dilatation and adhesion of the tendril tips; and their flowers are 5-merous, essentially hermaphrodite, and with expanding corolla. The second type of disk goes with 4-merous and some 5-merous flowers with corolla expanding in anthesis, that is, to the genus Cissus. We do not see the way to break this up into genera, certainly not on the number of parts, for this varies in some species, and while *C. stans* is 5-merous, the closely related *C. orientalis* is 4-merous. However it may be with some exotic groups, we must restore our two species, which formed part of Michaux's Ampelopsis, to the genus Cissus. Under that view the generic nomenclature is clear. The genus Ampelopsis (Michaux, p.p. and Torr. and Gray) is to be maintained on the lines long ago laid down in this country, and now reinforced, for those species which are popularly well known under this name. We do not feel obliged to defer to any work of Rafinesque as late as the year 1830. But, as to the present point, it seems to us that when Dr. Planchon followed him in the appropriation of one part of Michaux's Ampelopsis, he should also have adopted Rafinesque's name for the other part, namely, Quinaria, instead of making a new name, Parthenocissus, the former name being free for use. In our view both names are superfluous. As to true Vitis, it remains to be seen whether it will be at all possible to distinguish twenty or more North American species. Perhaps Engelmann allowed quite as many as can be defined. But Planchon's long and conscientious labors upon the genus and the family must be most helpful even where his conclusions are not at once accepted.[1]

[1] This is the last Review written by Professor Gray. — C. S. S.

EPPING FOREST.[1]

"EPPING FOREST, and How best to deal with it," is an article by Mr. Wallace in the "Fortnightly Review" for November, also separately issued, which should not be passed over as a matter of local concern. All Mr. Wallace's writings, even the most casual, will be found to touch and to illustrate some interesting question. A recent act of Parliament having decreed that one of the ancient woodland wastes, Epping Forest, which lies upon the very borders of London, shall be preserved forever as "an open space for the recreation and enjoyment of the public," the question what to do with it becomes a very practical one. It is not enough to say rejoicingly: "Here at length every one will have a right to roam unmolested, and to enjoy the beauties which nature so lavishly spreads around when left to her own wild luxuriance. Here we shall possess, close to our capital, one real forest, whose wildness and sylvan character is to be studiously maintained, and which will possess an ever-increasing interest as a sample of those broad tracts of woodland which once covered so much of our country, and which play so conspicuous a part in our early history and national folk-lore." Unfortunately much of it has been spoiled in all senses of the word. But Englishmen know how to plant, and the native trees which once covered the domain, with the undergrowth which of old accompanied them, could be made to flourish again. Probably the ancient forest could be essentially reproduced in all its former vigor, and former monotony. Mr. Wallace has something better than this in his mind, and his inspiration is caught from Professor Asa Gray's Harvard lecture on "Forest Geography and Archæology," which was published last summer in the "American Journal of Science," the ideas of which he adopts, happily summarizes, and applies to the case in hand. In re-foresting the open waste portions of Epping, he proposes to establish several distinct portions or broad tracts, each composed solely of trees and shrubs belonging to

[1] The Nation, No. 704, December 26, 1878.

some one of the great forest regions of the temperate zone. A climate of which it has, we believe truly, been said that it can grow treble the number of species of trees which the Atlantic United States can, and in which so many trees have been individually tested, offers favorable auspices for an undertaking of this kind upon a scale that may give a good idea of the features — not of this or that tree or shrub, but of a forest of the Alleghanies, of the Sierra Nevada, of British Columbia, and of Japan. Even the southern temperate zone may contribute from New Zealand its Kauri Pines and Beeches, under which Macaulay's overworked New Zealander may encamp on returning from his excursion to view the ruins of London bridge by moonlight.

When Mr. Wallace declares that "there is really no difficulty in producing in England an almost exact copy of a North American forest, with all its variety of foliage, with its succession of ornamental flowers, and with its glorious autumnal tints," we must agree that the experiment as a whole is hopeful, and much of it is already a success in piecemeal plantation. But we are not sure about autumnal tints under London skies, considering how much these differ between one season and another in New England. And, though every tree will grow in England, being put to no severe stress either in winter or summer, yet not every tree nurtured under our climate — so fierce in both seasons — will blossom in England, as witness our handsome leguminous tree, Cladrastis, or Yellow-wood. But a climate which will fairly nourish on one soil the trees of the Atlantic and the Pacific forests, those of Japan and Mantchuria, of Siberia, Himalaya, and the Caucasus, along with those to the manner born, deserves to possess them all. We, alas! can seldom grow on one side of our continent the trees and shrubs of the other. Moreover, there is very little forest east of the Rocky Mountains which an act of Congress could preserve; and, over that little, Congress and the Secretary of the Interior have lately been at loggerheads. Yet in California we have forests, still public domain, which are the veritable wonders of the world, which for the most part are doomed to irremediable destruction, but of which specimens ought to be preserved now when they may, now when it will cost nothing, and injuriously affect no man's interest. For the Redwood it is almost too late; yet a square mile, or half that area, of Redwood forest might still be reserved in Mendocino or Humboldt County. And from all accounts a square mile or two of true Big-tree forest, on or south of King's River, could well be set apart as a perpetual memorial. The Mariposa Grove is indeed such a reservation. But this is only a grove of a limited number of trees, many of them sadly injured by fires. Farther south this great tree is said to be the main constituent of extensive forests. A mile or two of Big-tree forest should be set apart before this district is invaded.

HOOKER AND BALL'S TOUR IN MAROCCO.

THE isolation of Marocco from the European world is strikingly shown in the fact that up to the publication of this goodly volume[1] the principal geographical authority for the interior of the country is Leo Africanus. This man was a Moor of Granada, who with his kinsfolk, at the time of the siege of Granada in 1492, fled to Fez, then the headquarters of Arabic culture, and was taken into favor by Mouley Ahmet, the founder of the dynasty still reigning in Marocco. He traveled throughout the empire, and wrote in Arabic a description of that part of Africa, which he must have had with him in manuscript when, in the year 1517, he was captured by Christian corsairs, and carried to Rome. The Pope, Leo X., hearing of this learned Moor, sent for him, treated him kindly, had him baptized, and gave him at the font his own names, Giovanni Leone; and so, as the writer of the narrative before us naively remarks, the Moor perhaps became as earnest a Christian as the pontiff himself. In Rome he translated his

[1] *Journal of a Tour in Marocco and the Great Atlas.* By Sir J. D. Hooker and John Ball. London and New York, 1878. (The Nation, No. 718, April 3, 1879.)

work into Italian; but it was not printed until 1550, in the first edition of Ramusio's "Collection of Voyages and Travels." Excepting Gerhard Rohlfs — who, by assuming the garb and professing the faith of a Mussulman, traversed districts where no Christian dare present himself, and who could make only surreptitious observations and trust them to memory — our British travelers were the first Europeans to travel in South Marocco and to reach the Great Atlas.

Their opportunity and their success were exceptional. It was a botanical expedition, in the spring of 1871, with the president of the Royal Society and director of Kew Gardens at its head, with Mr. Ball, a member, or at least an ex-member of Parliament, as his companion, and a third naturalist, Mr. George Maw, as their associate for half the journey. High influence was required for obtaining the Sultan's permission to penetrate the country at all; and utmost skill, determination, and no small assertion were called for to surmount the obstacles which were systematically interposed by the authorities, and to baffle attempts to lower the personal importance, and therefore frustrate the aims, of this scientific embassy. On reaching the city of Marocco the playing of the game began. The Sultan's letter instructed the governor of Mogador to "send the English *hakim* and his companions to the care of my slave El Graoui," said slave being the governor of the whole Great Atlas region. The city of Marocco was not in his province; and there the duty of providing for the sustenance and comfort of the travelers devolved upon Ben Daoud, governor of the city. Before entering the town our travelers had managed to learn that a very small house with only two rooms had been provided for them, and that Ben Daoud meant to make as little of them as possible. So a message was sent on that a larger house was needed, or else an enclosed garden in which to pitch their tents. In reply, a larger house with four rooms was offered. On reaching this house at nightfall it proved to be mean, dirty, and swarming; so when the *mona*, or present for the evening meal, came in, it was ordered back with scorn. It was felt that submission to any belittling at the outset would be the beginning of sorrows.

"Tell the Governor," said Hooker, "that my Sultana gives me a large house with a garden to live in; hospitality would require that the Governor of Marocco should provide me — the guest of his Sultan — with a better house; but, in any case, I shall not live in a worse one."

The messenger returned with the answer: "The Governor has no better house to give the Christians; but Marocco is large, and they are welcome to provide for themselves." Whereupon the cavalcade moved to the great square, or open space, beside the chief mosque and tower of the Kontoubia, sending at the same time a message to the Viceroy, son of the Sultan, that they should encamp in their own tents until a suitable house had been provided. The upshot was that on the following day the Viceroy installed them in the palace of Ben Dreis, with the adjoining garden, and soundly berated the city governor for his churlishness; but it leaked out that the poor governor had only obeyed the express orders of the Viceroy, who had directed him to begin by offering the mean house, then one somewhat larger, and to leave it to the Viceroy himself graciously to meet the higher demands if they should be insisted on. Fortunately, too, the row with the city governor threw the travelers into the hands and good graces of his rival, El Graoui, under whose protection and care they were to explore the Atlas range.

It was now full time to determine how this visit and the laborious journey they were to undertake were to be made intelligent and satisfactory to El Graoui and the other Moorish authorities. The gratification of a desire to learn something about the vegetation of the Great Atlas would have seemed a thin pretext for some sinister design. An insanity of this sort might possess the unaccountable soul of some one Christian; but that three should be simultaneously smitten with it would be thought to pass the bounds of probability. The pretense of collecting live plants for Kew Gardens could hardly be made plausible to the Moorish mind, except, haply, for that one use of plants that every one can understand. The Royal Gardens are literally the Queen's Gardens, and herbs that will cure diseases are among its most valuable collections.

Judicious representations upon this line would go far toward explaining the strange proceeding of the party of travelers, and would contain all the truth they were likely to get credit for. So "there is no doubt that the current belief among our own followers was, that the Sultana of England had heard that there was somewhere in Marocco a plant that would make her live forever, and that she had sent her own *hakim* to find it for her." And when it was seen what toil and hardship these botanical explorations entailed, the natural commentary upon the whole proceeding was: "The Sultana of England is a severe woman, and she has threatened to give them stick (the bastinado) if they do not find the herb she wants."

Those interested in the natural history of an almost unknown district lying on the very border of Europe; those interested in its geography and physical features; the statesman and philanthropist, who may here contemplate in a typical instance the decadence and depletion of a once prosperous people, and note the misery which two or three centuries of misrule may bring upon a smiling land; and those who enjoy the charm of fresh and graphic narrative, the sketch by a practised hand of keen and quick observations by practised eyes, — all these will find this volume attractive reading. To us it has been fascinating and full of suggestion. The main drawback to our travelers' enjoyment was the suffering which they unwittingly inflicted on the mountain villagers of the Atlas — the poor Berbers, who are altogether the worthier part of the population — through the insatiable rapacity of a large escort provided by the Moorish government for their protection. If the naturalists could have journeyed by themselves, with needful attendants under their complete control, their support would have entailed no hardship whatever; but the shameless extortion and abuse by the guards, which the travelers were unable to prevent, must have left a painful remembrance upon their minds as well as upon those of the despoiled mountaineers. It is hardly a consolation to the former that the latter are used to it.

The appendix to this charming volume is filled with scientific, philological, and geographical details. The most elaborate article is that "On the Geology of the Plain of Marocco and of the Great Atlas," by Mr. Maw.

As to the orthography of the name of the city and country, Morocco is peculiar to the English. In the adoption of Marocco the authors follow, as regards the first vowel, the universal continental usage.

BENTHAM ON EUPHORBIACEÆ.

THIS thoughtful essay [1] presents the general views attained to by Mr. Bentham on working up the genera of the great order *Euphorbiaceæ* for the ensuing volume of the "Genera Plantarum." We need not specify any of the results, except to indicate the author's decision in the case of the *Buxeæ*. He does not follow his predecessors, Baillon and J. Mueller, who, much as they differ in other respects, agreed in setting up the order *Buxaceæ*, taking their cue from Agardh, and making much of the dorsal rhaphe. Bentham concludes that this small group, however well defined, ought not in a general view to be regarded as of higher grade than one of the primary divisions, or tribes, of *Euphorbiaceæ*. We are not the less pleased with this that we quite expected it.

A wider interest will be felt in Mr. Bentham's *excursus* on nomenclature, or rather on some questions which the study of *Euphorbiaceæ* brought up, and which some recent discussions have made pertinent. The general laws of nomenclature of our day, and the principles on which they rest, are laid down in the code which was reported by Alphonse De Candolle to the Paris International Convention, in the year 1867, and, being approved, was published with a commentary in the autumn of that year, and in an English translation early in the following year. The laws, without the commentary, were printed in this Journal for July, 1868. The ten years

[1] *Notes on Euphorbiaceæ.* By George Bentham; Journal Linnæan Society, xvii. London, 1880. (American Journal of Science and Arts, 3 ser., xvii. 335.)

succeeding have tested, somewhat thoroughly, the questions (nearly all of minor moment) upon which differing usages prevailed; and though one or two points are still mooted, the great majority of phænogamous botanists are coming to be of one mind and practice. But, as Mr. Bentham remarks: "The result has not been quite effectual in checking the ever-increasing spread of confusion in synonymy. Besides the young liberal-minded botanists who scorn to submit to any rule but their own, there are others who differ materially in their interpretation of some of the laws, or who do not perceive that in following too strictly their letter instead of their spirit, they are only adding needlessly to the general disorder. In the application as well as in the interpretation of these rules they do not sufficiently bear in mind two general principles: first, that the object of the Linnæan nomenclature is the ready identification of species, genera, or other groups for study or reference, not the glorification of botanists; and secondly, that changing an established name is very different from giving a new name to a new plant."

It is to the latter point that this most experienced and even-minded botanist addresses himself. "The rule that long-established custom amounts to prescription, and may justify the maintenance of names which form exceptions to those laws which should be strictly adhered to in naming new plants, is unfortunately now frequently ignored. . . . The law of priority is an excellent one; and when a genus or species has been well defined by an early botanist in a generally accessible work, but has subsequently been neglected, and the plant became known under other names, it is well that the original one should be restored. . . . On the other hand, it creates nothing but confusion to suppress a generic name, well-characterized and universally adopted by long custom, in favor of a long-forgotten one, vaguely designated in an obscure work, out of the reach of the great majority of botanists. . . . The greater number of Necker's genera have been so imperfectly characterized, with so absurd a terminology, that they are quite indeterminable; and his names deserve to be absolutely ignored, except in the very few cases where Jussieu or other early French botanists have succeeded in identifying them, and corrected their characters; but even then it is doubtful whether these names should not bear the date of the correction, rather than of the original work. Adanson's 'Familles,' with all the inconveniences of its form and absurd orthography, is much more scientific, and many of his genera are well defined, and have therefore been properly adopted." . . .

Let us here interject a practical application. There is an old and well-established genus Smilacina of Desfontaines. There is a much older genus Tovaria of Ruiz and Pavon, founded in 1794, ever since accepted, and without a synonym. Recently Mr. Baker of Kew, finding that Necker has a Tovaria, published in 1790, and therefore four years earlier than that of Ruiz and Pavon, takes up this name in place of Smilacina, and leaves a new name to be made for the long-established homonymous genus. It will be said that the rule of priority demands the sacrifice, and that the identification of Necker's genus is sure, because the three Linnæan species of Convallaria which properly constitute Desfontaines' Smilacina are referred to it by name; and that, though it be a case of *summum jus summa injuria*, the injurious consequence is a necessity. But Mr. Bentham's characterization of Necker's work applies even to this instance. Twice over Necker's Tovaria is described as having a perianth of five sepals, and the berry is said to be one-celled. Desfontaines' Smilacina, on the other hand, is correctly characterized. Moreover, if we do not include this among those names of Necker which, Mr. Bentham says, "deserve to be absolutely ignored," we may yet find that the law of priority has another claim on it. In 1763 a much better botanist than Necker, namely, Adanson, founded a genus Tovara (essentially the same name as Tovaria) on *Polygonum Virginianum*, L., which is not unlikely to be taken up as a genus; and the name would supersede Necker's by the same rule that Necker's supersedes Desfontaines' Smilacina. All things considered, then, this is a case for the application of the homely but useful rule *Quieta non movere;* and much of Mr. Bentham's pertinent advice may be condensed into this maxim. But there remain nice questions to settle with regard to the names and extent of the Liliaceous genus.

"The representing the Greek aspirate by an *h* was generally neglected by early botanists; but now, ever since De Candolle altered Elichrysum into Helichrysum, modern purists have insisted upon inserting the *h* in all cases; and this has been so far acquiesced in that it is difficult now to object to it, though it has the effect of removing so many generic names to a distant part of all indexes, alphabetical catalogues, etc. Admitting the propriety of adding the aspirate in new names, I had long declined to alter old names on this account; now, however, I find myself compelled to follow the current." Which is, on the whole, regrettable, especially as Alphonse De Candolle would hold out with him. See the latter's comment on his Article 66, in which the remark is dropped, that "we do not see why we should be more rigorous than the Greeks themselves." Oddly enough, these same writers who must supply the aspirate to the *e* omit it from the *r*, and write *rachis* and *raphe*, instead of *rhachis* and *rhaphe*, — which is exasperating to lovers of uniformity.

It is unnecessary here to cite Mr. Bentham's appropriate illustration of the indivisibility of the two-worded name of a plant. The proper apprehension of this, and of the paramount rule that no unnecessary new names should be given to old plants, will go far to rid the science of a principal remaining ambiguity in nomenclature. For it clearly follows that when a plant has a rightful name under its proper genus, the specific half of it is not to be changed because of any earlier specific name under some other genus, to which the plant does not belong.

HENSLOW ON THE SELF-FERTILIZATION OF PLANTS.

THIS paper[1] is elaborate, mostly able as well as ingenious, in all respects considerable, and unconvincing. Its thesis is the Darwinian "Nature abhors perpetual self-fertilization," read backward. It concludes that, "not only are the majority of plants self-fertilizing, but that those which are exclusively so propagate abundantly and with extraordinary rapidity, are best able to establish themselves in foreign countries, as, being quite independent of insects, they run no risk of extermination on that score; . . . that, so far from there being any necessarily injurious or evil effects resulting from the self-fertilization of plants in a state of nature, they have proved themselves to be in every way the best fitted to survive in the great struggle for life." The hypothesis is also advanced "that they are all degraded forms," and that therefore "their ancestral life-history is a longer one than that of their more conspicuous and intercrossing relations." We fail to see how this follows, except upon the assumption that the earliest phænogamous plants had the most highly organized blossoms; and that would not accord with vegetable palæontology.

Mr. Henslow rejoices that he has one stanch supporter; "for, as has been seen, Mr. T. Meehan has arrived at the same conclusion;" and indeed he builds not a little upon facts supplied by Mr. Meehan's observations. He cites the latter's "admirable paper, which was reproduced in the 'Gardner's Chronicle' for September 11, 1875, and is in fact an 'apology' for self-fertilization." As he then marshals twenty reasons for believing particular plants to be normally self-fertilizing, and nineteen "chief facts which may be regarded as occurring correlatively with self-fertilization, some being actual causes which directly or indirectly bring it about," it would appear that it is no longer self-fertilization, but

[1] *On the Self-Fertilization of Plants.* By George Henslow; Transactions Linnæan Society, 2 ser., Bot. i. London, 1879. (American Journal of Science and Arts, 3 ser., xvii. 489.)

rather the existence and *raison d'être* of cross-fertilization that stands in need of apology, or of explanation.

He freely concedes that the flowers of many plants, and some whole orders, are so constructed that intercrossing is for them a necessity; also that most of those which are believed "to be normally self-fertilizing" because they can and do fertilize themselves habitually, yet "may in some cases be cross-fertilized by insects." It is admitted that the structure of the latter is adapted — most variously and wondrously adapted — to being fertilized by particular insects. As this comes to pass in plants and flowers of the highest organization and greatest specialization, Darwin and his school conclude that this is a most advantageous outcome, and means some real good to the species; that when this is accompanied with a loss of self-fertility, it is the loss of something no longer useful, something better than self-fertility having taken its place. But Mr. Henslow, reading this the other way, having determined "that self-fertilization is *per se* a decided advantage," and free from injurious liability, comes to regard intercrossing as merely "a compensatory process for the loss of self-fertility."

But how and why did this "compensatory process" come to pass? It is conceived on both sides that flowers were "primordially inconspicuous." (To this Henslow adds hermaphrodite and self-fertile, but that need not here come into account.) Both agree that insects have mainly determined their conspicuousness. Darwin says this has been determined through natural selection by the survival of the more and more conspicuous variations, correlated with their producing something good for the insect of which the coloration was a sign, and that the preferential survival of the more showy and attractive was a consequence of some benefit of the intercrossing. Henslow propounds the view that insects have determined the conspicuousness more directly, and not by benefiting but by irritating the flowers. "These, by being greatly stimulated by the repeated visits of insects, tend to become hypertrophied. Hence the corolla enlarges, becomes more brightly colored, the nectariferous organs increase the quantity of secretion, and the stamens develop more pollen. Such being the case, nourishment is withheld from the pistil, which is delayed in its development; consequently such a flower is very generally proterandrous." Mr. Darwin might accept this as an ingenious conception of the way the specialization comes about, still insisting on the advantage of the resulting intercrossing—"or else the thing would hardly come to pass," as the poet has it. And Mr. Henslow's hypothesis has to be supplemented to account for proterogyny, which is not much less common. But Henslow's supposed process works evil instead of good, and is therefore utterly anti-Darwinian and "dysteleological." For the result is a disturbance of the equilibrium and proper correlation between the andrœcium and gynœcium; and this, carried further, should upon this view result in the monœcious and diœcious states. So, accordingly, the cross-fertilization which comes into play in the case of separated sexes, and in that of self-sterile hermaphroditism, is not for any good there is in it *per se*, but because it may no better be. And all the elaborate, exquisite, and wonderfully various modes of adaptation of flowers to insects are only ways of repairing the damages inflicted upon blossoms by insects through their persistent visits! Did Mr. Henslow ever ask himself the question why the sexes are separate in animals?

The conclusion which Mr. Darwin has helped us to reach is, that intercrossing should be regarded as the aim in nature and on the whole most beneficial, and self-fertilization as a safeguard against the risks of crossing; that most hermaphrodite flowers have the advantage of both, the latter for immediate sureness, the former for ultimate benefit. Upon the new view, self-fertilization is the aim and the consummation, and cross-fertilization at best a succedaneum. By it insects may repair the damage they have caused to blossoms through endowing them with "the fatal gift of beauty," and stimulating their organs of secretion; and by it the winds may bring chance relief to those which, at length abandoned by their spoilers, have lost this attractiveness and fallen to the degradation of unisexuality. For these last, as has already been stated, are hypothetically regarded as degraded from higher floral types.

We are bound to glance at some of the considerations which are adduced in support of this thesis. They are multifarious and of unequal value. As has occurred in other cases, so here also, the weightiest objections to Mr. Darwin's view are those which he has himself brought out, namely, the fact that, as tested experimentally under cultivation, while some plants are much increased in vigor and fertility by artificial intercrossing, others are not sensibly benefited; and that the benefit derived in marked cases is not cumulative, but reaches its maximum in two or three generations. And even close breeding under cultivation occasionally gives rise to very vigorous and fully prolific self-fertile races. Then many plants are fully self-fertile in nature, and it is not proved that any such have lost or are in the way of losing either fertility or vigor through continued inter-breeding. But, before drawing from this the conclusion that cross-fertilization is of little or no account in nature, it should be remembered that bud-propagated races are in similar case. Races exist which have been propagated only from buds for hundreds of years, with seemingly undiminished vigor, and there is no proof that any one has succumbed under the process. But for all that we do not doubt that sexual reproduction contributes something to the wellbeing of the species, besides facilitating its dispersion. Again, no one questions the necessity of fertilization by pollen to the production of embryo in the seed; yet, even in this, the necessity is not so imminent but that some embryos may originate without it.

In short, the facts brought out by Darwin and others, and all the considerations of the present essays, are best harmonized by the conception which the former has consistently maintained, namely, that an occasional cross suffices to secure the benefit of intercrossing, whatever that may be. Nothing yet appears which seriously disturbs our conviction that just this is what nature generally provides for.

Mr. Henslow's proposition, "The majority of flowers are self-fertile," is doubtless true in the sense that they are capable of self-fertilization, and is not improbable in the sense that they "can and do fertilize themselves habitually." But his inference that the majority of flowers, or that any flowers, actually propagate for a series of generations by self-fecundation, or that a cross if it occur is "exceptional," and of no account, is surely unwarranted by the evidence which he has adduced.

Occasionally the reported facts will not bear scrutiny. *Gentiana Andrewsii*, it is said, never opens at all in America. It opens in sunshine in the middle of the day here in New England. And while looking at closed flowers we have seen a humble-bee emerge from one. We have in this Journal shown how it is that self-fertilization is impossible during the first three or four days of anthesis, but neatly practicable afterwards. It is rash to infer (as on p. 330) that papilionaceous flowers which shed their pollen early in proximity to the stigma are therefore self-fertilized. In most of the cases adduced the pollen is not lodged upon the stigma, but upon the style below it, and the adaptations for intercrossing, though the mechanism be different, are as explicit as in the analogous case of Campanula. "Fremont pathetically describes the solitary bee that rested on his shoulder at the top of Pike's Peak." The pathos is wasted as respects all but this particular bee; for the entomologists find the alpine region of the Rocky Mountains to be as well stocked with flying insects as are alpine regions in other parts of the world. They do not super-abound, but if from the alpine flora we subtract the evidently entomophilous and the anemophilous blossoms, the remainder will be nearly nil. And as to the correlation of this comparative scarcity of insects with the marked conspicuousness of blossoms, this is the way the lesson is read by a most eminent physiologist: "Even the glowing hue of alpine flowers is accounted for by the attraction which brighter-colored individuals exercise upon the insects, scarce in those heights and necessary for fertilization."

One or two of the author's own observations are perhaps to be revised. "*Gaura parviflora* . . . has no corolla and is cleistogamous, in that it is self-fertilizing in bud, as I found in specimens growing at Kew." Were they not imperfectly developed blossoms, perhaps late in the season? Here the flowers open freely, and have rose-colored petals. If he will examine fresh specimens of Scrophularia, it will soon be clear that his idea of their self-fertilization (p. 371) is a mistake. It is a mere slip in the "Genera Plantarum" through which abortive stamens are attributed to the cleistogamous flowers of Epiphegus. The authors evidently meant to describe the case just as Mr. Henslow found it to be, but used a wrong word.

"Weeds are probably all self-fertilizing or anemophilous. A weed is simply an unattractive plant, and possessing no feature worthy of cultivation." It may be as difficult to define "a weed" as to define "dirt." But, turning to the "Handbook of the British Flora," we find, as we expected, that the showy Corn Poppy, Cockle, and Larkspur are denominated weeds. Why weeds should possess the vigor and gain the predominance which they do is a large question, to which other solutions have been offered than that one which is in this essay very plausibly maintained. We cannot take up the topic here; but, without acceding to his general proposition, we are much disposed to agree with the author in this essay, as respects some of them, that aptitude for self-fertilization may have given them the advantage which has determined their wide dispersion.

The insistence upon the importance of self-fertilization is what gives this essay its value. As a whole it fortifies the proposition, well laid down by Herman Müller, which Mr. Henslow cites: "that, under certain conditions, the facility for self-fertilization is most advantageous to a plant, while, under other conditions, the inevitableness of cross-fertilization by the visits of insects is the more advantageous." But this is not our author's thesis. It comes to this: the plan of nature is either cross-fertilization supplemented by close-fertilization, or close-fertilization tempered by cross-fertilization. As restricted to plants the difference is not wide. Regarded generally, the Darwinian axiom is still best sustained.

PLANT ARCHÆOLOGY.

Investigations in fossil botany are recondite and technical, the materials generally unattractive, and the results unintelligible to the popular mind; but in Count Saporta's "Monde des Plantes,"[1] and under his happy exposition, the stony desert is made to rejoice and blossom as the Rose. The interest which we take in the vegetation of former periods is not so much geological as genealogical; and this interest diminishes with the distance from our own time and environment. We know nothing of the earliest plants — the beginnings of vegetable even more than of animal life are beyond our ken; no great satisfaction seems obtainable from the small acquaintance that has been made with the plants which flourished before the carboniferous period. And the botany of that age, notwithstanding its wealth of Ferns and its adumbrations of next higher types, impresses us as much with the sense of strangeness as of wonderful luxuriance. For even the fern-impressions, familiar as they may look to the unprofessional observer, are outlandish. The more the critical student knows of them the less likeness he finds in them, or in the coal-vegetation generally, to any species or genera now living.[2] But in the vegetation of cretaceous, and still more of tertiary times, familiar forms first come to view, and pedigrees may begin to be traced. Questions of ancestry touch us more nearly than those of history; so an enquiry into the source and parentage of the plants with which man is associated is more attractive than any question concerning the origin of the pristine vegetation of the earth. Moreover, our

[1] *Le Monde des Plantes avant l'Apparition de l'Homme.* Par le Comte de Saporta. Paris and New York, 1879. (The Nation, Nos. 742 and 743, September 18 and 25, 1879.)

[2] To those who wish to get a good coup d'œil of this vegetation from authentic records systematically arranged, we recommend the "Atlas to the Coal Flora of Pennsylvania and of the Carboniferous Formation throughout the United States," by Leo Lesquereux, an octavo volume of eighty-seven double plates, just issued by the Second Geological Survey of Pennsylvania. There is nothing else to be compared with it.

knowledge of the later fossil botany is comparatively full,—wonderfully so, considering how very recent this knowledge is,—and we are in a condition to apply it hopefully and confidently to the solution of problems which not long ago seemed to be beyond the reach of proper scientific enquiry, namely, to the explanation of the actual distribution of the species of plants over the earth. For the main data themselves, and for the clear exposition of them, we are most largely indebted to three men, who happily are still alive and active—Heer, Lesquereux, and Saporta.

The Linnæus and *facile princeps* of tertiary botany is Oswald Heer, of Zürich, now a septuagenarian, but still in harness. His "Recherches sur la Climat et la Vegetation du Pays Tertiare," rendered into French by C. T. Gaudin, was published nearly twenty years ago. It is a general and comparatively untechnical presentation of a long line of investigations, which have since been crowned by his several memoirs on arctic phyto-palæontology, now collected in the five volumes of his "Flora Fossilis Arctica." All these volumes, as well as others on the Swiss tertiary, have appeared within the last ten years, the latest only a year ago.

Leo Lesquereux, Heer's compatriot, and barely his junior, came to the United States fully thirty years ago, drawn hither from Neufchâtel by Agassiz. The greater part of his researches relate to the carboniferous flora, and he has recently thrown interesting light upon silurian botany, as has Dawson of Montreal upon the intermediate devonian. But those which at present concern us relate to the cretaceous and the tertiary of our own western regions. The most considerable of these works are the two notable quarto volumes, entitled "Contributions to the Fossil Flora of the Western Territories," published by the Geological and Geographical Survey of the Territories under Dr. Hayden, upon whom and whose survey they reflect high credit. One volume treats of the cretaceous, one of the tertiary flora.

Any proper enumeration of authorities upon the fossil botany of the later periods should include various other names, and especially that of Schimper, of Strassburg, who, like

Lesquereux, has divided his life between bryology and fossil botany, and whose classical "Traité de Paléontologie Végétale" is a systematic compendium of what was known of fossil plants up to the year 1874. But the volume now under notice is by a younger man, Gaston, Comte de Saporta, a Provençal, who has for fifteen years or more been investigating the rich tertiary deposits of Aix and vicinity, in the delta of the Rhone, the results of which have appeared from time to time in memoirs, mainly published in the "Annales des Sciences Naturelles." Besides these weightier and more technical publications, Count Saporta has contributed to the "Revue des Deux Mondes" and to "La Nature" subsidiary articles of a popular cast and of fine literary as well as scientific finish. These, now collected and re-edited, form a part of the volume before us, "Le Monde des Plantes avant l'Apparition de l'Homme," which has been published since the commencement of the current year. It is the most comprehensive and the most attractive, as well as the most recent, exposition of our subject, is a very readable book from beginning to end, of inviting typography, with abundant illustrations, both of woodcuts in the letter-press and intercalated plates. Although a popular, it is a truly scientific volume. The clear stream of the narrative is hardly at all troubled by the many technical terms which unavoidably strew its course, yet without obstructing its flow, for the author has the peculiarly French gift of happy exposition. As the volume is likely to be reproduced in English, let us hope that it may have a translator in whose hands it may lose nothing of its clearness, and as little as possible of its freshness and spirit.

To attempt a popular abstract of such a book would be like skimming the cream from the cream, and a critical review would cover too much or too technical ground. Still, we may give some general idea of the contents of the volume. The first part and the most discursive portion of the book is entitled "Phenomena and Theories." The introductory chapter discourses upon the introduction of life and the origin of the earliest terrestrial organisms; and the second chapter takes up the theory of Evolution or Transformism. We may skip these chapters, yet without advising the reader to follow the example, unless he is already familiar with the topic — now a little threadbare — for, as a popular presentation, it is neat and sensible, though not profound. It hardly need be said that Saporta is an evolutionist, using the term in its general sense, and apparently a thorough Darwinian. A vegetable palæontologist who studies the later geological deposits cannot be otherwise; at least, he must needs be a "transformist." Saporta concludes that palæontology, if it does not furnish demonstration, yet gives irresistible reasons for a belief in evolution. The ground and the nature of this conviction appear in his rounded statement, that there is not a tree or shrub in Europe, in North America, at the Canaries, in the Mediterranean region, the ancestry of which is not recognizable, more or less distinctly, in a fossil state. This is too absolutely stated, no doubt, but the qualifications it may need will not invalidate the conclusion.

The chapter on ancient climates which follows, and forms a proper introduction to the second part of the book, is worthy of particular attention. It is prefaced by an elementary but very graphic exposition of the phenomena and laws of climates and their diversities, from the regular succession of equal days and nights under the equator to the contrasted condition toward the pole of a year composed of a day and a night season, separated by a season of twilight; the change so rapid in the high latitudes that, while the summer day at the North Cape is two months long, at Spitzbergen seven additional degrees of latitude lengthen it to four months. Let the author point the contrast between the two extremes, as affecting man, in his own language, here somewhat exceptionally ornate: —

"Il est vrai que dans ce dernier pays [Spitzbergen] le soleil s'élève au plus de 37 degrés au-dessus de l'horizon ; il n'envoie que des rayons sans chaleur, *telum imbelle sine ictu ;* il éclaire de sa lueur pâle une terre glacée où frissonnent quelques plantes ensevelies sous les frimas, et qui ne sortent du sommeil qui les tient dix mois inertes que pour accomplir hâtivement leurs fonctions vitales et se rendormir de nouveau. Quel tableau, si l'on songe aux forêts vierges du Brésil et de Java, aux vallées profondes du Népaul, aux savanes noyées de l'Orénoque, où la vie surabonde, où une lumière ardente, vive et dorée, ondule de toutes parts, soulève de tièdes vapeurs, joue avec l'ombre, et fait resplendir les formes des plus merveilleux végétaux! Sous les tropiques, l'homme se sent écrasé par une vie exubérante, il lutte incessament pour maintenir sa place au milieu de la nature, dout il est dominé; ses plus fortes œuvres sont envahies en peu de temps; les arbres immenses reprennent possession du sol, dès que celui-ci est abandonné à lui-même. Dans l'extrême Nord, la faiblesse de l'homme est encore plus évidente, mais c'est du poids de la nature inerte qu'il est accablé. Les éléments règnent seuls dans ces régions dévastées, où l'atmosphère se trouve livrée à d'épouvantables tourmentes. La neige dérobe les aspérités du sol, la glace couvre la mer d'un sol factice, souvent mobile et toujours dangereux; la confusion est partout, le calme nulle part; chaque pas est pénible, la vie elle-même devient un effort que l'energie la mieux trempée ne peut soutenir longtemps sans succomber." (p. 212.)

The modifications of this contrast through the actual distribution of land and water, winds and currents, are then considered. As these have a fixity secondary only to the fundamental elements of climate, — namely, the heat of the sun, the inclination of the earth's axis to the plane of its orbit, and the relative density of the atmosphere according to the elevation of the land-surface, — the climate of any part of the world might be supposed to have been constant, oscillations excepted, through the long periods that have elapsed since existing species or their immediate ancestors were introduced. It is not very long ago that Arago demonstrated, to his own and the general satisfaction, that there has been no appreciable change of the earth's climate in man's time. Plants are the thermometers of the ages, by which climatic extremes and climates in general through long periods are best measured. For at least five or six thousand years the Vine and the Date-palm have grown in proximity, and have furnished grapes and dates to the inhabitants of the warmer shores of the Mediterranean. Yet a very moderate change either way in the temperature would have excluded the one or the other. So Arago concluded that man had witnessed no sensible changes in the climate of Europe; a good conclusion, if restricted within the limits of observation. But in Arago's lifetime the evidence was already accumulating which has now proved that in earlier times man and the reindeer lived together on the soil of southern France, when if grapes could ripen in Syria and northern Egypt, dates doubtless could not. Then, at a still earlier day, Palms flourished in Switzerland and Vines in Iceland. Then Maples, Lindens, Plane-trees, Spruces, and Pines formed forests in Greenland up at least to the eightieth parallel; and, indeed, our own southern Cypress, or Taxodium — which now barely maintains its existence between the mouths of Delaware Bay and the Chesapeake — flourished along with the Silver Fir of Europe, within two hundred leagues of the North Pole. A climate in Greenland in which Sequoias, now confined to California, Magnolias, Persimmons, and Grapevines were mixed with Maples, Oaks, and Poplars, could not have been colder than, or much unlike, that of Indiana and Kentucky now.

As we have seen, the vegetable world has had an eventful history, and this history Count Saporta undertakes to reconstruct from ancient and authentic documents.

These documents have settled one point with certainty, namely, that the great changes in the temperature of the polar regions have not resulted from any change in the earth's axis of rotation, such as certain physical geologists have supposed. That this has remained steady throughout the whole period in question is as good as proved by the identity of the miocene and other tertiary fossil plants in all longitudes, from the Mackenzie River and Alaska round to Spitzbergen, Iceland, and Greenland — in part the same species, in all the same or equivalent combinations. Not indeed the same species in the same latitude any more than now, but such latitudinal distribution as to show that the curvatures of the miocene isotherms were quite analogous to those of the present age. Moreover, the monotony which characterizes the sub-polar vegetation of the present day — when the most of the species and the combinations of species occur all round the world — equally characterized it then when clothed with forest trees of a temperate zone.

Upon the data in hand, now much extended, and on the supposition that the same species of tree had not appreciably altered meanwhile in its relations to temperature, Heer long ago elaborately compared the miocene climates with those of our time, and Saporta corroborates his conclusions. For the northern regions the difference is said to be equivalent to 25 or 30 degrees of latitude — that is, we have now, say in Europe, in latitude 40° and 45°, and in Atlantic America, in latitude 38° to 40°, the temperature and the vegetation which then flourished at latitude 70° in Greenland. Grinnell Land, in latitude 82°, only two hundred leagues from the pole, had a forest of coniferous trees; among them — associated with a Poplar-tree, a Hazel, and a Birch — was the Silver Fir of Europe, and the Bald Cypress of the swamps of the southern United States. The same combination, minus the Cypress, must now be sought in the more elevated parts of central and southern Germany. The Sequoias and Magnolias and Persimmons of Greenland, in latitude 70°, mingled with Maples, Oaks, and Grapevines, have their representatives partly in Virginia and on the Ohio River, partly in California. The miocene of the southern shore of the Baltic had Laurels, Oleanders, and Camphor-trees, but no Palms, so far as is known, thus answering to the Mediterranean flora, but miocene Palms reached to Belgium and Bohemia; and Provence, on similar data, had then about the climate of the coast of Zanzibar.

Another conclusion which Saporta confidently reaches — and which indeed is reached from all sides — is that of a very moist quaternary climate. Looking back to this comparatively recent period from our own, everywhere the streams have dwindled. Through great river-beds shrunken streamlets now meander in insignificant channels; springs reach the surface much lower down the valleys than of old; and the "rivers without water" of Egypt and Syria, and the reduced level of the Dead Sea, are so many evidences of a dryness supervening upon a general humidity greatly in excess of the present. These fuller watercourses of themselves indicate a more temperate or mean climate, a more equal distribution of heat and cold through the year. Was this equable climate frigid and rigorous or comparatively mild? Were Europe and the United States simply arctic and to be compared with the present Greenland and Spitzbergen, as the school of Agassiz maintains? Or should the comparison rather be made with southern New Zealand, where Tree-ferns almost overhang the terminal moraines of existing glaciers, in a climate which is neither cold nor warm? Saporta maintains the latter, and he is not alone. He insists that the high Alps and the Pyrenees are not the types of glacial Europe generally; that the arctic animals and plants, and the rigorous climate which we associate with these, belonged only to the close neighborhood of glaciers, but that the valleys below enjoyed a climate even milder than now, although vastly more humid. So, likewise, Mr. Ball (in a lecture recently delivered before the Royal Geographical Society of Great Britain) ventures to affirm that, even during the period of maximum cold, the highest ridges of the Alps were not completely covered with snow and ice; for we still see, by the appearance of the surface, the limit above which the ancient ice did not reach; and in the middle zone the slopes that rose above the ancient glaciers had a summer climate not very different from that which now prevails. And he concludes that the effect on the growth of plants in the Alps was to lower the vertical height of the zones of vegetation only one or two thousand feet.[1]

This would seriously affect the forests of Europe, but would not permanently disturb the alpine and sub-alpine vegetation.

Yet cold it must have been when the reindeer and musk-ox roamed over the plains of central Europe, and when the elephant or mammoth and even the rhinoceros which accompanied them, were equally clad with a thick coat of hair. But, says Saporta, with the remains of these very animals from which a frigid arctic climate is inferred, occur also, in the alluvia of the Somme and the Seine, those of an elephant nearly related to the Indian species, the hippopotamus of the African rivers, and the hyena of the Cape; and the vegetable

[1] But Mr. Ball is fairly astounding when he assumes that our arctic-alpine flora may have been the flora of high mountains at low latitudes in the carboniferous period.

remains include the Laurel of the Canaries along with the Vine. The trees of the same epoch farther north were Pines, Lindens, Maples, and Oaks. So, according to Saporta, even the glacial period formed only a seeming interruption to the general course, the steady and really unbroken diminution of terrestrial temperature from the earliest geological periods to the present. This must be admitted if the two classes of animals and plants — those adapted to cold and those to warmer climates — were really contemporaneous. Our geologists have maintained that they were not, but that climates have oscillated, and that warmer periods than ours intervened between the glacial epoch and the present, or were intercalated in the glacial period itself. But is not the distinction of periods an assumption for explaining the two kinds of fossil remains?

We need not enter here into the discussion of the cause of the higher temperature of ancient climates, and of that peculiar and temporary state of things attending and originating the glacial epoch, with which Saporta concludes his third chapter. And no space is left us in which to sketch even the outlines of the second part and main staple of his book, the history of the vegetable periods, beginning with the "primordial marine plants" of the Laurentian and closing with the pliocene, in which existing trees are everywhere identified. The general conclusion of these very rich, elaborate, and well-considered chapters is that the vegetation of the earth has been continuous through all ages, and that the explanation of the present is found in the past. The history of the genus Sequoia — of the two "big trees of California" — as recently sketched by Heer in a popular journal, "Das Ausland," is a fair illustration of this. The difference between these two trees is as notable as their resemblance and their isolation. They are the survivors of a numerous family, of wide distribution, which is first recognized in the cretaceous formation, in several species, and which reached its maximum in the middle tertiary, in fourteen recognizable species or forms. Almost from the first these separate into two groups, one foreshadowing the Coast, the other the Sierra, Redwood, yet with various intermediate forms. These intermediate species are extinct, the two extreme forms have survived. The likeness of these two trees is explained by their genealogy, their marked difference by the extinction of the connecting forms which in earlier times bridge the interval.

SERENO WATSON ON NORTH AMERICAN LILIACEÆ.[1]

MR. WATSON, in preparing the *Monocotyledoneæ* for the "Botany of California," came upon the order *Liliaceæ*, which is well represented in Pacific North America; and he had to consider how the genera and higher groups should be disposed. This led to a wide study of the order and a strict scrutiny of the American species; and the present "Revision of the North American Liliaceæ," occupying the greater part of the "Contribution" before us, is the result. It is generally agreed that this order is to have the wide extension which was given to it by the present writer a dozen and more years ago; and the proper collocation of its diversified forms, with interlaced affinities, has been a problem of no small difficulty. Mr. Baker, in England, has attempted the task for the order generally, and has sedulously elaborated some of the North American, but more of the Old World and the South American, genera and tribes. His arrangement and his systematic views are in many respects satisfactory, in some unsatisfactory as respects North American botany. Mr. Watson has the latter primarily in view, but still has to adjust the American genera into the general system. The arrangement he has planned consists of three series, the first of which parts into two subseries, and includes sixteen tribes, some of them divided into subtribes. The great endeavor has evidently been to make natural groups — and this endeavor has been really successful. The next thing is to assign characters, and here comes the difficulty. Absolute characters of the lead-

[1] *Contributions to American Botany*, IX. By Sereno Watson; Proc. American Academy of Arts and Sciences. Boston, 1879. (American Journal of Science and Arts, 3 ser., xviii. 313.)

ing groups are not to be had, even when North American forms only are considered. Those who imagine they could do better than Mr. Watson has done should make trial before they criticise. The character of the pericarp, whether baccate or capsular, the nature of the stock, whether bulbous, tuberous, or rhizomatous, the nature of the seed-coat, the inflorescence, direction of anthers, union or separation of styles, are all good characters to a certain extent, and all fail to furnish unexceptionable marks to distinguish the higher groups when natural associations are sought. It is not easy to ascertain what diagnostic characters in this monograph are most to be trusted. But the nature of the bracts (on the one hand scarious, on the other foliaceous or none) takes the lead in the first two series, and is followed by the persistence or deciduousness of the perianth, the insertion of the stamens whether on the perianth or at its base, the dehiscence of capsule, — all matters of little physiological importance, but for that reason perhaps surer guides to affinity than the more prominent adaptive characters. However, it may be said that the first series answers to the *Asphodeleœ*, with Yucca and *Hemerocallideœ* added; the second to the true *Liliaceœ*, with *Uvularieœ* and *Trillieœ* added; the third to *Melanthaceœ*, with the tribe *Tofieldieœ* appended. Thus disposed, it is doubtless judicious to designate the three primary groups as "series," and not as suborders, and to throw the stress upon the tribes.

The Melanthaceous series, which in our view best divides into the *Colchiceœ*, *Veratreœ*, and *Tofieldieœ*, — the first not American, — is here divided into the *Veratreœ*, *Helonieœ*, and *Xerophylleœ* (which two we should combine), into the midst of which the *Tofieldieœ* are intercalated. This last tribe, which should end the series, is quite exceptional, and is well composed of Tofieldia, Pleea, and Narthecium. Its marks are the equitant leaves, introrse anthers with parallel cells, and caudate seeds; but to bring Narthecium under the remaining character of "styles distinct or none," it is defined as destitute of style, but with "the slightly lobed stigma sessile upon the attenuated apex of the ovary." This is really much nearer the fact than would be supposed, as the cells of the ovary actually do taper up into the subulate style (as it has always and most naturally been termed), so that in the mature capsule the upper tails of the seeds reach up to within a short distance of the small stigma.

In a linear order it has not been practicable to approximate the *Convallarieæ* of the first series with the *Uvularieæ* of the second. The division of Uvularia gives a gratifying opportunity of dedicating a New England genus to the memory of one of the best of New England botanists, the late William Oakes (*Oakesia sessilifolia*, with its relative of the southern mountains, *O. puberula*); but he would not have relished the dismemberment of the Linnæan genus upon the characters, good as they are, neither in fact do we. The formation of the tribe *Yucceæ*, of Yucca and Hesperaloë, strikes us as excellent; and it seems right not to adopt the supposed second species of Hesperaloë until it is better known. Its principal distinctions (longer anthers and shorter style) may indicate heterogone dimorphism, which would be a novelty in the order.

Tribe *Nolineæ*, of the first series, must be regarded as an excellent group, composed of Dasylirion and Nolina; and it is gratifying to find that the outlying genus Nolina, founded long ago on a single Georgian species, is the northern representative of a considerable Texano-Mexican group, named Beaucarnea. It were to be wished that the plan of this Revision had allowed more citation of generic synonymy, and that it had been more explicitly stated that Beaucarnea is only Nolina. This union, indeed, is one of the happy hits of the present monograph.

As has been suspected, the Californian *Schœnolirion album* of Durand proves to be quite distinct from the Atlantic species on which that genus was founded. So Mr. Watson has embraced the opportunity, here offered, to dedicate a peculiar Californian genus to Judge Hastings — a judicious patron both of botanical and legal learning. Except for his exertions, his own liberality, and his direction of the liberality of others, we could not have had the "Botany of California," which Mr. Watson may now soon bring to a completion. The reader finds no mention of this under the genus Hastingsia, p. 217, nor under the species *H. alba*, p. 242. But an appropriate reference is made on p. 286.

Leucocrinum, Nutt., was conjectured by Endlicher to be the Mexican Weldenia, but it has just now been ascertained at Kew that Weldenia is a Commelynaceous genus.

Our species of Allium as now worked out by Mr. Watson with great painstaking, are thirty-six in number, exclusive of the introduced *A. vineale*. Some characters might be made more of in living plants, such especially as those furnished by the so-called "crests of the ovary." In *A. stellatum* these crests are remarkably developed, radiating from around the base of the style and recurving, the notch at the end of each fitting over the base of the alternate filaments, and the under side is nectariferous and attractive to bees. The flowers are proterandrous.

In separating the two species of Maianthemum we should have unhesitatingly referred the large Pacific coast form to *M. bifolium*. We should not have distinguished *Lillium Grayi* as more than a form of *L. Canadense*, one which extends northward to the central parts of New York. In view of geographical range, size, and general appearance, we should never have thought of *Uvularia flava* as a synonym of *U. grandiflora*. Mr. Watson finds good characters in the shape and markings of the capsule to separate *U. grandiflora* from *U. perfoliata*. Has any one ripe fruit of the small, yellow-flowered *U. flava?*

Chamælirium Carolinanum, Willd. This specific name is properly restored. It was the original name under this genus; and the name *luteum* is a false one (though the plant was *Veratrum luteum* of Linnæus), the blossoms being white without a tinge of yellow, duller white in the female plant, pure white in the male, the pedicels equally of this color.

No space is left in which to notice the Notes upon the Affinities and Geographical Distribution of *Liliaceæ*, nor the Descriptions of some New Species of North American Plants, about fifty in number, which make up the second part of this important "Contribution." Among them is a new Bolandra and a new Sullivantia from Oregon, both very much like (we fear too like) the original species. Here and in the Bibliographical Index, the name *Sullivantia Ohionis* is changed (perhaps accidentally) to *S. Ohioensis.* We knew of no law against genetive names of geographical more than of other places or stations, and such are not extremely uncommon. The name Ohionis was purposely chosen, and we hope may be retained.

The interesting new Erigoneous genus Hollisteria, discovered by the enthusiastic Mr. Lemmon (in San Luis Obispo County, east of the Coast Range), is of rather doubtful interpretation as to some points of structure. The inflorescence we suppose to be only seemingly axillary, the involucre is possibly a genuine trimerous one, and we take the two small stipule-like leaves to be real stipules, — a point which the published character does not decide, though it is implied in describing the leaves as alternate.

Being one of the most important of recent contributions to North American Botany, this publication deserves even a fuller notice than we can here give it.

DE CANDOLLE'S PHYTOGRAPHY.

VALUABLE as the present volume [1] is, it may very probably not be translated into English. So we propose to give a running account of its contents, adding here and there some brief comments, critical or otherwise. Treatises like this can be written only by botanists of long experience; and long experience, founded upon good training and accompanied by good judgment, gives the right to speak with a certain authority, particularly upon writing and publication in systematic botany, in which rules and method are most important. De

[1] *La Phytographie, ou l'Art de décrire les Végétaux considérés sous différents points de vue.* Par Alphonse De Candolle. Paris, 1880. (American Journal of Science and Arts, 3 ser., xx. 150, 241.)

Candolle is now one of our oldest systematists, one who as editor as well as author has had to consider every sort of phytographical question; and the volume he has here produced is a needful supplement to the "Philosophia Botanica" of Linnæus and the "Théorie Élémentaire" of the elder De Candolle, the two classical books which the serious botanical student should early and thoroughly master. Phytography has to do with form and method in botanical works; and natural history is nothing if not methodical. Its advancement by research and its educational value — which will be more and more appreciated as it is better taught — both depend upon correct morphology and upon well-settled method. Those who will not use its proper language and respect its customs, must not expect to be listened to, any more than is unavoidable. Observation and interpretation must go together, if either is to be of value; the naturalist must not only observe that he may describe, but describe if he would observe. In his preface De Candolle remarks upon the peculiar advantage of natural-history study in the combination of observation with judgment, and upon the importance to a student of acquiring a clear idea of what natural groups are, what a natural classification and the subordination of groups really mean, and how a naturalist arranges, names, and with precision defines the immensely numerous objects of his study. Men who have distinguished themselves in various professions and lines of life, have owed the advantage they have derived from this kind of training in youth, even though they never became naturalists.

De Candolle's book is in thirty chapters, many of them short and somewhat discursive, and generally abounding in recommendation and advice, rather than laying down positive rules.

The first chapter glances at "the evolution of botanical works" from Cesalpino, with whom scientific botany began in the middle of the sixteenth century, to Linnæus, whose rules and spirit still govern, and to our own times, noting the gradually increasing importance of herbaria as compared with botanic gardens. The second chapter touches upon the moral and intellectual dispositions necessary in botanical work, and asks the question what manner of men botanists are or ought to be. As their pursuits do not lead to fortune, and professorships are neither numerous nor well paid, he concludes that botany is just the science for disinterested people to prosecute from pure love of knowledge and the pleasure of discovering something new; that it does not deal with questions of a very high order, nor require very difficult or absolutely rigorous reasoning. The faculties which it brings into requisition are the spirit of observation and of order, sagacity, and a certain good sense in the appreciation of facts; that, if it does not shine with great éclat, at least the faults of its cultivators are not likely to harm any one; that, equally with the other sciences, it tends to elevation of character in that it requires an ardent love of truth, reposing as it does upon the idea that the veracity of its cultivators is absolutely complete. He concludes: "Les sciences jouent dans le monde le rôle d'une école practique de bonne foi. D'après ces réflexions, il est permis de penser que les botanistes sont ordinairement et devraient être toujours des hommes paisibles, inoffensifs, indulgent pour les erreurs de leur confrères, et occupés bien plus de l'avancement de la science que de leurs intérêts ou de leurs petites glorioles. . . . Ne nous arrêtons pas cependant sur de rares exceptions. La presque totalité des botanistes est pénétrée du sentiment de la justice et des convenances. On en trouverait difficilement un seul qui ne reconnût le principe fondamental de ne pas faire à autrui ce qu'on ne voudrait pas qui vous fût fait."

Still, our author continues, sometimes the perfectly honest and right-minded botanist may have failings. He may, for example, neglect to cite his predecessors, or cite them inexactly, either from negligence (not to speak of calculated omissions, which show want of honesty and soon bring down reprobation), or from the want of literary resources. The latter case may be deemed a misfortune, and no fault. But, our author rejoins, if he has not the necessary books within his reach, why not go where they are and consult them? Or if unable to do that, why need he publish?

Some good advice follows about polemics and captious criticism; which we pass over, as seemingly superfluous, so long as the botanists are almost without exception such peaceable and good people. Something is said of the need of a right appreciation of the extent of the science; of the danger of exclusive devotion to a single branch of botany, in which one may lose all just perspective; and, finally, of what accuracy means in natural history as distinguished from mathematical exactness. Everywhere the naturalist has to judge as well as to measure.

The third chapter discourses upon the manner of preparing and editing botanical works, and the most advantageous modes of publication, considers the different degrees of publicity: for instance, complete and durable publicity is attained when a Monograph of an order or genus, a Flora, a Species, or a Genera Plantarum is published and placed on sale by the booksellers; or when an article or memoir is contributed to any leading and well-known botanical journal, or to the bulletin of a purely botanical society, which publishes with some regularity and indexes its volumes; or when printed in the transactions or bulletins of any scientific society, if separate copies in sufficient number are printed and fairly distributed or placed on sale. The usage in some learned societies of paging each memoir separately and placing it separately on sale is referred to, with implied commendation. Let us add that in all such separate issues, the original pagination of the volume should be scrupulously preserved; and it were better that there should be no other. Less complete, but durable publication is that of *ouvrages de luxe*, so limited in number of copies, and so high in price, that only a few libraries can possess them; also articles in journals without full indexes, or with indexes only to a series of volumes.

Incomplete publicity is given when papers upon botany are inserted in the voluminous transactions of general learned societies, of which few individuals can possess the series or find room for them; also articles in reviews, encyclopædias, and the like, treating of many or of all sciences. Even journals of natural history alone fall under the ban, unless divided into separate parts for zoölogy and botany, — as is the long-continued "Annales des Sciences Naturelles" of Paris, and the "Journal of the Linnæan Society of London." A remedy or alleviation of these obstacles to publicity, and of others like them, is supplied by the catalogue of papers published by the Royal Society of London; which noble work was instigated by the late Professor Henry; yet, as this embraces all sciences and fills a goodly series of volumes, it can seldom be in the library of botanists.

In speaking of the obstacles to scientific publicity which are interposed by too limited editions, high prices caused by undue luxury in plates, and inopportune or inappropriate media of publication, De Candolle refers to customs in the book trade and in government patronage which need reform; and mentions incidentally what a botanical library costs. He says there should be by the side of every great herbarium and every considerable botanic garden, a special botanical library, without which it is impossible to determine exactly the plants of the one or the other, or to write any good Monograph or Flora. Such a library costs fifty or sixty thousand francs (ten or twelve thousand dollars), and needs about 12,000 francs for annual purchases. He asks how many such establishments there are in the world, and concludes that there may perhaps be between ten and twenty.

The section on the comparative superiority of certain kinds of works, sets forth the greater value of books or systematic works as compared with memoirs or articles.

The language to be employed in botanical publications is the topic of a special article. For descriptions, Latin, and the Latin of Linnæus. "Le Latin des botanistes n'est pas cette langue obscure et à réticences de Tacite, obscure et à périodes pompeuses de Cicéron, obscure et à grâces tortillées d'Horace, qu'on nous fait apprendre au collège. Ce n'est pas même le langage plus sobre et plus clair d'un naturaliste tel que Pline. C'est le Latin arrangé par Linné à l'usage des descriptions et, j'oserai dire, à l'usage de ceux qui n'aiment ni les complications grammaticales, ni les phrases disposées sens dessus dessous, ni les parenthèses enchâssées dans les phrases."

This for descriptions, except in local Floras, where popular use demands the vernacular; and we interpose the remark that English botanical language, freely incorporating as it does all Latin and Greek terms, comes next to Latin in convenience, compactness, and facility to all foreign botanists, who, being familiar with Latin, can seldom be at a loss. For discussions and reasonings, the botanists of each nation prefer their vernacular tongue; but De Candolle would restrict them to the four modern languages, one or two of which, beside his native tongue, every naturalist is nowadays supposed to be able fairly to read; English, German, French, and Italian. Indeed, De Candolle recommends Latin and the technically descriptive style even for *generalia*, on the ground of brevity; and he aptly suggests that the less capable botanists are of handling other than Linnæan Latin, the more brief, sententious, and strictly to the point their exposition will be.

Hints are given as to the best mode of collecting literary material, making and preserving notes (each upon separate slips of paper); upon the importance of adding clear explanations of drawings at the time they are made; and upon the desirability of refraining from publication until the work is thoroughly completed, but of then publishing as soon as possible. A manuscript work is said to have its maximum value at the moment of completion. Our author declares that second and third editions are seldom equal to the first. That depends. He objects also to posthumous publication, citing Roxburgh's "Flora Indica," published by Wallich, Plumier's plates, published by Burmann, and the wretched figures of Velloso; and he might have referred to the ill-advised printing of Griffith's rough notes and comments. But all depends upon the character of the manuscript and the length of time which has elapsed.

Chapters IV–XI traverse the whole subject of descriptions, under various aspects and a rather minute division of topics. As even a brief analysis would overpass available space, we will merely touch here and there upon certain points.

As to the relation of varieties to species, there are two modes of presentation, both of which have been followed by

Linnæus, and by most systematists, upon different occasions. Varieties are commonly designated by the small letters of the Greek alphabet, α, β, γ, etc., and also by names when they are pretty distinctly marked. Either the varieties, one or more, may be appended to the species (that is, to the form taken as the type, which usually must be the form originally described under the name), and therefore be treated as aberrant forms; or else the species is characterized as a group of forms, which forms are classified and defined just as species are under their genus. For instance, *Mentha Canadensis* is held to comprehend both a hairy and a smooth form, the two differing also somewhat in other respects. Linnæus founded the species on the former; and it is pretty well agreed that we are to refer the species back to him, however it be limited. Now we may either give a common character to the species, and then distinguish var. α. *villosa*, and var. β. *glabrata;* or we may characterize the species in general upon the originally named form, and append the variety β. *glabrata.* Either mode has its advantages and is likely to be employed in certain cases. The former classifies the varieties under the species, perhaps more naturally, and exhibits the polymorphous character of what we call a variable species; and De Candolle considers that it will prevail in proportion as the forms of a species come to be well known; the latter holds closer to the bibliography. There is danger of some misunderstanding when the two modes are used in the same work. In the "Synoptical Flora of North America," the former mode is invariably adopted, partly on the score of brevity. Either the originally described form, or a medium or common form is taken as the type, and the varieties are treated as departures from this. Even when the specific character is drawn so as generally to cover the varieties (as should be done as far as possible), some form, and the history of the species generally indicates what form, is kept in view as the norm or alpha. Of course, except for cultivated plants, there is no knowing and no pretence of determining which was the parent form, or in what order the several varieties may have diverged from a pristine stock.

As De Candolle points out, there is much ambiguity and looseness in the use of this word "type," which it would be well to avoid. Properly the type of a species is the species or genus, or the full idea of it, which no one individual or species may embody, which in the case of a group no single representative or member can fully exemplify. To apply the term to a form which well exemplifies the essential characters of the species or genus is quite natural, and hardly involves any confusion. But the term is also used in a historical sense, as referring to the particular form on which a species was founded, or the species on which the genus was characterized or which its founder had mainly in view, but which very often proves not to be the best representative of the group, sometimes not even a fair one. Finally a particular specimen which the original author described, or an authentic specimen, is said to be a type, or a typical specimen; and this De Candolle objects to. But after all, such terms can hardly he held to a single sense in technical any more than in ordinary language. Something must be left for the context to determine.

In drawing up the characters of groups, such especially as orders and genera are exceptions or what we call exceptions to be indicated in the character; or shall this express only what is generally true? De Candolle discusses the question, but leaves it, as must needs be, for practical judgment to determine. On the one hand the point or the usefulness of a character is blunted or dissipated by the intercalation of alternatives and exceptions, yet characters must be somehow made to correspond with the facts. The method of Bentham and Hooker, of a separate specification of the principal known exceptions, is commended.

Should outlying or anomalous groups be incorporated with the orders they most resemble, or be merely appended as "genera affinia," and the like? The latter was inevitable in the earlier days of the natural system; but increasing knowledge, as well as considerations of symmetry and convenience, more and more fix the place of these floating groups; so that their general incorporation into the orders by Bentham and Hooker in the Genera Plantarum of our day is in the natural course of things. But botanists have to remember that many of them are still riddles.

De Candolle classes descriptions under the two general heads of *developed* and *abridged*. A developed description is a detailed account of the whole conformation, without regard to differentiæ. The type of an abridged description is the diagnosis, such as the specific phrase, or as Linnæus called it, the *nomen specificum;* what we now universally term the specific name being his *nomen triviale.* In the course of phytography both these have become rare or of special use as regards species, and a hybrid between the two has been engendered which is more serviceable than either. The long and independent descriptions of the olden time are now seldom written. Except for special cases, the development of the natural system in its subordination of groups in ever increasing numbers and definiteness, has rendered them superfluous. What was once stated in the developed description of a species in one formula, and a vast deal more, is now parceled out among the ordinal, tribal, generic, sub-generic, or sectional, sub-sectional, and other characters, each of which deals primarily, if not wholly, with *differentiæ.* The characters of each grade, being diagnostic, may be comparatively short; but taken together they become almost exhaustive. But to avoid going again over the same ground, subsidiary matters not diagnostical, yet needful or useful, are not rarely intercalated among the more essential points, instead of being collected in a separate paragraph. Consequently the specific diagnosis may be prolonged and get to partake of the nature of a developed description. The remedy for over-length is to multiply divisions and sub-divisions between the genus and the species. To do this well, to arrange the species group within group most definably as well as most naturally, tasks the powers and the patience of a systematic botanist, and tests his aptitude for discerning affinities, and solving practical difficulties.

Developed descriptions are in place in such general works as De Candolle's "Systema" (which was soon overweighted and crushed by them), and above all in monographs of orders or genera. In his sixth chapter, devoted to this topic, the author cites, in order of date, the principal monographs of orders or tribes (excluding those of a single genus) which may be taken as models (about two dozen only), and points out some of their merits or defects. The subject of abridged descriptions is taken up in chapter VII.; and this connects itself with a great number of subsidiary questions and particular details, running on through twenty chapters more (individually short), and forming the most practically useful part of the book. There are so many points which it were well to call attention to, or sometimes to comment upon, for which space is now wanting, that we must defer the remainder of this critical notice to the next issue of the Journal.

It is to be regretted that, for the completeness of this work, the author did not comprehend in it the subject of nomenclature of groups — an important part of phytography — and reprint in it his opuscula, entitled "Lois de la Nomenclature Botanique," along with some further commentaries, such as his experience and some adverse criticisms from an opposing school may have suggested. This may still be desired, although the little treatise has already been widely disseminated in three languages, and although, as the author incidentally remarks, his own view is shared by an immense majority of descriptive botanists.

We proceed with our remarks upon this interesting volume, taking up certain points as they strike attention in turning the pages, but passing over many others of equal or superior importance.

There is a short chapter upon enigmatical descriptions or botanical riddles, and how they come about. The author has taken the pains to collect and tabulate the "species dubiæ" of the last four volumes of the "Prodromus," to see who is accountable for them, taking into account only botanical authors no longer living, and excluding those who have contributed no more than three. So good a botanist as Blume heads the list, one so indifferent as Siebold is accountable for the fewest; so not much comes from such a tabulation. The practical point is that Blume, as well as Miquel and Kunth, who stand high on the list, have fallen much into the habit of founding species on fragmentary or quite insufficient herbarium specimens; instead of passing them over without mention, or at least without naming them. One is apt to suppose that a description of an incomplete specimen, say without flowers, may be readily eked out later by another hand supplied with the missing parts. This, as De Candolle says, is a mistake. The succeeding botanist is hindered more than he is helped by such work. And it is the same with species founded on figures, such, for instance, as those of Moçino and Sesse, upon which the elder De Candolle established species and some genera in the earlier volumes of the "Prodromus." As to "genera dubia vel non satis nota," very few can be laid at the door of first-class botanists. In the list of names of deceased botanists which are notable for their absence, the name of Torrey is inserted between that of the elder De Candolle and that of the elder Hooker.

In the chapter on the description of groups superior to species, the author enumerates and sketches the character of the six "Genera Plantarum" which have appeared within the 180 years of modern botany; the immortal works of Tournefort ("Institutiones," 1700), Linnæus (the first edition of whose "Genera" was published in 1737), A. L. Jussieu (1789), Endlicher (1836–1840), Meisner (1836–1843, which is much less known), and finally of Bentham and Hooker, which began in 1862, and is now three quarters finished. Tournefort fixed the rank and character of genera, Linnæus tersely and clearly defined them, Jussieu arranged them under natural orders, defining these, Endlicher and Lindley developed the hierarchy of groups superior to orders, also the tribes inferior to them, and the latter is deservedly praised for his sagacity in discerning affinities, the former for the perfection of his style; and to Bentham and Hooker is justly awarded the crowning merit of having, far beyond their predecessors in this century, verified or developed the characters of the genera by a wide and direct study of the herbarium materials.

Floras, or descriptions of natural groups, not in their entirety but so far only as represented in a particular country or region, are discussed in chapter X. In Floras, as in more general works, abridged descriptions or diagnoses suffice, indeed are preferable in all cases where the region has been pretty well explored, and where the materials can be thoroughly elaborated. Formerly all considerable Floras were written in Latin, at least the characters. So they would continue to be if the convenience of botanists and the advancement of science only were to be considered. But Floras are used by many to whom even Linnæan Latin would be a stumbling-block. Fortunately the difference between good botanical English and botanical Latin is not wide, and will not seriously trouble a French or German botanist. The converse hardly holds. The greatest Flora written in English, we might say the best great Flora in any language which has ever been produced and completed, is Bentham's "Flora Australiensis," in seven octavo volumes. Touching upon works of special illustration, the "Botanical Magazine" is justly singled out for praise, for sustained botanical correctness under difficulties, and for its great influence upon the science.

De Candolle insists much on the importance of describing and well classifying the varieties of a species, and of distinguishing them as much as possible into grades, such as subspecies or races, varieties, subvarieties, etc. We suggest that this can be done with great advantage only when the forms are comparatively definite, or have been described as species. We think that only the more salient and definite varieties should be distinguished by names; otherwise the names and the groups will be limitless.

In the eleventh chapter, on partial descriptions of groups from the point of view of organography (a term which our author prefers to morphology), of physiology, botanical geography, etc., our author has some pertinent remarks upon the helps which all such studies are offering to phytography, which will gradually extend its domain over them; and upon the obvious advantage and great need of having results of histological researches expressed descriptively, under something like a common terminology, and with due regard to rules which have governed the more matured branches of botany, — rules and practices which eliminate a deal of verbiage, facilitate comparison of views, and ensure mutual intelligibility. Of botanical descriptions for the purposes of systematic botany, it could be said that whatever is not clear is not botany. May such clearness be hoped for in the future of histological botany?

Chapter XII. treats of the unavoidable mixture of artificial with natural grouping. Truly natural groups are often artificially defined, that is, are indicated by single characters; or truly artificial characters are used for the sake of convenience in the division of natural groups. Of the latter sort are the divisions *Polypetalæ*, *Gamopetalæ*, and *Apetalæ* in Dicotyledons; also those founded on the mode of curvature of the embryo in *Cruciferæ*, introduced by Brown, who cautiously used them for genera, but raised to the rank of primary or subordinal characters by De Candolle. Hypogyny, perigyny, and epigyny are in the same category, and probably no one was more sensible of it than Jussieu himself, whose point and forte was the constitution of orders, not their collocation under these artificial heads. De Candolle suggests that, while to the more natural divisions are appropriated the terms of Class, Cohort, Orders, Tribes, Genera, and Sections; such names as Division, Subdivision, Series, etc., might be restricted to artificial divisions, and that these should take adjective names not of generic origin, such as *Liguliflore*, *Polypetalæ*, and the like.

Chapter XIII. relates to difficulties in phytography which have grown out of various methods or absence of method in the nomenclature of organs, and from the want of consideration of the law of priority in such matters. The result of which in some departments, such as histological morphology, is a state of anarchy not unlike that which prevailed in the names of groups before the days of Tournefort and Linnæus. We may hope that order and lucidity will some day dawn upon this chaos and a common language replace this confusion of tongues. Meanwhile De Candolle offers certain counsels, the utility of which, he says, is not doubtful nor the application very difficult.

(1) Hold fast to common and universally known names, whether in Latin or in modern languages. *Radix*, *caulis*, *folium*, *flos*, etc., with their vernacular equivalents, are not to give place to new-fangled substitutes. This, he thinks, will rid us of "such useless terms as *caulome*, *phyllome*, etc." Now these terms, along with *trichome*, seem to us legitimate and useful, as succinct expressions of a morphological idea; they are annoying only when pedantically ridden as hobbies over ground on which they are not wanted.

(2) Do not entertain the idea that a change in the mode of considering or defining an organ requires a change of name. Although Linnæus did take the leaf-blade for the leaf, and define it accordingly, that did not much hinder the coming in of a truer view, involving merely a change of the definition. But one may intimate that De Candolle here comes into conflict with another rule he insists on, namely, that terms should have unmistakably one meaning. When we say — as we ever shall — that leaves are ovate, we speak according to the Linnæan definition; when we say that their insertion is alternate, we use the word in a more comprehensive sense; when we have occasion to declare that cotyledons, bracts, petals, etc., are leaves, we use the word in the most comprehensive sense. All this involves considerable ambiguity; and the endeavor to keep the new wine in the old bottles causes no little strain. It is borne because it has been applied gradually. If Linnæus had started with, or even reached our ideas, we should happily have had a nomenclature to match. Now we must be content, for descriptive purposes, to employ some words both in a restricted and in a comprehensive sense, and let the context fix the sense, just as it must in ordinary language. Technical precision is only a matter of degree. But it is clear that the excellent rule here laid down need not forbid the introduction of terms to express our conceptions, such as *rhizome*, *caulome*, *trichome*, and the like. Yet these are ill-chosen terms, except the last. In particular, *rhizoma* has long ago been appropriated for something which is not of root nature, but the contrary.

(3) The third counsel is to change the name of an organ,

as we do that of a genus or species, only when it is positively contrary to the truth, or when it has been pre-occupied.

(4) Avoid giving special names for rare or ill-definable cases of structure. An epithet or short periphrasis is vastly preferable to a new and strange term, which will be seldom used and may be hardly understood. De Candolle truly remarks that after a great multiplication of terms and distinctions generally comes some good generalization, which does away with a crowd of particular names; that what has happened in carpology is likely to occur for microscopic organs; and he adds: "Nous assistons au 'feu d'artifice' d'une trentaine de noms de ces états des cellules" [in our vernacular, we have seen them "go up"]; "il en restera seulement quelques-uns généraux ou fréquents, qui seront toujours nécessaires."

(5) Between two or more names choose, not the most agreeable, or even the most significant, but the one best known and most widely recognized.

(6) Between names equally known and used, adopt the oldest. Which are the older names is not difficult to know in the case of common organs, but is very much so in modern histology.

(7) In this matter of priority or of usage, consider only names taken from [or in conformity with] Latin or Greek. As in systematic botany, scientific and not vulgar names are to be accounted in this regard. Those who like *spaltöffnung* for *stoma* or *stomate*, and *scheitelzelle*, must needs follow their own fashion; but the genius of our own and the French language resists their importation, while it adopts or adapts with ease technical terms from classical sources.

(8) Not to admit names contrary to these rules.

Chapter XIV. surveys some difficulties in phytography which arise from the variant, changed, or contradictory use of certain botanical terms, and from the employment of vernacular terms which cannot be latinized. The latter has just been referred to incidentally. Even the French describe the dehiscence of a certain kind of capsule as "en boîte à savonette." In English we do not attempt to say "in soapbox fashion," and should not be understood if we did, but we adopt the Linnæan Latin "circumcissile." In general, De Candolle concludes that a vernacular term, whether the name of an organ or of a botanical group, which will not enter into a Latin text by a modification of its termination, is not scientific, and may give place to one which is.

A few terms are mentioned which have been more or less changed in meaning since the time of Linnæus; such as *lanceolate*, which has gradually varied more or less, and for a part of the change the present writer is held to account; also *glaucus*, which classically means sea-green in hue, but which has been generally used in botany to designate sometimes a certain whitishness, and sometimes a whitishness caused by a minute waxy exudation in the form of a powder: the latter is the same as *pruinosus*. Others may be as surprised as we were to learn that neither *glaucus* nor *pruinosus* are Linnæan terms.

Among the terms used ambiguously, it is surprising that De Candolle does not refer to *pistillum*, first introduced into botany by Tournefort, and used in the sense of the modern term *gynœcium*, therefore only one to a flower; modified by Ludwig to denote a female member of the flower (having ovary, stigma, and commonly a style), of which there may be several or many in a flower; and adopted in the latter sense by Linnæus, yet generally with a use that avoids contradicting the sense of Tournefort. Mirbel, Moquin-Tandon, and St. Hilaire among the French, have openly departed from Tournefort's use, and speak freely of pistils in the plural. Brown and De Candolle have used the word in the manner of Ludwig and Linnæus when they have used it at all, but have generally evaded its use; other botanists, especially British, have gone back to the Tournefortian sense of gynœcium. The present writer has a note on the subject in the new edition of his "Structural Botany" (1879 and 1880), p. 166.

Sinistrorse and *dextrorse* in the direction of ascent of climbing stems or the overlapping of parts in a bud, etc.— De Candolle had formerly insisted upon the desirability of following what he takes to be the authority and practice of

Linnæus in the use of these terms; and he here returns to the subject, reinforcing his former arguments. It is most desirable that these terms should not continue to be employed in contradictory senses, one party calling that sinistrorse which the other calls dextrorse; it is also fitting that the principle of priority should prevail and that the authority of Linnæus should be respected. Let us, therefore, in the first place give an abstract of the points which De Candolle here makes.

But first, we take it for granted that a stem or such organ, having no front or back, can have no right or left of its own: so when we say that it twines to the left or right, we can mean nothing else than the right or left of the observer. The contradiction comes from the different position which the observer is conceived to occupy. De Candolle supposes the observer to be placed within the coil or ascending helix, and that this is the more natural position. The other party supposes the observer to face the object from without; and from this position the Hop twines to the left, *i. e.*, turns in ascending from the observer's right to his left, while the Convolvulus turns from his left to his right; the first is sinistrorse, the second dextrorse; while to De Candolle, standing within the coil, the first is dextrorse, the second sinistrorse. Now, says De Candolle, Linnæus in the first edition (1751) of the "Philosophia Botanica," § 163, page 103, says: "Sinistrorsum hoc est quod respicit sinistrum, si ponas te ipsum in centro constitutum, meridiem adspicere; dextrum itaque contrarium."

De Candolle remarks that the phrase "meridiem adspicere" is of no account [but it indicates a certain confusion in Linnæus's mind], for it matters not in what direction you look. He adds — what we had all overlooked — that in the errata, on p. 360, Linnæus corrected the word sinistrum into dextram. But, inasmuch as two editions of the "Philosophia Botanica" were printed at Vienna in Linnæus's lifetime, and this correction was not introduced into them, he concludes that the correction was cancelled by the author of it. And he notes that the expression "sinistrorsum hoc est quod respicit dextram" is a most awkward one for denoting the right-about change which the erratum had in view. Nevertheless the correction was so made in the edition of the "Philosophia Botanica" by Gleditsch in 1780, two years after the death of Linnæus, also in that of Willdenow, published ten years later. But De Candolle the elder, in the "Flore Française," and in all his writings, followed the original text, as also has the present De Candolle, who cites as maintaining the same view, Braun (who for a time gave way to the opposite), Bischoff, Mohl, Palm, Dutrochet, Nægeli, and even Darwin. But we should say that Darwin, noting the conflict of views, had carefully evaded both, using instead the expression "with the sun, and against the sun"; yet sometimes saying "from left to right," as equivalent to "against the sun" (as on p. 34), showing that he took the external position to be the natural one.

Among those who have used the terms sinistrorse and dextrorse and defined them in the way which supposes the observer to stand outside of the helix, are Aug. St. Hilaire, Duchartre, Bentham and Hooker, Eichler; and the present writer may be added, although our author appears not to be aware of it. While trusting that the younger botanists will follow the example of Linnæus and the majority of authors, De Candolle recommends that those who depart from it, and even those who adopt it, shall state their point of view by some convenient abbreviation, such as *extus vis.* or *intus vis.;* and thus lessen the danger of a misunderstanding. This is indeed essential.

De Candolle remarks that he can discover no reason for the *ab extra* point of view except a tacit but perhaps nowhere expressed assumption that it requires some effort to suppose one's self in the centre of a helix or spire. He thinks a moderate effort will accomplish this. The reply may be that, in the case of a stem climbing a hop-pole, or of the scales imbricated on the axis of a pine-cone, or of a flower-bud on the stage of a dissecting microscope, the contemplation of the object from without calls for no effort at all. So natural does this extraneous position appear to be that we found ourselves describing these objects from that point of view without thinking of any other, — so natural, as we shall see, that Linnæus fell into it himself, and there remained. Yet, that the opposing view has also its fitness is obvious from the fact that the physicists and mathematicians are divided in usage, no less than the naturalists.

In the actual state of the case, the question which view ought to prevail in botany must be determined therefore on a balance of considerations: 1, Priority and authority, such as that of Linnæus; 2, Naturalness; 3, Preponderant actual usage. We had maintained in this Journal (for March and for May, 1877) and in "Structural Botany" (6th ed., note on pp. 51, 52) that the *externe visum* view has decidedly the best case on the second ground, and, except in botany, on the third also. And now that De Candolle has drawn our attention to the matter, we are going to claim the remaining ground likewise, and to contend that the contrary usage in botany came in from non-attention to the teaching and practice of Linnæus himself.

On p. 39 of Linnæus's only own edition of the "Philosophia Botanica" he defines and illustrates the directions of twining thus: "*Sinistrorsum*, secundum solem vulgo: Humulus, Helxine, Lonicera, Tamus. *Dextrorsum*, contra motum solis vulgi; Convolvulus, Basella, Phaseolus, Cynanche, Euphorbia, Eupatorium."

Nothing is said about the position of the observer. But in every one of the examples of sinistrorse (Helxine being *Polygonum convolvulus*), the stem winds around the support passing from right to left of the observer confronting the coil; and in every one of the dextrorse examples (Eupatorium being Mikania) it winds in the opposite direction. That is, dextrorse and sinistrorse are used in the *externe visum* sense. On p. 103 the same is repeated, except that reference to the sun's apparent course is omitted and additional examples are added, most (but not all) of them accordant with the preceding. So far, it would seem that Wichura was not mistaken in his statement that De Candolle had followed a different method from that of Linnæus. And this appears to be the whole case as respects direction of twining.

But on the same page, to "Corolla sinistrorsum" is appended the foot-note which has made so much trouble, namely, "Sinistrorsum hoc est, quod respicit sinistrum, si ponas Te ipsum in centro constitutum, meridiem adspicere; Dextrorsum itaque contrarium." That is to say, in defining the direction of overlapping of the parts of a perianth, Linnæus took the open flower instead of the bud, and proposed to look down upon it from above or within. Now it may well be that Linnæus subsequently perceived the contradiction between his terminology for overlapping and that for twining; and that his brief erratum, on p. 310, "pro sinistrum lege dextram," was intended to bring the former into congruity with the latter, which it does, but in an awkward way. Perhaps he saw the incompatibility of the cited examples; in fact, about as many of them accord with the outside as with the inside point of view. Any way, the erratum is his own; it seems unlikely that he authorized its omission from the Vienna editions; and Gleditsch and Willdenow should not be blamed for heeding his behest in their editions. For, so far as it goes, it tends to render their author consistent with himself. If Linnæus had revised the page himself, he would have left out the "meridiem adspicere," which has nothing to do with the matter, and doubtless he would have completed his assimilation of the direction of petal-obliquity or overlapping with that of stem-winding; and so the whole confusion from which we are endeavoring to escape would have been avoided.

In adopting the external point of view — now fortified by original authority — it is well to note that we shall be in accord with the modern physicists and mathematicians, and also with common people. The ordinary screw, on which the thread winds from left to right of the confronting observer, and which is driven home by the semi-rotation of the hand and fore-arm from left to right, is everywhere known as the right-handed screw; and this, with the corkscrew, is taken as the norm and exponent of right-handed rotation by Clerk-Maxwell ("Treatise on Electricity and Magnetism," i. 23), and by Sir Wm. Thomson.

The analogies which have been adduced in favor of the

inside position are mostly drawn from objects which have a right and left of their own; a building, for instance, has a right and left side or wing because it has a front and a rear. The right side of an assembly presided over by an officer who faces the members is quite arbitrarily, but naturally, taken to be that on the right of the chairman. But the right-hand figures on a drawing or engraved plate are taken to be those on the right hand of the observer, notwithstanding that the plate, having face and back, has a right and left of its own.

Chapter XV. refers to certain difficulties which grow out of ambiguous terms of ordinary language; for example, the various meanings of the word (fin) end or purpose, and the ambiguities in the use of the terms nature, natural, supernatural (which lead off into philosophy, but are here treated rather in reference to style of exposition); also the change which has occurred in the scope of the word history in natural science.

Chapter XVI. is an interesting and pertinent one, upon the manner in which facts observed under the microscope are described, and on the great saving of space and advantage in clearness which would be gained by the adoption, for all matters perfectly capable of it, of the Linnæan descriptive style, and of Linnæan Latin. Extracts from the German of Schacht, the French of Payer, and the Italian of Gasparrini are given, and by their side a rendering in descriptive Latin; and the words and letters are counted. The German specimen so treated is diminished to considerably less than half the number of words and a little less than half the number of letters. The French simmers down to one third the number of Latin words and less than half the number of letters; and in the French of descriptive botany to less than one half. The Italian extract of 51 words and 256 letters is expressed in Latin of the Linnæan form by 21 words and 127 letters.

Style in botanical works is discussed in chapter XVIII., which all young botanists should study, especially the portion which treats of the admirable style of Linnæus. In speaking of botanical style in the modern languages, the author notices the great advantage which the languages of Latin stock have inherited, and which the English-writing botanists have acquired, of ready and free use of Latin and latinized technical words by direct transference. Botanical French, English, and Italian, are contrasted with the German in this respect. Noting that the German of conversation inclines to be clear and sententious, while in botanical writings the words lengthen more and more and the sentences become badly involved, our author remarks that recently having read a couple of pages of Vegetable Anatomy, and feeling his brain somewhat fatigued with the frequency of such words as Sclerenchymfäsergruppen, Gefässbundentwicklung, and Entwicklungseigenthümlichkeit, he asked himself if that was good German style. He then recollected that Goethe, one of the very greatest of German literary writers, was also a profound naturalist. He opened his "Metamorphose der Pflanzen," read a page or so, and experienced a relief which he likens to that felt by a sea-tossed ocean voyager when the vessel suddenly glides into a quiet harbor.

Chapter XIX. discusses the propositions to employ letters and figures, chosen arbitrarily or otherwise, to represent specific and generic characters, — repulsive contrivances, to which our author lends no countenance.

Chapter XX. treats questions of orthography, abbreviations and signs, pagination, typography; the twenty-first chapter, of titles and indexes; both full of interesting details upon which we cannot touch, although we are longing to put in our oar.

Chapter XXII. animadverts upon the tendency of certain modern cryptogamists to set all botanical rules at naught. The next gives advice about articles in journals, dissertations, and the like; the next treats of translations; another, of figures, and has many noteworthy remarks; chapter XXVI., of auxiliary and bibliographical works; and chapter XXVII. is a chronological table of the progress of phytography, beginning with a Chinese encyclopedia 1000 years before Christ, and ending with Sachs's "Lehrbuch," 1868–1877. Botanical students will find it very interesting and instructive.

The remaining chapter begins the second part of the volume, "Preuves des Descriptions;" which is principally devoted to herbaria, their history, formation, and management; — a most important chapter, the analysis of which would form an article by itself. Last in order and not least in importance, a full enumeration is given of botanical collectors and authors who have formed herbaria, with an indication of the place where their herbaria or collections are preserved.

DARWIN'S POWER OF MOVEMENT IN PLANTS.

FIRST let us congratulate the scientific community, no less than the author, that Mr. Darwin's experimental researches are seconded, and are we hope long to be continued, by the son whose name appears upon the title-page, and whose independent papers already published approve his worthiness for that honor. This volume[1] is from beginning to end a record of a series of researches and of the inferences which they directly warrant. Naturally it will not fascinate the general reader after the manner of "The Origin of Species" and some of the volumes which succeeded that epoch-marking production; nor has it the fresh charms of the treatises "On the Movements and Habits of Climbing Plants," and of "Insectivorous Plants," of which it is the proper continuation and supplement.

The organs of plants take certain determinate positions and execute certain movements, some of them universal or general, some of them special, some of them very striking and seemingly strange, most — but not quite all of them — evidently advantageous to the plant or essential to its wellbeing. Roots point toward the earth; stems point away from it; young stems bend toward the light, and the upper face of leaves is presented to it. Stems that twine "*circumnutate*" (a capital term), *i. e.*, bend successively to all points of the

[1] *The Power of Movement in Plants.* By Charles Darwin, assisted by Francis Darwin. With illustrations. London and New York, 1880. (American Journal of Science and Arts, 3 ser., xxi. 245.)

compass, and this wholly irrespective of external influences; and the twining around a support is the direct consequence of the circumnutation. Most tendrils freely circumnutate, and thereby are enabled to reach the object which they grasp. Most tendrils (and in certain cases some other parts) are very obviously sensitive to external contact or irritation, to which they respond by movement and change of form, and thus they grasp or do other advantageous acts. Some movements, especially of leaves, occur with regularity upon the access of light, others with its withdrawal; a few, such as the small leaflets of *Desmodium gyrans*, proceed irrespective of night and day. The specification need not be extended. The general facts in all their great variety are familiar to scientific readers. The inquiry of this volume is as to their ground and origin, or, as in this connection we should rather say, their development and history. For instance, circumnutation gives rise to twining and gives efficiency to other ways of climbing. But Darwin is bound to suspect, and even to show, that circumnutation is not a special endowment of the stems and tendrils of climbing plants, but rather a more developed manifestation of a general faculty. And the same is to be said of the movements of tendrils and leaves, or their appendages, whether automatic or in response to external irritation or stimulus. All this is what the experimental researches detailed in this volume undertake to ascertain and have satisfactorily made out.

An abstract of the volume might be somewhat tedious, and is certainly unnecessary for biological readers, who are sure to possess and study it. But the gist is readily to be gathered, without running through the iterated details or scanning many of the illustrative and curious figures which record the movements under investigation, by the simple perusal of the introduction and of the concluding chapter, in which the matter of the volume is summed up.

The sum and substance of the case is, that all these powers and faculties are manifested in the seedling immediately upon germination, and most of them are then remarkably exemplified. The caulicle or initial portion of stem below the cotyledons (with the elongation and protrusion of which the germination of dicotyledonous seeds usually begins) circumnutates as soon as it comes out into the open air, and even earlier: this is the earliest manifestation of an automatic movement which is shared by all the succeeding portions of stem developed from it, in the early life of most plants, whether climbers or not. In the latter, and especially in twining plants, we see this general faculty at its maximum and in beneficial exercise. More remarkable and novel it is to learn that the initial root, growing from the lower end of the caulicle (not inaptly called by Darwin the *hypocotyl*), also shares in this faculty of circumnutation. As it penetrates the soil in its downward course, it cannot largely manifest this faculty, and indeed its power of circumnutation is always small; "but the circumnutating movement will facilitate the tip entering any lateral or oblique fissure in the earth or a burrow made by an earth-worm or larva; and it is certain that roots often run down the old burrows of worms. The tip, however, in endeavoring to circumnutate will (successively) press against the earth on all sides, and this can hardly fail to be of the highest importance to the plant" (being supplemented by another faculty, that of sensitiveness at the tip presently to be mentioned); for "when the tip encounters a stone or other obstacle in the ground, or even earth on one side more compact than on the other, the root will bend away as much as it can from the obstacle or the more resisting earth, and will thus follow with unerring skill the line of least resistance." Then, beside the almost universal heliotropic movement, by which each leaf or leaflet presents its superior surface to the direction of the greater light, Mr. Darwin shows that these organs also circumnutate, beginning even with the cotyledons or seed-leaves; although their sweeps generally form so narrow an ellipse that they move up and down in nearly the same vertical plane, a movement describing a circle being converted into one up and down.

These circumnutatory movements are of the most fundamental and therefore mysterious character. Although most commonly connected with growth, they are at bottom independent of it. This — contrary to some German physiologists — we must conclude from both DeVries' and Darwin's investigations. They are produced by the changing turgescence of the cells on different sides of a stem or footstalk, which may or may not be fixed by some consequent growth or solidification. This Mr. Darwin, we presume rightly, concludes to be the faculty or susceptibility upon which heliotropism, geotropism, and the like (not to speak of *apheliotropism*, *apogeotropism*, *paraheliotropism*, *diaheliotropism*, *hyponasty*, *nyctotropism*, and other terms which the incautious student may take to be powers instead of abbreviated expressions) — in other words, upon which the solar rays and some occult influence of the earth — act, modifying the sweeps or converting them into forth and back or other special movements. Among these, that which has been termed the sleep of leaves, better and briefly designated by the word nyctotropism, is thoroughly investigated in this volume, and is shown to be far more general than has been supposed; and the conclusion is that the end subserved is a needful protection of the surfaces, mainly of the superior surface against cold from nocturnal radiation. *A priori*, looking at the structure of the leaf, one would have thought that the under surface had the greater need of such protection.

Not only are all these movements incipient in the seedling, but some of them are manifested more rapidly and extensively than in most mature plants. This should needs be, since, as Mr. Darwin states it, "Seedlings are subjected to a severe struggle for life, and it appears to be highly important to them that they should adapt themselves as quickly and as perfectly as possible to their conditions." Very properly, therefore, no small part of this volume is devoted to the seedling and to the behavior of its several parts. The most novel and unexpected results relate to the young root. Judging from its simplicity and from the medium in which it is developed, one would not look there for the endowment which Mr. Darwin finds in it. But this root-tip and the vegetable cells which compose it conspire to teach us that the most simple structures may be wonderfully gifted. The tiny root exhibits three kinds of movement: first that of circumnutation, in which, endeavoring to bend in all directions, its tip "will press on all sides, and thus be able to discriminate between the harder and softer adjoining surfaces, . . . and to bend from the harder soil and follow the lines of the least resistance," so modifying advantageously its course from that to which geotropism constantly tends to give it. Moreover, the growing end of the root is sensitive to contact, and in a complex manner. If pressed above the tip, it bends there toward or around the impinging body, much as the end of a tendril bends around a support: thus it may follow, as roots do, along the unequal surface of a solid body. But, thirdly, if the tip itself be locally pressed, it exhibits different and more surprising sensitiveness, for it transmits an influence to an upper adjoining part, causing it to bend away from the affected side. This sensitiveness to contact is confined to a little more than one millimeter of the tip; the part which bends is 6 or 7 or even 12 millimeters above. So, when the sensitive tip in its downward growth strikes obliquely upon a stone or other obstacle, the part above at this distance, to which some influence must be transmitted, bends and carries the point away from the obstacle. Yet later, when a new portion of the side impinges upon the stone or other body, it will bend at that part toward instead of away from it, and so follow along its surface. It is the tip, likewise, that can discern that the air is moister on one side than on the other, and which thence "transmits an influence to an upper adjoining part, which bends toward the source of moisture." It is the tip only which is sensitive to gravitation. Well may Mr. Darwin affirm that there is no structure in plants more wonderful, as far as its functions are concerned, than the tip of the radicle. Also, that "it is impossible not to be struck with the resemblance between the foregoing movements of plants and many of the actions performed unconsciously by the lower animals." "But the most striking resemblance is the localization of their sensitiveness and the transmission of an influence to an excited part which consequently moves. Yet plants do not of course possess nerves or a central nervous system; and we may infer that with animals such structures serve only for the more perfect transmission of impressions and for the more complete intercommunication of the several parts." The closing sentence of the book may be appended to this. "It is hardly an exaggeration to say that the tip of the radicle, thus endowed and having the power of directing the movements of the adjoining parts, acts like the brain of one of the lower animals."

The movements "excited by light and gravitation," as well as the nyctotropic or sleep-movements so called, are (as we have already stated) all referred by Mr. Darwin to modified circumnutation, "which is omnipresent while growth lasts, and after growth has ceased whenever pulvini are present," as in several classes of leaves. As respects the relation of external agents to the movements, note Mr. Darwin's remark: "When we speak of modified circumnutation we mean that light, or the alternations of light and darkness, gravitation, slight pressure or other irritants and certain innate or constitutional states of the plant, do not directly cause the movement; they merely lead to a temporary increase or diminution of those spontaneous changes in the turgescence of the cells which are already in progress."

Certain parts of plants turn or grow earthward. When this is attributed to gravitation, as it commonly is, the physicists have opportunity to complain of the misuse of a term. Although Mr. Darwin, like other writers, speaks of the influence of light and of gravitation in the same breath, without discrimination, we note with satisfaction his disagreement with those who "look at the bending of a radicle towards the centre of the earth as the direct result of gravitation," and note especially the closing dictum. "Gravity does not appear to act in a more direct manner on a radicle than it does on any lowly organized animal, which moves away when it feels some weight or pressure." Why, we would ask, need the word gravity or gravitation be used at all in this connection?

The introduction to this volume contains a short article upon the terminology which is adopted in it, chiefly as regards such words as epinasty and hyponasty, geotropism and related terms, which is most convenient to employ, and also the names of these several parts of the embryo and seedling. This is, we believe, almost the first English book in which the axial part of the dicotyledonous embryo below the cotyledons (the radicle of the systematic botanist even of the present day) is distinctly recognized as hypocotyledonous or initial stem, although on the continent and in America this has long been taught and accepted. None the less so although the term radicle has been retained for it (until recently by the present writer, at least), in order not to break with the terminology of systematic works. Mr. Darwin, in this volume, shortens the expression of "hypocotyledonous stem" into the term "hypocotyl,"—a fairly good English term, certainly better than the French *tigelle.* The objection to them both is that the words will not take the substantive Latin form, as all such terms should. Wherefore the better name—an old one which we have reverted to in the last edition of the Botanical Text-book (Structural Botany)—is *caulicle* or *cauliculus.* The initial root, which grows from the lower end of the caulicle (or "hypocotyl"), Mr. Darwin calls the radicle, following in this the ordinary English usage, except in very definitely distinguishing it from the cauline part above it. Being simply root, we have preferred uniformly to call it so, thus avoiding a word which the systematists have all along applied to the caulicle. Although initial stem and initial root are most clearly discriminated in the present volume, yet in the accounts of the germination or the ordinary Dicotyledons, it appears to be implied or stated, either that it is the root-part which first projects from the seed-coats and that the stem-part begins its development later, or that the axial part of the embryo conspicuously preëxisting in the seed is root and not stem. We take it to be quite otherwise, namely, that this axial part in the seed is cauline, and that ordinarily it protrudes or makes some growth in length before root-formation begins.

A few misprints of names of plants will in no wise mislead or trouble any botanist, except possibly in the case of *Apium-graveolens*, which on p. 422 and 424, and in the index, is printed Apios.

DE CANDOLLE'S ORIGIN OF CULTIVATED PLANTS.

M. Alphonse De Candolle's "Géographie Botanique Raisonnée," in two volumes of nearly 700 pages each, was published in the year 1855, and has been for several years out of print. It is not surprising that the now venerable but still well-busied author should decline the labor of preparing a new edition, involving, as it would, the re-discussion of certain questions under changed points of view, and the collocation of a vast amount of widely scattered new materials which the last quarter of a century has brought to us.

Happily, the chapter on the geographical origin of the species of plants generally cultivated for food, and for other economical uses, could be detached. This, the author has sedulously studied anew; and the present volume[1] is the result. As yet we have it only in the original French; but it is said that an English translation is in preparation. So, if the work is not already in the hands of botanists and other scholars generally, we may expect that it soon will be; and, contenting ourselves with a mere mention of its plan and scope, we may proceed to remark, here and there, upon points which strike our attention.[2] We may expect this to be for

[1] *Origine des Plantes Cultivées.* Par Alphonse De Candolle. Paris, 1883. (Bibl. Scientifique Internationale, XLIII.) (American Journal of Science and Arts, 3 ser., xxv. 241, 370; xxvi. 128; with J. Hammond Trumbull.)

[2] To avoid repetition, it may be mentioned here that, in the following annotations, the "Relations of the Voyages" of Columbus are cited from Navarrete's "Coleccion de los Viajes," etc. (Madrid, 1858, and 1827–37); references to Peter Martyr d'Anghiera's first three Decades "De Rebus Oceanicis et Novo Orbe" (written before 1517) are to the Cologne edition of 1574; references to Oviedo's "Historia General y Natural de las Indias" — of which the first nineteen books, published in 1535, included a revised and enlarged edition of his "Relacio sumaria de la Nat. Historia de las Indias," printed in 1526 — are to the edition published by the Royal Academy of History of Madrid, 1851–55; Jean de Lery's "Histoire d'un Voyage faict en la terre du Brasil" (in 1557–8) is referred to in his revised edition in Latin, "Historia Navigationis in Brasilian" (Genevæ, 1586); Fr. Hernandez, "Nova Plantarum, etc., Historia," in the edition of

many years the standard work upon the subject, and to undergo revision in successive editions; and we are sure that the excellent author will welcome every presentation of discussion which may chance to throw any new light upon the sources or the aboriginal cultivation of certain plants which the Old World has drawn from the New.

The first part of the volume, of only 22 pages, is mainly occupied with a consideration of the means employed for the determination of the sources whence the various cultivated plants have been derived. The botanist enquires where a given cultivated plant grows spontaneously, or what was its wild original; and he has to judge, as well as he can, where it is truly indigenous or where a reversion from a cultivated to a wild condition. This, as respects weeds and the like, is a difficult matter, even in a newly settled country like North America, much more so in the Old World; but as respects the plants of agriculture, the case is usually simpler. The botanists resident in a country are not likely to be far misled by the occurrence of wilderings; but, in the case of travelers and collectors, perhaps too much has been made, even in this volume, of plants only once met with growing spontaneously, and inferred to be indigenous. Plentifulness is of no account, else the Century plant and Opuntia would be thought indigenous to the Mediterranean region, the Ox-eye Daisy to the United States, and certainly the Cardoon to the Pampas, where there is now probably more of it than anywhere in the Old World. Archæology and palæontology are often helpful, as by the identification of fruits and seeds in ancient Egyptian tombs, or of paintings upon their walls, or of fragments in ancient bricks; or the débris of lake-dwellings rescued from lacustrine deposits, as in Switzer-

Rome, 1651; "Rariorum Stirpium Historia" by L'Ecluse (Clusius), in the first edition, Antwerp, 1576; his "Exotica," including his translations of Monardes and Acosta, Antwerp, 1605, with his "Curæ Posteriores" (posthumous), 1611. — J. H. T.

Il va sans dire — yet should explicitly be said — that all the historical and philological lore, which gives this article its value, is contributed by my associate. — A. G.

land; and from the tufas of southern France, the kiökenmöddings of Scandinavia, the mounds of North America, and the ancient monuments and tombs of Mexico and Peru. Historical documents are also important for the date of certain cultures in particular countries; and here it is stated that the principal cultures have come from three great regions, namely: China, southeastern Asia and Egypt, and intertropical America. De Candolle also remarks that in the Old World agriculture was developed along rivers, in the New, upon plateaux,—a fact which he attributes to the primitive situation of certain plants worth cultivating. But this is not quite obvious. Linguistic learning may be turned to much account; as in tracing a plant toward its home by the name which has gone forth with it in all its migrations. Like other instruments this must be used with some knowledge and judgment. Blé de Turquie (maize) did not come from, and probably not by way of, Turkey, any more than did the animal of that name. Jerusalem Artichoke has naught to do with Jerusalem, but came from North America, and is no artichoke. Pomme d'Acajou, anglice Mahogany-apple, is neither an apple nor a pomaceous plant, nor has it anything to do with mahogany. New Zealand Flax came indeed from New Zealand, but is not a flax. Among errors from the careless transference of names from one plant to another, that of Potato, which belongs to the Batatas or Sweet Potato, is familiar. Of mistakes which have been made in the transference of a popular name from one language to another, De Candolle mentions the Arbre de Judée of the French, which in English has become Judas-tree. We may add that of Bois fidèle, of the French West Indians, which, taken up by their English successors as Fiddle-wood, has been perpetuated in the generic name Citharexylum.

The several lines of evidence—botanical, archæological, palæontological, historical, and linguistic—may be used to supplement or correct each other. How they may be brought to bear, and how their combination may give satisfactory results, is practically shown in Part II,—a study of the species as regards their origin, their earliest culture, and the principal facts of their dispersion, — which makes up the principal bulk of the volume, namely, from page 23 to page 350.

This part is divided into chapters, *e. g.*, Plants cultivated for their subterranean parts, such as roots, bulbs, tubers, etc. Those cultivated for their herbage, whether for human food, for forage, for fibres, for stimulation, etc.; but the proper medical plants are left wholly out of view, as likewise plants cultivated for ornament. So the chapter on plants cultivated for their blossoms, or parts connected with these, is brief enough, treating as it does only of the Clove, Hop, Safflower, and Saffron. For the Rose, *Acacia Farnesiana*, and all plants however largely cultivated for perfume or for essential oils are left out of view. So also are the sweet-herbs of the kitchen-garden, and all condiments, except Horse-radish. Plants cultivated for their fruits and seeds occupy the closing chapters. Among the latter the Cotton-plant is placed. The arrangement matters little, and that adopted may be the most convenient. A good index makes ready reference to any topic.

In the order of the book we come first to *Helianthus tuberosus*, the Topinambour of the French, Jerusalem Artichoke of the English; in the United States the tubers simply called artichokes. Almost all we know of the origin and source of these esculent tubers has been recovered since the publication of De Candolle's earlier work, in 1855. Although the contemporary accounts specified its introduction from Canada, and Linnæus so cites it in the "Hortus Cliffortianus," the subsequent reference to Brazil was followed without question down to De Candolle's "Prodromus"; and the present author, in the work above mentioned, doubted the Canadian as well as the Brazilian origin. It now appears that Schlechtendal (in Botanische Zeitung, 1858) was the first to recover a part of the documentary history. Our own article on the subject — to which there is nothing of importance to add — was contributed to this Journal for May, 1877.[1] Singularly, it has

[1] In it reference was made to Lescarbot's mention of roots "grosses comme naveaux . . . ayans un goût retirant aux cardes," etc., and cited his "Histoire de la Nouv. France," in the edition of 1612 (p. 840). In a subsequent edition (1618), cited by M. De Candolle, Lescarbot adds that he

remained unknown to De Candolle, although it is referred to at the close of Decaisne's independent and exhaustive article, in the "Flore des Serres," 1881.

It can now be said that the wild plant to which *Helianthus tuberosus* has been traced is not *H. doronicoides*, Lam., although it was confounded with that species in Torrey and Gray's "Flora." Lamarck's plant is a sessile-leaved species. Decaisne's remark that *H. tuberosus* is the only species of the genus which is at all tuberiferous may be qualified. A form of what appears to be *H. giganteus*, but is not yet very well known, grows in Minnesota and the Saskatchewan region, has been mentioned by Douglas under the name of "Indian Potato" of the Assiniboine tribe, by Bourgeau as "*H. subtuberosus*," in herb. Kew, and by Dr. C. C. Parry in Owen's Minnesota Report, page 614, under the name of *H. tuberosus*. The scanty tubers which we have seen in dried specimens do not compare well with those of *H. tuberosus;* and that species has never been found wild so far north (that we know of), not even in the most southern parts of Canada West. The aborigines who cultivated it must have obtained it from the valleys of the Ohio and Mississippi and their tributaries, where it abounds.

had brought these roots into France, where they began to be sold under the name of Topinamboux, and that their Indian name was Chiquebi. On this last point, Lescarbot was wrong. Chiquebi was an eastern Algonkin name for the tubers of *Apios tuberosa,* the common "ground nuts,"—not for those of *Helianthus tuberosus.* It is easy to see how Lescarbot was misled. Father Biard's "Relation de la Nouv. France" was printed in 1616, and in it (chap. 22) there is mention of certain "racines, appellées en Sauvage Chiquebi," which grow spontaneously under oaks: "elles sont comme des truffes, mais meilleures, et croissent sous terre *enfilées l'une à l'autre en forme de chapelet,*" etc. Lescarbot doubtless caught the name from Biard, and misapplied it. Father Paul Le Jeune (Relation, 1634, chap. 7) mentions these ground-nuts, "une racine que nos François appellent des *chapelets,* pource qu'elle est distinguée par nœuds en forme de grains." Lescarbot's "Topinamboux" indicates a popular belief, in France, in the Brazilian origin of *H. tuberosus.* The Tupinamba Indians of Brazil — a division of the Tupi-Guarani family — had been allies of the French in the sixteenth century, and their name was probably well known in France through the relations of J. de Lery and other voyagers. Lescarbot (Histoire de la Nouvelle France, 1612, p. 178) follows Lery in writing the name "Tououpinambaoult." — J. H. T.

Helianthus annuus, L. — the history of which was almost equally confused, and which we had identified with a wide-spread species of the western United States — is omitted by De Candolle, yet might claim a place: for Decaisne, who has treated it at length in the paper above cited, informs us that a form of it (called "Russian Sunflower") is cultivated in Russia for the oil of its large seeds, and, if we mistake not, for fattening poultry. Our Indians also cultivated it for the oil of the seeds, which they used for greasing their hair, also for eating and other purposes. Champlain noted this (in 1610?), and Sagard about a dozen years later.[1] The latter says (Histoire du Canada, 1736, page 785): "Ils font estat du tournesol, qu'ils sement en quantité, en plusieurs endroits à cause de l'huyle qu'ils tirent de la graine," etc., piously adding: "Mais comment est-ce que ce peuple sauvage a pû trouver l'invention de tirer d'une huyle que nous ignorons, sinon à l'ayde de la divine Providence." The wild original of this Sunflower must

[1] Champlain's earlier record of the cultivation and use of the Sunflower is essentially like that of Sagard, and both relate to the same stations, namely, the Huron towns near the southeastern point of Georgian Bay. This Champlain reached by way of the Ottawa (R. des Prairies) and Lake Nipissing. The lamented Decaisne has here introduced some confusion into the history, which we hasten to rectify. In his article in the "Flore des Serres" (xxiii. p. 108, p. 2 of the pamphlet), he says, "Je trouve dans Champlain l'observation suivante (Voyage Nouv. France, réimpress. 1830, tom. i. p. 110):"

"En remontant le St. Laurent et avant l'arriver au Lac Ontario, je visitai cinq des principaux villages fermés de palisades de bois, jusqu'à Cahiagué," etc., and so on to the mention of the "grande quantité de bled l'Inde (Maïs) qui y vient très beau, comme aussi des citrouilles. Herbe des soleil, dont ils font de l'huile, de la graine de laquelle ils se frottent la tête."

This, the latitude of 44.50° being stated, would refer Cahiagué and the Sunflower cultivation to the neighborhood of Ogdensburgh and Prescott, far away from the actual place (the Indian town mentioned being the Huron name of the mission station of San Jean Baptiste, in what is now Simcoe Co.), and it introduces a palpable anachronism, Ontario having been an unknown name in Champlain's time. In fact, there is nothing answering to the early part of this pretended quotation, either in the original of Champlain or in the edition here cited by name and page. The excellent Decaisne could never have tampered with the quotation himself. He must have taken it at second hand and neglected to verify it.

have been obtained by the Canadian Indians from beyond the Mississippi, and some degrees farther south. Judging from the breadth of the flower-heads soon after its introduction into Europe, it must in aboriginal hands have assumed much of the abnormal development which distinguishes the cultivated Sunflower from its wild original of the western plains.

Solanum tuberosum, L. — The question of the Potato was fully discussed by De Candolle in 1855; and the present review of it only confirms the now generally admitted conclusions. These are summed up in the statements, that the plant is spontaneous in Chili under a form quite identical with the cultivated species, that its aboriginal cultivation had extended as far north as New Granada, but apparently no farther; that allied tuberiferous species, which our author regards as distinct (though others partly doubt it) are found along the Andes and through Mexico, and within the borders of the United States; that when known in Virginia and North Carolina in the second half of the sixteenth century, it was not derived from our Indians; and that it was carried to Europe first by the Spaniards between 1580 and 1585, and afterwards by the English.

Batatas vulgaris, Choisy, *Convolvulus Batatas*, L., the Sweet Potato, is one of a few cultivated plants which have attained to a very wide distribution over the warmer parts of the world in early times; and it is one which no botanist pretends to have seen in a truly wild state. The evidence inclines to an American origin; but it had reached the Pacific islands in prehistoric times, and was cultivated in China in the second or third century of our era. De Candolle states that —

"Clusius, one of the first to speak of the Batatas, says that he had eaten it in the south of Spain, where it was said to have come from the New World. He indicates the names of 'Batatas,' 'Amotes,' 'Ajes.'"

The testimony of Clusius (L'Ecluse) to the American origin of the Sweet Potato, though not of the highest value, might be more strongly stated. He visited Spain and Portugal in 1566. The first edition of his "Historia Rariorum Stirpium" was printed in 1576, and contains the description of Batatas, which M. De Candolle cites from the edition of 1601. He gives a figure of the plant, of which, he says, he had observed three varieties growing in the south of Spain. He states their American origin, not as a doubtful matter or with a "l'on pretendait," but as a well-established fact: "Spontè nascitur in novo orbe, vicinisque insulis, unde primum in Hispaniam delata est." "Now," he adds, "it is planted in many places near the coast of Andalusia; but those grown at Malaga are preferred, and are transported to Cadiz and Seville. We sometimes have them fresh in Belgium, but they will not germinate here, the country being too cold." As to the name — he was as undecided as have been some botanists since his time: "the Spaniards call them 'Batatas,' and also 'Camotes' or 'Amotes'; some also 'Ajes'; yet, as they say, they differ among themselves, and the root of Batatas may be much the sweeter and the more tender."

This confusion of names dates from the time of Columbus, for Clusius was not, by half a century, the first to speak of the Batata. (It may be worth noting, in parenthesis, that Batatas, the scientific name adopted by Linnæus, and as the name of a genus by Choisy, is the Spanish plural of Batata, the aboriginal name.) Even Peter Martyr and Oviedo do not agree in all particulars as to the distinction between "Ajes" and "Batatas" — a distinction which both recognize. In the 9th book of his 2d Decade, written about 1514, Peter Martyr (ed. 1574, p. 191), describing the fruits, etc., of the province of Uraba, Darien, names, for the first time, Batatæ. "They dig from the earth," he says, "roots that grow spontaneously (*suapte natura nascentes*); the natives call them 'Batatas' [accus. plural], which when I saw I thought to be rapes of Lombardy [? 'Insubres napos'] or great earth-tubers [*Cyclamen Europæum?* *Rapum terræ* and *Tuber terræ* of the old botanists]. In whatever way they are cooked, roasted or boiled, they yield in delicate sweetness[1] to no confectionery or other eatable whatsoever." They are, he adds, "also planted and cultivated in gardens." In his 3d Decade (lib.

[1] The sweet potato was an inspiration to Peter Martyr, who rarely indulged himself in such a flight as "dulcorata mollities."

4, p. 240) he mentions "Maize, Yucca, Ages, and Battatæ" as plants that grew in Honduras when Columbus landed on that coast in 1502; and in the same Decade (lib. 5, p. 261) he names the same four plants as the ordinary food of the people of Caramaira (east of Darien) "as of the others," and again takes occasion to name the battatas, as surpassing all else "mirâ quâdam dulci mollitie — especially if one falls on the better sort (nobiliores) of them."

Oviedo gives a good description of the Batata, which, when he wrote (1525–35), was commonly cultivated by the Indians in Hispaniola and elsewhere, and highly prized (Historia, lib. vii. c. 4). It resembles the "Ajes," he says, in appearance, but tastes better and is far more delicate. The leaf is more notched (*harpada*) than that of the "Age," in nearly the same fashion. Some varieties are better than others, and he gives the names of the five kinds which are most highly esteemed. [Peter Martyr (dec. iii. lib. 9, p. 302) included the same five names among the nine varieties of "Ajes" that he mentioned as distinct; but in this, as in other matters pertaining to natural history, Oviedo is the better authority.] "When the Batatas are well cured, they have often been carried to Spain, where the ships happened to make a quick passage, but more often they are lost on the voyage. Yet," adds Oviedo, "I have carried them from this city of Saint Domingo, in Hispaniola, to the city of Avila, in Old Castile."

The "Gentleman of Elvas" who wrote the "True Relation" of De Soto's expedition to Florida, in 1538, mentions Batatas, then growing in the Island of Terceira (belonging to Portugal).

Cieça de Leon, who was in Peru in 1547, speaking of the fertility of the valleys near the Pacific coast, and the plants cultivated by the Indians, names among these, Sweet potatoes (Chron. del Peru, c. 66). In the Quichuan language they were called "apichu"; in the dialect of Quito, "cumar." Mr. Markham, in a note to his translation (Hakluyt Soc., 1864, p. 234) mentions, on the authority of Dr. Seemann, "the curious and interesting fact that 'kumara' is also the word for Sweet potato in Tahiti, the Fiji Islands, and New Zealand." Garcilasso says these roots "which the Spaniards call 'batatas' and the Indians of Peru 'apichu,'" are of four or five different colors, etc. "The least good are those that have been brought from Spain."

Jean De Lery found them in Brazil in 1557, and described them under their Tupi name — "Hetich," as he wrote it — of which "the soil of Brazil is as prolific as that of Limousin or Savoy is of rapes." He describes the Indian method of planting; yet, "since these roots are the principal article of food of this country, and are met with by travelers in various places, I judge that they grow spontaneously" (Hist. Navig. in Brazil, p. 165). Montoya (Tesoro, 1639) gives the Tupi-Guarani name, "*Yetĭ*," and mentions numerous varieties.[1]

Monardes, in the third part of his "Simpl. Medic. ex Novo Orbe," published in 1574 (translated by Clusius, ed. 1593, p. 439), states that Battatæ "are now so common in Spain, that ten or twelve caravel loads are sent annually from Velez-Malaga to Seville."

De Candolle (who has elsewhere printed a short article upon the subject) calls attention to the fact, which ought to be familiar, that sweet potatoes are roots, not tubers, and that Turpin long ago published good figures illustrating this; also that while these roots are free from acrid or noxious qualities, all the *Convolvulaceæ* with tubers, of which there are many, and not a few of large size, are inedible and acrid, — mostly, as we know, violently purgative.

Manihot utilissima, Manioc, Cassava-plant. — De Candolle assigns good reasons for concluding (as did Robert Brown, without giving his reasons) that this important food-plant of the tropics is American, not African. But he leaves unnoticed the convincing fact that "Manioc" and "Manihot" are Brazilian names, slightly corrupted, of a plant cultivated in St. Domingo and Cuba before the landing of Columbus, and which became known to Spanish and Portuguese discoverers before 1500, by its Haytian name, "yuca," or "hiucca."

[1] Hans Stade, who was a captive in eastern Brazil in 1549, briefly mentions these "roots called '*Jettiki*,' of pleasant taste." (Captivity, Hakl. Soc. ed. p. 166.)

Peter Martyr (1493) describing the food of the islanders, names "Iucca, from which they make bread" (Dec. i. lib. 1, p. 7; ed. 1574); in the third book of his second decade (p. 148) he mentions Iucca, Ajes, and Maiz, as the three plants used by the natives for bread; in the third decade (lib. 5, p. 262) he describes the mode of the propagation by cuttings, of cultivation, and of the preparation of "Cazabbi" from the root; and he states that "there are many kinds of 'iucca'" (p. 263). Oviedo (Historia, lib. 7, c. 2) describes "the bread of the Indians that is called 'caçabi,'" which is "made from a plant they call 'yuca,'" and he distinguishes two species of the plant. Acosta (Hist. of the Indies, transl. by E. G.; Hakluyt Soc. ed., p. 232, 1588–90) gave a good account of the plant "yuca," and the kind of bread made from it, called "caçavi."

Peter Martyr (Dec. iii. lib. 9, p. 301) relates the Haytian tradition of the origin of the cultivation of "yuca" in their island. "They say that a 'Boitius' [*i. e.*, magus, or diviner], a wise old man, after the lapse of many years, saw, on the banks of a river, a plant that was like a cane; pulling it from the earth, he made this wild plant a cultivated one. He who first ate the 'Iucca' raw, quickly died. But because its taste was sweet, they determined that a way of using it should be diligently sought for. When roasted or boiled, it was less hurtful. At last they came to the knowledge of the latent poison in its juice," etc.

Gomara (Hist. gen., c. 71), Acosta (Hist. nat. y moral de las Indias, 1588–90; lib. 4, c. 17), Monardes (De Simplicibus medic., transl. by L'Ecluse, 1593, p. 437), and other writers of the 16th century gave good descriptions of the plant "Yuca," and of the "caçavi" or "cazabi" prepared from the root. By the blunder of European editors, in the last half of the 16th century, the Haytian name was transferred from the plant to which it belonged to one of another order, the Yucca of Linnæus and of modern botany. The mistake was pointed out by Lobel.

Jean de Lery (Hist. Navig. in Brasil. c. 9) describes the two species that were cultivated in Brazil in 1557—under their Tupi names, "Aypi" [*M. aypi*, Pohl] and "Maniot" [*M. utilissima*]. Marcgrav (Hist. plant. Bras., p. 65) mentions many varieties of both species, and gives "Mandioca" as the name of the root; "Mandiiba" or "Maniiba" for the plant. Of the products of the root, Cassava retains its Haytian name ("caçavi") nearly; Tapioca is a corruption of the Brazilian (Tupi) "tipioca" or "tipiocui."

Dioscorea sativa, alata, etc. Yam. — De Candolle informs us that these species, or their allies, are wholly unknown to botanists in a wild state; that, although cultivated in the East Indies, they have no Sanscrit names; that they seem not to have been widely cultivated in Africa, but that the authors of the 17th and 18th centuries speak of them as widely diffused over the south Pacific islands, from Tahiti to New Caledonia and the Moluccas. In the summary they are assigned to southern Asia (Malabar? Ceylon? Java?), and to the eastern Asiatic archipelago. Although a large part of the genus is indigenous to tropical America, it is thought that the cultivated species were probably introduced from the Old World. The following presentation of the evidence, as concerns America, may set the question in a different light: —

The natives of Cuba and St. Domingo, when Columbus discovered those islands, cultivated two kinds of plants for their roots. These were called in the language of the islanders of St. Domingo, "Ages" or "Ajes," and "Yuca." Neither of these plants was known to the Spaniards. About "Yuca" there is no question; it was the "Manihot," or "Manioc," of which we have already spoken. It is nearly as certain that the "Ages" was a species of Dioscorea, to which, in their ignorance of the language of the islands, the Spaniards at first gave the name of "Ñame," "Niame," "Inhame," or other corruptions of a foreign (probably African) name; and this name seems to have been occasionally misapplied both to the "Yuca" and the "Batata."

L'Ecluse, who had traveled in the south of Spain and in Portugal, in 1563, says that the Colocasia (*C. antiquorum*) "first brought from Africa, was common in many places in Portugal, near streams of water, that it was sought for by negro slaves in Portugal, who ate it both raw and cooked," and that it was "called by the Portuguese, following the Moors, 'Inhame,' — by the Andalusians, 'Alcolcaz,'" etc. (Rarior Stirpium Hist., p. 299.) In a note to his translation of "Garcia ab Horto" (1574, p. 217), he says that "the plant called 'Inhame' by the Portuguese has very broad leaves, and grows near the water, or in water, — not spontaneously, but when once planted it propagates itself from the roots," etc.

Some of the companions of Columbus had seen the "Inhame" (or "Ñame") in Africa, and were ready to transfer its name to the first cultivated roots they saw in America. A few days after the discovery of Cuba (Nov. 4, 1492), Columbus saw fertile fields "full of 'Mames' ["these are 'Ajes' or 'batatas,'" notes Las Casas], which are like carrots ('Zanahorias'), and other plants, including Kidney beans and Beans (faxones y fabas) much unlike ours." (Navarrete, Colec., i. 200.) These "mames" are mentioned again Nov. 6 (id., 203) — in both places, probably by an error of the copyist, for "niames"; for, the next month, some natives of Hispaniola brought "bread of 'niames,' which are roots that grow as large as rapes ('rábanos'), which they plant and cultivate in all their fields, and on which they live; and they make bread of them, boil them, and roast them; and they have the taste of chestnuts, and no one eating them would believe they were not chestnuts" (id., 238). A few days later, the Spaniards learned the names of these roots — or of others with which they were at first confounded. The Admiral sent a present to a friendly cacique. The officer who carried it reported, on his return, that "all this island (St. Domingo) and Tortuga are cultivated like the country about Cordova. The lands are planted with 'Ajes' — which are little shoots (ramillos) that are planted, and at the bottom of each grow roots like 'zanahorias' which they use for bread," and these roots "are very savory, and taste like chestnuts." They have them here larger and better than he had seen in any place; for, he said, he had "[seen] such also in Guinea" (id., p. 242). Again, the natives "brought bread made of 'niames,' which they call 'Ajes'" (id., 251); and, Dec. 26, they gave the Admiral a "collation, of two or three kinds of 'Ajes,' and of their bread that they call 'cazavi,'" etc. (id., 263). After this the name of "niames" gives place to "ajes" (or "ages"). On the second voyage of Columbus, the natives, near Isabella (in St. Domingo), brought great quantities of "ages which are like rapes (nabos) very excellent eating," and "this age, the natives of Caribi (the Caribbean Islands) call nabi, and the Indians [of Hispaniola?] hage" (id., 368, 369).

In two or three of the passages to which reference has been made — particularly those in which bread is mentioned — the Spaniards seem to have confounded the "ages" with the product of the "yuca" (Manihot), or to have included both under the general name of "niame" (or its equivalents, "Ñame, Igname," etc.) Amerigo Vespucci — or some one of the several translators through whom the relation of his first voyage comes to us — says, that in 1497, "the common food of the natives of Paria was the root of a certain tree (arborea radix quædam), which they reduce to a good enough flour, and that some call this root 'Iucha,' others 'Cambi,' but others 'Ignami'" (Navarrete, Colec., iii. 216).[1]

This confusion of names, in the first decade of discovery in America, was natural and unavoidable. The foreign name, "niame, igname," was applied without much discrimination to roots cultivated by the natives of the islands and the mainland — primarily, to "ajes," occasionally to "yuca" (Manihot), and perhaps to "batatas." In the relations of the voyages of Columbus only two cultivated roots are named — "Ages" and "Yuca." The first book of Peter Martyr's first decade (dated 1493, but probably revised before its publication in 1511) names only these two; and in the third book of his second decade he mentions the use of the same two roots by the natives

[1] It is to this passage that Humboldt refers, in "Nouv. Esp.," 2d ed., ii. 468 (cited by M. De Candolle, p. 63), as evidence that the name "Igname" was heard on the continent of America by Vespucci in 1497; but, as will be seen, Vespucci (or his copyist) does not say that this name was used by the natives.

of Comagra, in Darien (p. 148); but in a subsequent chapter (dec. ii. c. 9., p. 191) he adds — as has been mentioned in a preceding note — a third kind of roots, which the natives of the province of Darien call "Batatas," that grow in their country spontaneously. From this date to the middle of the 16th century the distinction between these roots, though occasionally lost sight of, is generally observed. Oviedo (Historia, l. vii. cc. 2, 3, 4, pp. 268–73) describes the "caçabi" and two species of the plant ("yuca") that yields it; "ajes"; and "batatas." The "ajes," he says, were cultivated in Hispaniola and in all the other islands, and on the continent; they were of various colors — white, reddish, inclining to mulberry, and tawny, but all white within, for the most part; the stem of the plant extends itself like that of "correhuela" (Convolvulus or Bindweed), but stouter; the leaves cover the ground, and are shaped much like "correhuela" and nearly like Ivy or panela, with some delicate veins ("unas venas delgadas"), and the little stems ("astilejos"), on which the leaves hang, are long and slender, etc. The leaf of the "Batata," he says (p. 274), is more toothed or notched (harpada) than that of the "Aje," but of nearly the same fashion; and the two plants are much alike, but the "Batatas" are sweeter and more delicate, etc.: some of the "Ajes" weigh four pounds each, or more. In some parts of Castilla del Oro (in Darien) there are "Ajes" that are small and yellow, etc. (p. 273). His description of the two plants permits no reasonable doubt that his "Ajes" were of the genus Dioscorea. Moreover, they were not identical with — though they resembled — the imported "ñame" or "yam": for Oviedo states (Historia, lib. vii. c. 19, p. 286), "that 'name' (called 'names') is a foreign fruit, not natural to these Indies, which has been brought to Hispaniola and other places, and is suited to this evil race of negroes, and a profitable and good subsistence for them. . . . These 'names' seem to be 'ajes,' but are not the same, and generally are larger than 'ajes.'" They had already multiplied greatly in the islands and on the mainland.

The distinction between "Ajes" and "Batatas," though

clearly apprehended, was sometimes lost sight of. Peter Martyr (dec. iii. lib. 9, p. 302) says that "the species of 'Ages' are innumerable — the varieties being distinguished by their leaves and flowers;" and he gives the American names of nine of the varieties; but five of these nine are named by Oviedo (p. 274) as varieties of "Batatas." [See Batatas, ante.]

The "Gentleman of Elvas," who wrote the narrative of DeSoto's expedition, mentions a fruit, at Santiago, Cuba, called "batata," the subsistence of a multitude of people, principally slaves, and which now (1538) grows in the island of Terceira, belonging to Portugal. . . . "It looks like the 'ynhame,' with nearly the taste of chestnuts" (Relaçam Verdadeira, ch. 5).[1]

Jean de Lery, who was in Brazil in 1557, though he gives a good description of the "Batata," does not mention the Yam; but it is figured and described by Piso (Hist. Nat. Brazil., 1648, p. 93), as "Inhame" of St. Thomas, called "Cara" by the natives of Brazil, and "Quiquoaquecongo" by the Congo negroes. Ruiz de Montoya has the name "Cará" in his Tupi dictionary, 1639, and mentions five varieties. As the Tupi name for the Virginia Potato (*Solanum tuberosum*), "Carati" (*i. e.*, white Yam), is formed from that of the "Inhame," it would seem that the latter was of earlier introduction. So, in the Mpongwe — a language of the Congo group — the Potato is called "mongotanga," "white-man's yam."

Portulaca oleracea, Purslain. — Botanists have taken it for granted that this weed of gardens and other cultivated grounds was transported to America from the Old World. But Nuttall found it apparently indigenous on the upper Missouri forty years ago, and Dr. James in Long's Expedition, along the eastern base of the Rocky Mountains in what is now the State of Colorado. From thence to Texas it grows wild

[1] In one Indian language of the south, the Choctaw, the Sweet Potato is now called "ahe"; while the Virginia Potato (*S. tuberosum*) takes the adopted prefix of "Irish," "Ilish ahe," or is sometimes called "ahe lumbo," "round ahe."

along with two other nearly related species. Moreover, the following evidence tends to show that its introduction, if introduced by human agency, took place before the landing of Columbus.

On their first sight of the New World, the Spaniards were much impressed by the strangeness of all forms of animal and vegetable life: "all the trees are as unlike ours, as day is to night" — wrote Columbus, Oct. 17th, 1492, six days after landing at San Salvador: "and so are the fruits, and so the plants, and the stones, and all things" (Navarrete, i. 183). On the 28th, on the north shore of Cuba, he saw — apparently for the first time — a familiar plant: "halló verdolagas muchas y bledos," — he found much "purslane" and "bletum" (id., 192). It seems hardly possible that the Admiral and his companions could mistake a strange plant for a salad herb so well known as "verdolagas" to Spanish eyes and palates. Again, Oviedo, writing about 1526, in a list of "plants in the island of Hispaniola which are like those of Spain, and which were before the Christians came to these parts, and are natives of this land, and were not brought from Spain," mentions "verdolagas or pertulaca," and "bledos or bletum" (Blitum).

In his description of "perebeneçuc," written in 1525, he says that plant grew in great abundance in Saint Domingo and in many places on the continent, in the woods and fields; even "purslane ('verdolagas') is not more abundant here" (id., lib. xi. c. 5, p. 378).

Jean de Lery, in Brazil in 1557, was as much impressed by the novelty of the flora as Columbus had been in the West Indies. "I declare," he wrote (Hist. Navig. Brazil., 168), "as far as it was permitted me to discover in wanderings through the woods and fields, that there are no trees or plants, or any fruits, that are not unlike ours, these three excepted, portulaca, ocymum, and filex" (in the original French edition, 1578, p. 217, "pourpier, basilic, et fougière").

Capt. John Smith, in Virginia in 1606, found "many herbes in the spring, commonly dispersed throughout the woods, good for broths and sallets, as Violets, Purslain, Sorrell, etc.; besides many we used whose names we know not" (Smith's Gen. History, 1632, p. 26; and repeated by Strachey, Travaile into Virginia, p. 120). Smith's purslain was probably *Sedum ternatum.*

Sagard-Theodat, in the relation of his "Grand Voyage du Pays des Hurons," in 1624 (p. 331), says that the Hurons make little use of herbs, "although the pourpier or pourcelaine is very common there, and grows spontaneously in their fields of corn and pumpkins."

W. Wood, who was in New England from 1629 to 1633, names "Purselane" among plants growing "in the woods, without either the art or the help of man" (N. E. Prospect, pt. 1, c. 5). We doubt its growing literally in the woods, as unlike its natural habit, and place more confidence in the statement of Champlain, who, in his earlier voyages, 1604–11, found plenty of excellent "pourpier," for his salads, on the coast of New England, growing among the Indian Corn; "the savages making no more account of it than if it were a noxious weed" (Voyages, ed. 1632, p. 80).

Humulus Lupulus, Hops. — Although the matter has nothing to do with the introduction of Hops into cultivation, it is noticeable that De Candolle assigns the home of the plant only to Europe and western Asia. It is undoubtedly indigenous to North America also, and is mentioned as such in the American works. In Gray's "Manual," besides the printing of the name in the type appropriate to indigenous species, the plant is expressly stated to be "clearly indigenous." But, through some oversight, in the "Prodromus" (xvi. 29), it is stated, in connection with this very reference, that the plant was introduced.

Oca. — Considering that "Maté" and "Coca" find place in this volume, although perhaps rather employed than cultivated (at least the former), the absence of Oca (*Oxalis tuberosa* and *O. crenata*) is noticeable. This esculent root deserves mention, if only for the antiquity of its culture in Peru. The name, which is Quichuan, appears to have belonged, specially, to *Oxalis tuberosa.* Another root "like the oca in shape, but not in taste," called in Quichua "añus," was less esteemed.

Both were cultivated in Peru in the time of the Incas, and in the districts where no Maize grew, the crop of these tubers was of much importance (Garcillaso, Comment., b. v. c. 1; b. viii. c. 10). J. de Acosta, 1588–90, says "there are an infinite number" of roots used for food in the Indies, "but the Papas (Potatoes) and Ocas be the chief for nourishment and substance" (Nat. and Moral Hist. of the Indies, lib. iv. c. 18).

Our notes upon plants cultivated for their herbage, tubers, roots, etc., have run to such a length that the remainder, concerning some plants cultivated for their fruits and seeds, must be left for another article.

PART II.

THE fourth chapter relates to plants cultivated for their fruits; the fifth, to those cultivated for their seeds. Our present annotations concern a few species or forms of *Cucurbitaceæ*, the history of which has been involved in some obscurity and confusion.

A word, in passing, upon the Peach, upon the history of which this volume throws some new light. De Candolle had formerly suggested China as its home, and he has brought together additional evidence in favor of that view. He shows how this conclusion goes against an old idea that the Peach is a derivative of the Almond, which is indigenous to western Asia, and was unknown to the Chinese anterior to the Christian era, while they had peaches of various sorts long before. Upon Pyrus there is a note relating to botanical orthography, p. 183, which we append, as it has an application to a few other words.[1]

[1] "L'Orthographe Pyrus, adopté par Linné, se trouve dans Pline, 'Historia,' ed. 1631, p. 301. Quelques botanistes ont voulu raffiner en écrivant Pirus, et il en result, pour une recherche dans un livre moderne, il faut consulter l'index dans deux endroits, ou risquer de croire que les Poiriers ne sont pas dans l'ouvrage. En tous cas le nom des anciens est un nom vulgaire, mais le nom vraiment botanique est celui de Linné, fondateur de la nomenclature adoptée, et Linné a ecrit Pyrus."

Pears and apples were prehistoric in Europe, both wild and cultivated.

Lagenaria vulgaris. Bottle Gourd. — Although doubtless an Old-World plant (De Candolle attributes it to India, Molucca, Abyssinia), yet it is not quite certain that it had not reached the New World before Columbus. At least the following notes may be put upon record.

M. De Candolle mentions the case of the name gourd for pumpkin ("potiron") by English writers, as "an example of the confusion of popular names and the greater precision of scientific names." Such confusion becomes more perplexing when we have to deal with popular names of the 15th and 16th centuries. Parkinson — a good observer and a respectable botanist — complained, in 1640, of "our modern writers who confound Pepo, Melopepo, and Cucurbita so promiscuously that it is not possible to find out the distinct certainty of them all; for some make that Pepo that others call Melopepo, and others, Cucurbita." (Theater of Plants, p. 770.) Scientific names of the 16th century are as obsolete as popular names of the same period. They do not help us to distinguish Lagenaria from Cucurbita, or Pepo from Melopepo; or Citrouille from Citrullus. Early voyagers to America wrote cucurbita, calabaça, courge, or zucca, as a name for any gourd or pumpkin, and occasionally for a calabash which was not even a cucurbit. The relation of the first voyage of Columbus repeatedly mentions the "calabazas" used by the natives of St. Domingo and other islands for carrying water (Navarrete, Collec., i. 180, 188, and Dec. 3, 1492). Columbus saw, near the east end of Cuba, fields planted with "calabazas" and other productions of the country (id., p. 225). Yet we know from Peter Martyr that some of the gourds ("cucurbitæ") used in the islands grew on "cucurbiteas arbores" as tall as elms (Dec. i. lib. 3, and iii. lib. 4; pp. 38, 246). This tree, *Crescentia Cujete*, is described by Oviedo (Historia, lib. viii. c. 4) under its Haytian name, "Higuëro"; in Nicaragua it was called "Guacal"; and in Brazil, "Cuiete"[1] (Marcgrav, Hist. Nat. Brasil., 123). J. de

[1] Not "Cujete" — unless *j* has the German sound. The Tupi name is formed from *cuï* (*cou-in*, Lery) "the shell" or hard rind of a nut or fruit (and the bowl or calabash made from it) and *eté* "good, precious."

Lery (Hist. Navig. in Brasil., 154) describes the tree under the Brazilian name of "Choyne"; but elsewhere (p. 246) he says "the natives have *Cucurbitæ* (courges) and other kinds of fruits," from which "they make their bowls, called 'coui,' and other vessels."

It is certain that "calabaças," which were not arboreal, but genuine cucurbits, were abundant — and were believed to grow spontaneously — in the islands and on the main land, before 1526. Oviedo (Historia, lib. vii. c. 8) observes that "calabaças, in the Indias, were as common as in Spain, and of the same kinds (de las mismas), long and round, or banded (çeñidas), and of all the shapes they usually have [in Spain]." They were much used "in all parts of these Indias, both the Islands and the Main," and "are one of the common things that the Indians cultivate in their gardens." They were not cultivated for food — "for they do not eat them" — but for carrying water; "and they have other calabaças that are in all respects like the aforesaid, except that they are bitter to the taste; and there are many of these that grow of themselves without cultivation."[1] The same author (lib. xi., e. 1), in a list of plants introduced from Spain, names Melons and Cucumbers (pepinos), but not Gourds.

The relation of the voyage of Amerigo Vespucci, 1489, in a description of the Indians of Trinidad and the coast of Paria, says that "each carried, hanging at his neck, two small dried gourds (*cucurbitas*), one containing the plant that they were accustomed to chew, the other, a certain whitish flour," etc., and that each woman carried a "cucurbita" of water (Navarrete, iii. 252, 254).

The "Cucurbita lagenæ formâ," which Marcgrav found in Brazil, 1637–8 (Hist. Nat. Brasiliæ, 44), though "very probably *Lagenaria vulgaris*," yet, as M. De Candolle observes, "does not prove that the species was in that country before the voyage of Amerigo Vespucci in 1504;" but we

[1] M. De Candolle, p. 198, citing this passage from Ramusio's Italian translation of Oviedo's "Historia," has "zucche" for "calabaças" of the Spanish original, and takes no notice of what is said of their spontaneous growth.

know from Lery, above cited, that the natives of Brazil used *Cucurbitæ*, for bowls and drinking-vessels, at least as early as 1557. Moreover, the richness of the Tupi vocabulary in gourd-names suggests — if it does not absolutely prove — that several varieties of Lagenaria were known to the Brazilians long before the visit of Piso and Marcgrav. The "Tesoro de la Lengua Guarani" (*o Tupi*) of Father Ruiz de Montoya was first printed in 1639. It gives for gourd ("calabaço"), the Tupi general name, "Ia" [which is a compound of *ĭ* "water," and *yá* or *á* "fruit"], and for the varieties — among others — it names *ĭaçĭ* "round gourd"; *ĭáyurumî* "narrow-mouthed gourd"; *ĭátĭ* "long-necked gourd"; *ĭáobá* "wide-mouthed gourd"; *ĭáquatiâ* "painted gourd"; *ĭácuîpé* "spoon gourd" (used for making spoons); *ĭáapê* "small gourd, used for drinking"; *ĭáquâ* "great gourd"; *ĭácuî* "gourd like a great dish" or bowl, etc.: not including the derivatives of *cuî*, or the edible "calabaças" — to be mentioned hereafter.

"Acosta, too," says M. De Candolle, "speaks of Calebasses which the Peruvians used for cups or vases, but the Spanish edition of his book is of 1591, more than a hundred years after the conquest." (?) Acosta says more than this. After mention of the "Calebasses or Indian Pompions . . . especially those which are proper to the country" [Peru], he adds: "There are a thousand kinds of Calebasses; some are so deformed in their bigness that of the rind cut in the midst and cleansed, they make, as it were, baskets to put in all their meat for their dinner. Of the lesser, they make vessels to eat and drink in," etc. (Hist. nat. y moral de las Indias; translation, revised by Markham, lib. iv., c. 19, p. 238.)

Cucurbita maxima, *C. Pepo*, *C. moschata*. Pumpkin, Squash, etc. — In the "Géographie Botanique" not one of the cultivated *Cucurbitæ* is attributed to America, and a reference to Nuttall's record that the warted squash was grown by the Indians on the upper Missouri is the only mention of any aboriginal cultivation of squashes in North America. In the present volume there is merely a reference, in this respect, to Dr. Harris's article in this Journal (xxiv., 1857), and to Mr. Trumbull's note in the "Bulletin of the Torrey

Club" (1876), with the comment that: "Cela nous apprend seulement que les indigènes, un siècle après la découverte de la Virginie, 20 à 40 ans après la colonisation par W. Raleigh, faisaient usage de certains fruit de Cucurbitacées." Nevertheless *Cucurbita Pepo*, upon botanical indications solely, is attributed to temperate North America in the general table, to a Mexican or Texan origin in the body of the work. This rests upon the collection by Lindheimer, in Texas, of a form of this species "apparently indigenous." That was between thirty and forty years ago; no wild specimen has since been received from all that region (nor from any other); and it is wellnigh certain that the species was commonly cultivated in all that country by the aborigines. If ever found truly indigenous, it will probably be farther south than Texas. *C. maxima* is now set down as from Guinea, on the strength of a single finding of it "apparently indigenous" on the banks of the Niger. *C. moschata* (to which Vilmorin refers the Canada Crook-neck Squash) is in the list of species of completely unknown or uncertain origin.

In this state of the case, it is certainly worth while to present the evidence — gathered with much care and pains — which assures us that one or two, and perhaps all three, of these species, and many varieties, were largely cultivated throughout America, from the tropics to Canada, before the voyages of Columbus.

Allusion has already been made (under Lagenaria) to the difficulty of distinguishing the genera of *Cucurbitaceæ* under the names by which they are mentioned by voyagers and explorers of the first century after the discovery of America; and the question of species is particularly difficult. Yet we find abundant evidence — especially as respects North America — (1) that, in various parts of the country remote from each other, the cultivation of one or more species of Cucurbits by the Indians was established before those places are known to have been visited by Europeans; (2) that these species or varieties were novel to Europeans, and were regarded by botanists of the sixteenth and seventeenth centuries, as well as by the voyagers and first colonists, as natives or denizens of the region in which they were found; and (3) that they became known only under American names; one of these names (Squash) becoming, in popular use, generic, and two others (Macock and Cushaw) surviving, as names of varieties, into the present century.

To present this evidence as nearly as possible in the order of time, we refer, first, to the relation of the first voyage of Columbus. December 3, 1492, entering a small river [the Rio Boma], near the eastern end of the island of Cuba, he found near it a populous Indian village, and saw large cultivated fields "planted with many things of the country, and calabazas, a glorious sight (que era gloria vella)!" See Navarrete, Colec., i. 225. It is not certain that these "calabazas" were not bottle-gourds (Lagenaria), but it is, to say the least, highly improbable that the enthusiasm of Columbus would have been so kindled by the promise of a harvest of little value to Europeans.

Oviedo (Historia, l. xi. c. 1) names among plants and seeds brought from Spain to Hispaniola "melones" and "pepinos" — of which imported varieties were already abundant in the island before 1535; the seed of "cogombros" brought from Castile had not succeeded so well.

In July, 1528, Cabeça de Vaca found near Tampa Bay, in Florida, "maize, beans, and pumpkins in great plenty, and beginning to be fit for gathering." In 1535–6, when passing through Texas, the Indians supplied him with prickly pears and, occasionally, maize; but after crossing "a great river coming from the north" — probably the Rio Grande — he and his companions came to a region having "fine dwellings of civilization, whose inhabitants lived on beans and pumpkins" — and, when the season was not too dry for raising it, maize (Relacion, 1542; translated by B. Smith, 1871).

In the summer and autumn of 1539, De Soto found the Appalachian country, in western Florida, well supplied with "maize, beans (fésoles) and pumpkins (calabaças);" the pumpkins of Uzachil were "better and more savory than those of Spain;" there were "fields of maize, beans, and pumpkins," not far from Tampa Bay, where he first landed from Cuba; at Pacaha, on the Mississippi, the northernmost point he reached (1541), he found again "many pumpkins and much maize and beans"; and, still westward, at Coligoa, "beans and pumpkins were in great plenty; both were larger and better than those of Spain; the pumpkins when roasted had nearly the taste of chestnuts" (Oviedo, lib. xvii. cc. 24, 28; True Relation, etc., by a Fidalgo of Elvas; translated by Buckingham Smith, pp. 45, 47, 122, 285). Oviedo writes "calabaças," but the author of the Portuguese "Relaçam Verdadeira" (1557) has, in one or more of the places cited, "aboboras."

In 1535, Jacques Cartier, the first explorer of the St. Lawrence, found among the Indians of Canada "grand quantité de gros melons, concombres and courges" (Bref Recit de la Navigation, etc., 1545; reimpr. Tross, 1863, ff. 24, 31).

Sagard, whose "Grand Voyage du Pays des Hurons" was made in 1642, makes repeated mention of the native squashes ("citrouilles du pays") which the Hurons raised in abundance, and which he found very good, boiled or baked (pp. 85, 105, 140, 331). In his "Histoire du Canada" (283) he describes the method by which the Indians hastened the germination of the seeds of these "citrouilles du pays," and "raise them with great ease."

Lahontan (Nouv. Voyages, 1703, ii. 61) describes the "Citrouilles" of (southern) Canada — "sweet, and of a different kind from those of Europe, where," as several persons assured him, these would not grow. "They are of the size of our melons; the flesh yellow as saffron. They usually bake them in the oven, but they are better roasted under the embers, Indian fashion," etc. Lahontan had as little doubt as Sagard had, that these "citrouilles" (cultivated by the Indians of Canada from the time of Cartier, at least) were genuinely "du pays."

As to the *Cucurbitaceæ* of Virginia, M. De Candolle admits, "only, that the natives, a century after the discovery of Virginia, twenty to forty years after the colonization by W. Raleigh, made use of certain fruits of *Cucurbitaceæ*" (p. 201). Let us reëxamine the evidence. Captains Amidas and Barlow, in the first vessels sent by Sir Walter Raleigh to the New World, landed on an island in Ocracoke Inlet (now within North Carolina) in 1584. While the vessels remained there, and while they were at Roanoke Island near by, the Indians entertained them kindly, and "sent them, commonly every day, a brace of bucks, conies, etc., sometimes melons, walnuts, cucumbers, pease, and divers roots" (J. Smith's Gen. Hist. p. 3).

What these "melons," or some of them, were, we learn from later explorers and the first colonists of Virginia (proper).

Captain John Smith says that the Indians of Virginia (1606–8) "plant amongst their corn Pumpions, and a fruit like unto a musk-melon, but less and worse, which they call Macocks," etc. (Gen. Hist. p. 29). Strachey, who was in Virginia in 1610, describes these "macock gourds" in nearly the same words (Trav. into Virginia, p. 72); elsewhere, he says the "macokos is of the form of our pumpions — I must confess, nothing so good, — 't is of a more waterish taste," and he mentions also the "pumpions" planted by the Indians, and "a kind of million" which they "seeth and put into their walnut-milk, and so make a kind of toothsome meat" (p. 119). "The Indian Pumpion, the Water-melon, Musk-melon," etc., are named among fruits introduced into Bermuda, by the English, before 1623 (Smith's Gen. Hist. p. 171).[1]

[1] L'Ecluse (Clusius) heard of these Macocks in 1591 or earlier. In his "Exotica" (1605; lib. iii. c. 2) he describes a fruit — "Macocqwer Virginiansium, forte" — which had been sent him from London by James Garet, brought from "the province of Wingandecaow, which the English call Virginia." He conjectured that this might be "the fruit which the natives of that region call Macocqwer" — but his figure and description do not favor this identification. The fruit, he says, is nearly orbicular; four inches in diameter; with a hard rind, yellowish on the outside; many seeds, flat and heart-shaped ("cordis, ut vulgo pingitur, formam referentia"). L'Ecluse thought it might be one of the gourds which the natives used for rattles, as the Brazilians used their Tamaraca, etc. His specimen was old and dried, the pulp blackened, the rind covered with a dark membrane, "per quam sparsæ quædam fibræ à pediculo ad summum." This must have been a fruit of *Crescentia cucurbitina*, a calabash, which is a native not only of the West Indies, but also of southern Florida.

Among Johnson's additions to Gerarde's "Herball," 1636, there is a description of "Macock Virginiani, sive Pepo Virginianus; the Virginian Macock or Pompion" (pp. 919, 921). The description is dated 1621, and signed by John Goodyer. The plant has "great broad shrivelled yellow flowers, like those of the common Pompion." The fruit, "somewhat round, not extending in length, but flat like a bowl, but not so big as an ordinary bowl, being seldom four inches broad and three inches long; of a blackish green color when it is ripe. The substance or eatable part, of a yellowish white color. . . . Seeds like the common Pompion, but smaller." The "small round Indian Pompion," and "the cornered Indian Pompion" — the latter resembling our common "scolloped Squash" ("*Pepones lati*, Broad Melons or Pepons" of Lyte's Dodoens, p. 588) — are described and figured in "Johnson's Gerarde," p. 920.

Beverley's "History of Virginia," 1705, p. 124, mentions the Macocks, "a sort of Melopepones, or lesser sort of Pompion or Cashaw," which he identifies with the "Squash or Squonter Squash" of New England. "The Indian name," he says, "is still retained by them." Professor Schele de Vere (of Virginia) states that it still "survives in its anglicized form of Maycock" (Americanisms, 1871, p. 60).

The "Cushaw" ("Ecushaw," Hariot) is described by Beverley (Hist. of Virg., p. 124) as "a kind of Pumpion, of a bluish green color, streaked with white when they are fit for use. They are larger than the Pompions, and have a long narrow neck. . . . The Cushaws and Pompions they lay by, which will keep several months good, after they are gathered" (p. 152). Bartlett, "Dict. of Americanisms," notes the name Cushaw, "sometimes spelled Kershaw," as "Western" for a pumpkin. Beverley's description makes it nearly certain that the variety so named was the (New England) winter "crookneck" squash — which, five and twenty years ago, might have been seen hanging, by its necklace of flannel "list," in every New England farmer's kitchen, from early harvest time till wanted for Thanksgiving or Christmas pumpkin-pies.

The Rev. Francis Higginson, who came to New England in

1629, wrote from Salem, a few weeks after his arrival: "Here are stores of pompions, cowcumbers, and other things of that nature which I know not" (N. E. Plantation, 1630); and, again: "We abound with . . . sundry sorts of fruits, as musk-melons, water-melons, Indian pompions, Indian pease, beans, and many other odd fruits that I cannot name" (Young's Chron. of Mass., 265). William Wood, who was in New England from 1629 to 1633, says of the Indians of Massachusetts: "In summer, when their corne is spent, 'Isquoutersquashes' is their best bread, a fruit like a young Pumpion" (N. E. Prospect, p. 76). Roger Williams, 1643, names these "*Askútasquash*, their vine-apples, which the English from them call squashes, about the bignesse of apples, of several colors, a sweet, light, wholesome refreshing" (Key to the Language of America, 103). Again, Josselyn (1638–71, N. E. Rarities, 57) mentions these "squashes . . . more truly Squontersquashes, a kind of melon or rather gourd, for they oftentimes degenerate into gourds; some of these are green, some yellow, some longish like a gourd, others round like an apple, all of them pleasant food boiled and buttered, etc. But the best yellow squash, called an Apple squash, because like an apple, and about the bigness of a Pome-water, is the best kind: they are much eaten by the Indians and the English." But he distinguishes these from the "Pompions [of which] there be several kinds, some proper to the country; they are dryer than our English Pompions, and better tasted; you may eat them green" (p. 91). The last words (here italicized) give a nearly literal translation of the Algonkin-Indian name of Cucurbits, — in the dialect of New England, "asq," plural "asquash," "green things," or (to be eaten) "immature." Eliot, in his version of the Bible (1663) names three kinds of asquash: askoot-asquash [= Askùt-asquash, R. Williams, Isquoutersquash-es of Wood, Squontersquash-es of Josselyn, ut supra], for "cucumbers"; quonooasquash "gourds" [literally, "long *asquash*"]; and monaskoot-asquash "melons."

Squashes were first known to the Dutch by their Algonkin name. Van der Donck, after speaking of the pumpkins of New Netherland (1642–53), adds: "The natives have another species of this vegetable peculiar to themselves, called by our people 'quaasiens,' a name derived from the aborigines, as the plant was not known to us before our intercourse with them. It is a delightful fruit, as well to the eye on account of its fine variety of colors, as to the mouth for its agreeable taste. . . . It is gathered early in summer, and when it is planted in the middle of April, the fruit is fit for eating by the first of June. They do not wait for it to ripen before making use of the fruit, but only until it has attained a certain size. They gather the squashes and immediately place them on the fire without any further trouble. . . . The natives make great account of this vegetable." (Descript. of N. Netherlands, 1656; transl. in N. Y. Hist. Soc. Coll., 2 ser., i. 186.)

Thus far we have cited, with one or two exceptions, American authorities. M. De Candolle, after mentioning "the three forms of Pepones figured by Dodoens, edition of 1557, to which a fourth, *P. rotundus major*, was added in the edition of 1616," and a figure of *P. oblongus*, in Lobel. "Icones," 641, observes, that "the names given to these plants indicate a foreign origin; but the authors can affirm nothing in this regard; the less so, because the name Indian signifies, either, of southern Asia or of America" (p. 204). A collation of the descriptions of Pepones or Cucurbitæ, given by European botanists of the sixteenth century, does away with this ambiguity.

Tragus (Hieron. Bock), "De Stirpium Nomenclaturis," etc., 1552, p. 830, described and figured Melo, Pepo, Cucumis, and Citreolus; and (p. 832) named, also, *Cucumis sylvestris*. In the next chapter (p. 834) he wrote "De Cucumere seu, ut vulgo loquuntur, Zucco marino" — with a figure. "Many kinds of strange plants," he says, "have been brought from remote parts into Germany, in the last few years." Among others, these "poma æstiva," of which some are large, some small, some round, some oblong, some sweet, others bitter, of various colors. "Some call these 'Cucumeri,' and assert that they are 'Turkish Cucumeres,' with which opinion I cannot agree. . . . I call them *Mala æstiva & Indica*," of which he distinguishes four kinds, *M. Indica crocea, lutea, citrina*, and *nigra.* "Commonly," he says, "they are called Zucco marina, because they first came to us from parts beyond the sea, some from Syria, some from India, which the names given them attest; for they are commonly called Zucco de Syria and Zucco de Peru."

The figure of *Cucumer marinus*, Ital. *Cocomere marino*, etc., in the "Effigies Plantarum" of Fuchs, 1549, is a reduced copy of Bock's, and substantially agrees with that of *Pepo rotundus* in Lyte's "Dodoens," p. 587, which was "called also *Cucumis marinus;* of some, *Zucco marino;* in French, *Concombre marin, Pompons Turquins,*" etc.

Matthioli of Padua (Comm. in Dioscor., ed. 1559, p. 292) is more explicit. "There are," he says, "various kinds of cucurbits foreign to Italy, which can be kept fresh far into the winter. They say that these came into Italy from the West Indies, whence they are called by many Indian. Their taste is sweetish, not so insipid as ours," etc.; and his figure of *Cucurbita Indica* agrees with that of Bock's Zucco marinus (or "Zucco de Peru") and with Lyte's *Pepo rotundus.*

It is certain, then, that the botanists of the 16th century to whom M. De Candolle refers, used Indian — when applied to varieties of Cucurbita — in the sense of American. In the 17th century, the evidence is not less direct. Parkinson (Theatrum Botanicum, 1640, pp. 769, 770) figures and describes (1) *Cucurbita lagenaria major*, the greater Bottle Gourd; (2) *C. longa*, the long Gourd; (3, 4) *C. clypeiformis & verrucosa*, and *Anguria Ægyptiaca*, the Simmel [Scallop Squash], and the rugged Gourd [warted Squash, orbiculate depressed], and the Egyptian Citruell or Watery Million; (5) *Cucurbita Indica, ovalis, pyriformis, & fere rotundus*, Indian Gourds, oval, pear-fashioned, and almost round. Of these Indian Gourds he says: "There is very great variety of these Gourds (or Millions, as some call them, or Pompions, as I may call them) that came out of America or the West Indies, from sundry places, both farther south among the Spanish colonies, and nearer hand, in our own of Virginia, New England, etc." He notes the great variety of size, shape, and color, "some as great as our pompions, some as small as an apple, some discolored on the outside, green with whitish or yellowish stripes, . . . some also reddish, spotted, or striped, and some of a deep yellow."

Piso and Marcgrav (Hist. Nat. Brasil., 1648, p. 44) describe and figure a plant called "Iurumu" [=Yurumu] by the Brazilians, and by the Portuguese, "Bobora." M. De Candolle, p. 201, is inclined to agree with modern botanists in referring this to *C. maxima;* but, as he remarks, it appears to have been a cultivated plant. If introduced from abroad, the name given it by the Tupis was probably formed, by prefix or affix, from that of some native (or naturalized) species to which it had some resemblance. In Montoya's "Tesoro," 1639, we find "Yuruá" "calabaçillos silvestres," small wild calabazas; but the name "Yurumu" did not yet appear. Almost a century before the visit of Piso and Marcgrav, Jean de Lery saw in Brazil (1557) "certains citrouilles rondes, fort douce à manger," called by the natives "Maurongaus" (Voyage, ed. 1578, p. 217). The Tupi name "morangá" (the first two vowels nasal) denotes a "handsome fruit."

Lycopersicum esculentum, Tomato. — We have only to note an oversight in respect to the Mexican cultivation of the *Mala Peruviana*, as it was named by some botanists of the 16th century. De Candolle refers to Humboldt's statement that the cultivation of this esculent was ancient in Mexico, but adds that there is no mention of it in the earliest work on the plants of that country, namely: Hernandez, "Historia." But Hernandez (ed. 1651, p. 295 ff.) actually has a chapter "De Tomatl, seu planta acinosa vel Solano," and describes several sorts under their Mexican names.[1]

[1] I find only one writer in the 16th century who gave the Tomato a name indicating a Peruvian origin — namely, Anguillara, whose treatise "De Simplicibus" was first printed in Italian, at Venice, 1561. On his authority the name *Poma Peruviana* is introduced in the synonymy by C. Bauhin, in his annotations on Matthioli, 1598 (p. 761), and the "Hortus Eystettensis" (attributed to Besler), published in 1613, is referred to by the same writer (Pinax, 1771, p. 167) an authority for the names, *inter alia*, "Poma amoris fructu rubro & Mala Peruviana." The "Hortus Eystettensis" is the only authority cited for *Mala Peruviana* in the work to

Persea gratissima, Alligator Pear of the English, l'Avocat of the French; a singular corruption of a native name, as De Candolle remarks, which had no more to do with an alligator than with a lawyer. Our author does not carry back the native Mexican name quite to its original, which was "Ahuacahuitl" corrupted by the Spaniards into "Aquacate," "Avogade," etc. Champlain, who saw it in Mexico in 1599 or 1600, calls the fruit "Accoiates" and "Acoyates"; "Voy. to the W. Indies" [Hakluyt Soc., 1859], p. 28.

Oviedo described "the wild pear-tree of the main land," in 1526. It grew "in the province of Castilo del Ora (Panama), in the sierras of Capira and the country of the cacique of Juanaga," etc. In the revision of his first work, in 1535, he adds, that he had, some years before, seen these trees cultivated by the Indians in Nicaragua (Historia, lib. ix. c. 22). It was still a tree of "Terra firma" — not yet introduced into the Islands. Clusius saw it in a garden in Valencia — "said to be brought from America" — thirty-five years earlier than the date (1601) mentioned by De Candolle. He described the Persea in the first edition of his "Historia rariorum Stirpium," 1576 (lib. i. c. 2), published five years after his journey in Spain.

Passiflora. — This genus is wholly omitted by De Candolle; unaccountably so, considering how much Granadillas have been cultivated and prized in tropical countries. A note on the subject may not be out of place, as a species was cultivated by our own Indians.

which M. De Candolle refers — the "Historia Stirpium," attributed to J. Bauhin, but published long after his death (in 1551), with large additions by his son-in-law Cherler, and by Chabræus and Graffenreid. Guillandinus, of Padua, in a treatise "De Papyro," 1572, named the "Tumatle Americanorum" as a species of "*Pomum Amoris* or *Solanum pomiferum*"; and earlier, Matthioli had described it (Comment. in Dioscor., ed. 1559, p. 537) as a "kind of Mala insana" [*Solanum Melongena*], which was "beginning to be imported" into Italy, and which was "popularly called *Pomi d'oro*, that is, *Mala aurea.*"

Anguillara may have confounded the Tomato with another of the American *Solanaceæ*, introduced at about the same period — the Thorn Apple (*Datura Stramonium*) which Guillandinus (1572) named *Mala Peruviana*, and the French called "Pomme de Perou." — J. H. T.

As to early history and aboriginal nomenclature, Monardes (De Simplicibus Medic., c. 66) says that the Granadilla was spontaneous in Peru, and that the fruit was highly esteemed by the Indians and by the Spaniards. Cieça de Leon (Chron. del Peru, c. 28), about 1550, saw it growing in the valley of the Rio Lile, near Cali (now in Colombia), and in the country about Pasto. The flower and fruit are mentioned by J. de Acosta (Nat. and Moral Hist. of the Indies, b. ii. c. 38) — "the fruit sweet, and too sweet, in the opinion of some." Lery (1557–8) does not appear to have found it in Brazil, but it was common there before the middle of the 17th century. Piso and Marcgrav (Hist. Nat. Brasil., 1648, pp. 70, 106) reckon nine species or more, of which four were cultivated — two especially for their fruits. The Tupi (Brazilian) name was "Mburucuîa" ("Montoya," 1639), which Piso and Marcgrav, *l. c.*, wrote "Murbcuia"; the species which was generally cultivated for its fruit was "Mburucuîa-guacu" (*i. e.*, great "Murucuya"). Father R. Breton (Dict. Caraibe, 1665) gives "Merécoya" as the Carib name of the fruit; but this seems to have been adopted from the Tupi — for in that language "Mburucuîa" denotes the "fruit of a vine."

It is remarkable that the Tupi (and Carib) name went with the plant and its fruit to the country of the Algonkins before the coming of Europeans. One species (*Passiflora incarnata*) was cultivated by the Indians of Virginia. "They plant also Maracocks, a wild fruit like a lemon, which also increase infinitely" (Capt. John Smith, Gen. Hist., p. 29): and, again (p. 25), Smith mentions the "fruit which the inhabitants call 'Maracocks' — pleasant, wholesome fruit" — among "things which are naturally in Virginia." Strachey (Travaile into Virginia, 72, 119) describes the "fruit called by the natives a 'maracock,' which the Indians plant," etc.

Although no longer planted, the fruit of the spontaneous plant is still eaten in the southern Atlantic States; and its popular name, "May-pop," is probably the last stage of the Tupi original.

Now our *Passiflora incarnata* is so like *P. edulis* (well

known in cultivation), the home of which is in Brazil,[1] that botanists have been unable clearly to distinguish the two, except by the fact that ours, dying down to the ground at approach of winter, remains herbaceous. It occurs in a rather narrow geographical range; and Dr. Masters, in his elaborate study of the order (Trans. Linn. Soc., xxvii. 641; see also Flora Brasiliensis), says that "being so far separate from the remainder of its allies of the same subgenus, [it] may be considered as an outlier." Altogether we may infer that the fruit and the name were originally derived from the same South American source.

Musa, Banana. — The author concludes, as did Robert Brown, that Banana and Plantain are varieties of one species; also that this species is of the Old World; that in all probability it was not known in the West Indies when discovered by Columbus; but that in respect to the western side of South America there is some evidence which is not easily ruled out, especially the statement of Garcilasso that the Peruvians had the Banana before the conquest, and of Stevenson (Trav. in S. Amer.), who is said to have seen in the ancient Peruvian tombs beds made of Banana leaves. This is discredited because the author found beans in the same tombs, "et que la fêve est certainement de l'ancien monde." But if Stevenson wrote beans, without doubt he meant the seeds of Phaseolus, not of Faba. It would rather seem that the Banana, like the Sweet Potato and Cocoa-nut, had early been transported over the Pacific.

Phaseolus vulgaris, Kidney Bean. — Three weeks after his first landing in the New World Columbus saw, near Nuevitas in Cuba, fields planted with "faxones and fabas very different from those of Spain," and two days afterwards, following the north coast of Cuba, he again found "land well cultivated with these fexoes and habas much unlike ours." "Faxones"

[1] Not Mexico, although indeed said to have been brought from New Spain to the garden of the Farnese palace, in Rome, as early as 1619. It is described, under the name of "Maracot" — with excellent plates showing the plant, the flower, and the fruit — by Tobias Aldinus, in "Rarior. Plant. Horti Farnesiani" (Romæ, 1625), pp. 49–59.

or "fexoes" were — as Navarrete notes, "Colec." i. 200, 203 — "the same as frejoles or judias," Spanish names for Kidney beans, which the Portuguese call "Feijaos." Oviedo (1525–35) speaks of the "fésoles, as the Spaniards call them, of which there are many kinds in the [West] Indias." These fésoles, he says (lib. vii. c. 18), "are called by Pliny 'fagivoles': in Aragon we call them 'judias,' and the seeds of those of Spain and of this country are properly the same." The natives of Hispaniola raise these fésoles, but they are much more abundant on the main land, especially in New Spain and Nicaragua. "I have, in the province of Nagrando in Nicaragua, seen them gather a hundred hanegas (bushels, nearly) of these fésoles: and they also, in that country and other parts of that coast, have many other kinds of fésoles, besides the common sort: some have yellow seeds, others spotted," etc. In another place (lib. xi. c. 1), Oviedo, mentioning plants that had been brought from Spain to Hispaniola and other parts of America, "in the beginning," names "Fésoles, called in Aragon 'Judias,' and in my country [Castile] 'Arvejas luengas':" but "of these, there is no need of bringing more seed, for in this island and on the main many bushels are harvested every year, and in the province of Nicaragua they are indigenous (naturales de la misma tierra), and a great number of bushels are produced yearly of these and of other fésoles of other sorts and different colors," etc.

From this time (1535) onward, nearly every writer who mentioned plants cultivated by the Indians named, together or in close connection, maize, beans, and pumpkins. Reference to several of these writers has been made in our notes on *Cucurbitæ.* Cabeça de Vaca found beans cultivated by the Indians of Florida in 1528, and again near the western limit of his wanderings (in New Mexico or Sonora) in 1535. De Soto, at his landing in 1539, found "fields of maize, beans, and pumpkins," near Tampa Bay; and at Coligoa (west of the Mississippi) "beans and pumpkins were in great plenty; both were larger and better than those of Spain;" and so, at other places, on his travels to the west and north.

Jacques Cartier, the discoverer of the St. Lawrence, on his

first voyage, 1534, found that the Indians near the mouth of that river on the Bay of Gaspé had abundance of maize, and had "beans (febues) which they name 'Sahu,'" or (as spelled in the vocabulary printed with his Discourse du Voyage) "Sahe."[1] The "Bref Recit" of his second voyage, 1535–36, mentions the use of corn and beans by the Indians of the St. Lawrence — "bled & febues & poix, desquels ilz ont assez" (f. 24).

Father Sagard in his "History of Canada" and in the account of his journey to the country of the Hurons, 1625, mentions the cultivation and use of "fezolles" by the Indians. The Hurons used in their succotash ("neintahouy") "a third or a quarter part of their fezoles, called ogaressa" (Grand Voyage, 83, 138).

Lescarbot, 1608, says that the Indians of Maine, like those of Virginia and Florida, plant their corn in hills, "and between the kernels of corn they plant beans marked (féves riolées) with various colors, which are very delicate; these, because they are not so high as the corn, grow very well among it" (Hist. Nouv. France, ed. 1612, p. 835; see also p. 744).

The relation of the voyage of Captains Amidas and Barlow to Virginia, 1584, mentions pease, melons, etc., at Roanoke Island, but does not name beans; but Harriot, who accompanied them on this voyage, includes both "Wickonzour, called by us pease," and "Okindjier, called by us beans," among the productions of that country. Capt. John Smith, who was in Virginia in 1607, and Strachey, who was there in 1610, describe (in nearly the same words) the Indian manner of planting corn and beans: "they plant also pease they call assentamens, which are the same they call in Italy fagioli: their beans are the same the Turks call garvances, but these they much esteem for dainties" (Smith's Gen. Hist., 28; Strachey, Trav. in Virginia, 117). Evidently, these names

[1] The language spoken by these Indians was a dialect of the Huron-Iroquois group, and we trace the name sahe (as Cartier caught it) in the Mohawk *osahe-ta*, "fésoles" of Bruyas (17th century), and the Onondaga *ousahèta* and *hôsahèta*, "poix, féve" (Shea's Onondaga Dictionary).

are confounded. Garvance was the French name of the Chick Pea (*Cicer arietinum*), the Spanish garbanzo; and it is not probable that the Turks gave this name to any kind of beans; while fagiuoli was the Italian equivalent of Latin phaseoli. Strachey's Virginian vocabulary gives assentamens (and otassentamens) for "pease," and peccatoas, peketawes, for "beans."

It must be remembered that at the beginning of the 17th century kidney beans — as well as vetchlings (Lathyrus) — were popularly regarded as a kind of pease, or "peason." Turner, in his "Names of Herbes," 1548, says that "Phasiolus otherwyse called Dolichos, may be called in English 'long peasen or faselles;' . . . in French phaseoles:" and "*Smilax hortensis*, . . . in French, as some wryte Phaseole . . . may be called in English Kydney beane," etc. (Eng. Dial. Soc., ed. 1881, pp. 62, 74). Lyte's "Dodoens," 1578, follows Turner for the English names of Phaseolus, "Kidney beane and Sperage; of some they are called Faselles, or Long Peason," etc. (p. 474): his "common Peason" and "middle Peason" are Ervilia (*Ervum Ervilia* L?) and *Pisum arvense* L.; while *P. sativum* is distinguished as "Great Peason, Garden Peason, and Branche Peason, because, as I thinke, they must be holpen or stayed up with branches" (id. 476).

So, on the continent, the Spanish names for fésoles was "arvejas luengas" (Oviedo), *i. e.*, long vetches, and the garden pea (*P. sativum*) was "un corto genero de Arvejas" (Calepin's Diction., ed. 1616), a short kind of vetches. The confusion of names is frequent in writers of the 16th and first half of the 17th centuries, in French as well as English and Spanish. "Magno sane labore" — as Tragus found — "Phasioli, Orobi, et Pisa, nec non Cicer arietina, maxime agrestia, secernuntur," (Stirp. Hist., 1552, p. 613).

Champlain uses "poix" and "fèves," interchangeably, as names of the American Phaseoli. In Breton's "Diction. Fr.-Caraibe" (1666) the Carib name, "caláoüana," stands for "pois de Brésil" and "febue de Brésil"; "pois rouges, dit Anglois, míbipi;" and "Pois, mancónti" — this being, probably, the (introduced) *P. sativum*.

Sagard (1624–5) says that the Hurons called the coarser part of their pounded maize — after the meal had been sifted from it — "Acointa, c'est à dire Pois (car ils lui donnent le mesme nom qu'a nos pois);" and in his "Dictionnaire Huronne," he has "Pois, Acointa," "Fezolles, Ogaressa;" whence we infer that French pease [*P. sativum*] were already cultivated by or known to the Hurons. The Abnakis of western Maine, in the 17th century, called pease, "awen-nootsi-minar," *i. e.* "French (or foreign) seeds." Tanner, 1830, gives as the Chippeway name of the "Wild Pea-vine" [*Phaseolus diversifolius ?*] "Anishemin," *i. e.* "Indian (or native) seeds." In nearly all North American languages, the names for kidney-beans (Phaseoli) are of earlier formation than those for garden pease. The latter are usually formed on the former: *e. g.* Chahta, *tobi*, "bean"; *tobi hullọ*, [wild] "pea"; *tobi hikint ŭni*, "garden pea" (Byington); Dakota, *o*^{n}*mnicha*, "bean"; *o*^{n}*mnicha hmiya*^{n}*ya*^{n} [*i. e.* "round bean"], "pea."

Without multiplying citations — we may assume that the "pease" and "poix" which early voyagers found cultivated by the American Indian were species of Phaseolus — not Pisum.

Five and twenty years before the settlement of Virginia, the Indians of Carolina and Florida had "fine citroüilles and very good beans" (Lescarbot, Nouv. France, 778). Lawson, 1700–1708, gives a more particular description of the southern beans cultivated by the Indians. "The Kidney-beans," he says, "were here before the English came, being very plentiful in Indian corn-fields." "The Bushel bean," a spontaneous growth, very flat, white and mottled with a purple figure, was trained on poles [*P. multiflorus ?*]: "Indian Rounceval, or Miraculous Pulse, so called from their long pods and great increase; they are very good, and so are the 'Bonavies, Calavancies [= Garvances ?], Nanticokes,' and abundance of other pulse, too tedious to mention, which we found the Indians possessed of when we settled in America"(Voyage to Carolina, pp. 76, 77).

In the northern States, we find little difficulty in establishing the identity of *Phaseolus vulgaris* with the beans cultivated by the Indians at the first coming of Europeans. These were from the first distinguished as "Indian beans," from the Garden beans (*Vicia Faba*) introduced by the English. In 1609, Hudson, exploring the river which bears his name, saw at an Indian village — in the vicinity of Schodac and Castleton, Rensselaer County, N. Y. — "a great quantity of maize or Indian corn, and beans of the last year's growth" (Hudson's Journal, in De Laet, 1625, b. iii. ch. 10, and Juet's, in Purchas: N. Y. Hist. Soc. Coll., 2 Ser., i. 300, 325).

1631–42. The Indians of New Netherland "make use of French beans of different colors, which they plant among their Maize. . . . The Maize stalks serve, instead of the poles which we use in our Fatherland, for the beans to grow upon" (De Vries, Voyages, transl. in 2 N. Y. Hist. Soc., iii. 107).

1653. Van der Donck, in his "Description of the New Netherlands," distinguishes the beans cultivated by the Indians before the coming of the Dutch, and the Turkish beans which had been introduced: "Of Beans there are several kinds; but the large Windsor bean [*Vicia Faba*] . . . and the Horse-bean will not fill out their pods. . . . The Turkish beans which our people have introduced there grow wonderfully. . . . Before the arrival of the Netherlanders [1614] the Indians raised beans of various kinds and colours, but generally too coarse to be eaten green or to be pickled, except the blue sort, which are abundant," etc. He then describes the Indian mode of planting beans with maize, *ut supra* (N. Y. Hist. Soc. Coll., 2 Ser., i. 188–9). This is the only reference we have found to the introduction of any species of Phaseolus into North America. Van der Donck's book was written more than forty years after Hudson's coming, and the author first arrived in New Netherland in 1642. His statement as to the introduction by the Dutch of the best kind of "Turkish beans" for "snaps," salad, or pickling, is not to be accepted without reserve; but the fact that Turkish beans "grow wonderfully, fill out remarkably well, and are much cultivated," while the imported Windsor beans [*Vicia Faba*] and horse-beans proved failures, is to be noted.

Wood, who was in Massachusetts from 1629 to 1633, says that the Indians "in winter-time have all manner of fowles, Indian beanes, and clams" (N. E. Prospect, pt. 2, ch. 6). Roger Williams, 1643, gives the Indian name of these beans in the Narragansett dialect: "Manusqussed-ash" (plural); Cotton's Massachusetts vocabulary (1727–8) has (sing.) "Mônasquisset, an Indian bean;" President Stiles, about 1760, heard the name in the Pequot dialect as "Mushquissedes" (MS. Vocab.); Zeisberger, 1776 and 1803, wrote it in the Delaware, with dialectic modification, "Malachxit;" and we can trace it in the modern Shyenne "Mónisk" (Hayden's Vocab., 1862) and "Monchka." In the Chippeway, the kidney-bean has received — probably from some local variety — a different name: "Miskodissimin," *i. e.* "red-dyed seed (or fruit);" and this name, modified as "M'skochī-tha," was used by the Shawanees of Ohio.

To return to New England, Josselyn, who was in this country, 1638–9, and again, 1663–71, in his catalogue of "plants proper to the country," names "Indian beans, falsely called French beans:" "the herbalists call them kidney-beans, from their shape and effects. . . . They are variegated much [in size and color]; besides your Bonivis and Calavances, and the kidney-bean that is proper to Roanoke: but these are brought into the country: the others are natural to the climate" (N. E. Rarities, p. 56; Voyages, p. 73–4). Here is reference to at least two species of American beans, one "proper to New England," the other from Roanoke — perhaps *P. multiflorus.*

Besides the names already mentioned — "Mônasquisset," with its variants — there is another, in northern Algonkin languages, for kidney-beans, which must have originally belonged to some high-twining variety. Eliot used it, in the plural, for "beans" in 2 Samuel, xvii. 28, *tuppŭhquam-ash* — which literally signifies "twiners;" and Rasles (1691–1700) gave, in the "Kennebec-Abnaki" of Maine, for "faséole," *a'teba'kwé* — from the same root. A modern Abnaki vocabulary shows that this name is still in use — as "ad-ba-kwa."

As to the American origin of *P. lunatus*, the Lima bean and its varieties, there seems now to be no question. "It is evidently," says M. De Candolle (p. 276), "a Brazilian species, dispersed by cultivation, and perhaps long ago naturalized, here and there, in tropical America."

But as to the origin of *P. vulgaris* he is not free from doubt. He finds "(1) that this species was not cultivated in ancient times in the East Indies, the southwest of Asia, or Egypt; (2) that we are not absolutely certain that it was known in Europe before the discovery of America; (3) that at that epoch the number of varieties suddenly increased in the gardens of Europe, and all authors began to speak of it; (4) that the majority of species of this genus are found in South America; and (5) that seeds which apparently belong to this species have been found in Peruvian tombs [at Ancon] of a date somewhat uncertain, mixed with many other species, all of which are American." (p. 275.)

The proof that *P. vulgaris* (and *P. nanus*), in varieties almost innumerable, were cultivated by the natives of America before the coming of Europeans, seems to be conclusive. The resolution of M. De Candolle's doubts as to the American origin of the species must depend chiefly on the identification of the species known as Phaseolí (Phaselí, Fagiuolí, Fésoles, etc.), in Europe, before the discovery of America. This identification may not be impossible, but the space at our disposal will not permit us to attempt it in this article, or even to reëxamine the authorities on which M. De Candolle admits the probability "that the Dolichos of Theophrastus was our pole bean (*haricot à rames*), and the Fasiolos our cultivated bush bean (*haricot nain*)," p. 271. At present, we have only to offer one or two notes.

1. The distinction indicated by Galen (De Alimentis, lib. i. cc. 25, 28) between the Phasiolos (φασίολος) of Dioscorides and Phaselus (φάσηλος) — presumably the "vilis faselus" of Virgil — if well founded, seems to have been lost sight of in the middle ages. In Italy, the Greek and Latin names Phasiolos, Faseolus, Faselus, Fasillus, etc., passed into the modern Fagiuoli. Piero de' Crescenzi, of Bologna, whose

treatise on agriculture was written near the beginning of the 14th century, in Latin, and translated into Italian about 1350, mentions, among field plants, Faseoli (Fagiuoli), as well known; "some of them are red, some white. . . . They are planted conveniently among panick, millet, and chick pease; they are also planted in gardens, among cabbages and onions."[1] It is not certain that the red and the white were of the same species, or genus, or that either was a species of Phaseolus, L. In the first half of the sixteenth century the white Phaseoli were the more common and less esteemed. The young and tender pods were eaten, with the included seeds, in salads, or boiled with other vegetables.

Two other early figures show that the Faseoli were not so "well known" to the herbalists of the 15th and beginning of the 16th century as to Crescenzi in the 14th. One is from a Venice edition of the "Hortus Sanitatis," 1511; the other from the "Tacuini Sanitatis" of Elluchasem Elimithar, Strasburg, 1531, p. 49. They are equally unlike the modern Phaseolus, the earlier figure in Crescenzi, and each other. The second may have originally been intended for a Teasel (*Dipsacus sylvestris*), the *Virga pastoris* of the herbalists. Calepin's Dictionary (ed. 1616) says, s. v. Faseolus, that the name Fasilli is now given by the common people to "a species of Cicercula." This "vulgaris Phaseolus" of Matthioli[2] and other Italian botanists of the period is figured and described in the later editions of his commentary on Dioscorides,[3]

[1] "De Agricultura," lib. iii. c. 10 (Italian, Ed. Venice, 1504). The Latin text was first printed at Strasburg in 1471, and with figures, 1486; the Italian version was printed at Florence, 1478. The figure of Faseolus in the earliest (Latin) edition we have seen, without date, but probably of Louvain, about 1480, has little resemblance to the Phaseolus of modern botany.

M. De Candolle remarks (p. 272) that "authors of the 15th century say nothing of Faseolus, or any analogous name," and that "this is the case with P. Crescenzi," — referring to a French translation of Crescenzi, printed in 1539, which we have not seen.

[2] "Vulgares Phaseoli, quibus passim in cibis vescimur, dum satis in campis virent, non repant," etc., Matth. "Apologia adv. Amathum," 1559, p. 33; "Vulgaris usus Phaseolus." *Ibid.* 31.

[3] Ed. C. Bauhin, 1598, p. 341. In the earlier edition (Venice, 1559, p.

as having an erect stem, ternate leaves, white flowers, kidney-shaped seeds, white, "except the umbilicus, which grows black." It may be worth noting that "a black spot in the place of the cotyledon" was a characteristic of the Faseoli described by Albertus Magnus (13th cent.), which appears to M. De Candolle "to be the dwarf Haricot (*P. nanus*) of our epoch." This black spot is more strongly suggestive of Dolichos than of any known variety of *P. vulgaris: e. g. Dolichos unguiculatus*, L. (French, *D. Mongette*, *Banette*, *Haricot cornille*) — not mentioned by M. De Candolle, but much cultivated in Italy, and of which there are a great number of varieties — which has seeds "marked by a prominent black spot about the umbilicus."[1]

2. M. De Candolle (p. 272), with a reference to Delile and to Piddington's "Index," remarks that though "no Hebrew name corresponding to the Dolichos or Phaseolus of the botanists" is known, yet "a name less ancient, because it is Arabic, namely 'Loubia,' is found in Egypt for the *Dolichos Lubia;* and in Hindustani, under the form 'Loba,' for *Phaseolus vulgaris.*" This name seems to be clearly referable to the Greek. It has not been traced earlier than to Jahia ebn Serapion — an Arabian physician of the 9th or 10th century — whose work "De Simplicibus," compiled chiefly from Dioscorides and Galen, was translated into Latin in the 15th century.[2] In a chapter (lxxxi) on "Lubia," *i. e.* "Faseoli," he quotes from Dioscorides the description of *Smilax hortensis* (*χηπαία σμίλαξ*) "whose seeds some call Lobia;" and it is evident that the name Lubia (as it was transliterated from the Arabic text by the translatór) was transferred to the

264), Phaseolus is figured as a low, bushy, and spreading, but not twining plant.

1 Several other species of Dolichos (*e. g.*, *D. sesquipedalis* L., Ital. "Fagiuolo Sparagio," Engl. "Asparagus bean") are similarly marked. Vilmorin-Andrieux et Cie., "Les Plantes Potagères," 1883, p. 280. Other names for this species are: Germ. "Ostindische Riesel-Spargel Bohne," "Nagelische Fasel;" Ital. "Fagiuolo dell' occhio;" Span. "Garrubia," "Moncheta," "Judia de Careta."

2 Milan, 1473, and Venice, 1479; but better known to botanists of the 16th century in the Strasburg edition of 1531, edited by Otho Brunfels.

Arabic from the Greek of Dioscorides. It is probable, to say the least, that it has been rightly appropriated to *Dolichos Lubia*, Forskal (De Candolle, 278), rather than to any species of Phaseolus.

The length to which our annotations have extended forbids all notice of the third part of this book. This, however, is very brief. It contains a tabulation of the plants of cultivation, and of the results of the preceding discussion of them; also an article on the regions in which the principal species have originated or have been brought into cultivation, in which it is stated that of the 247 species under investigation the Old World has furnished 199, and America 45, leaving three which are doubtful in this regard; and the extreme poverty of the southern hemisphere beyond the tropics is a striking feature. An article on the number and nature of species cultivated at different periods is noteworthy. So, also, is the enumeration of the cultivated plants which are unknown in a wild state; from which it is gathered that 27 species have never been found wild by any botanist, 27 more are doubtful in this respect, while 193 are of recognizable origin.

Of the "Reflexions diverses," at the close, we note only the final one, that "In the history of cultivated plants I have found no indication of communications between the inhabitants of the Old and New World anterior to the discovery of America by Columbus. The Scandinavians, who had carried their expeditions to the northern United States, and the Basques of the Middle Ages, who had extended their whaling voyages perhaps to America, would appear not to have transported a single cultivated species. The Gulf-stream has equally been without effect. Between America and Asia two transportations may have been effected, one by man (the Batatas), the other either by man or by the sea (Cocoa-nut)."

Perhaps the Banana should be ranked with the Sweet Potato in this regard. And we may merely conjecture that the Purslain came to our eastern coast with the Scandinavians or the Basques.

BENTHAM AND HOOKER'S GENERA PLANTARUM.

THE completion of this great work[1] marks an era in systematic botany. As the senior author is nearly eighty-three years of age, we may say that, essentially, it brings to a worthy close the long course of scientific labors of the most accomplished, sagacious, and continuously industrious phytologist of our day, almost the only survivor of those who personally knew Antoine Laurent Jussieu, and were associated with the elder De Candolle; one whose line of authorship, begun in the year 1826 (antedated only by that of the venerable Roeper), comes down to the present year, apparently, in almost unabated vigor, and is as remarkable for sustained importance as it is unparalleled in length. Upon the junior author, who should not yet feel the burden of years, however weighted with official cares, and upon such specialists as can be mustered, may devolve the responsibility of such *addimenta* as may be needed to bring the earlier volumes up to the date of the last, and even to supplement the work from time to time, as the "Genera Plantarum" of Linnæus was supplemented by successive editors. But it is not probable that the task which has now been happily accomplished will be undertaken again, still less executed, before the twentieth century is well entered upon. Useful compilations we may expect, and monographs which may here and there better represent advancing knowledge of particular groups; but a production like this, covering the whole field of phænogamous botany, ordinally and generically, and with the uniformity of treatment and scale thus secured by two close associates in one continuous work, cannot be looked for again for a long while.

It has few predecessors. The first "Genera Plantarum" in fact (though not exactly in name) was that of Tournefort, published in the year 1700. Then and there, as Linnæus de-

[1] *Genera Plantarum ad Exemplaria imprimis in herbariis Kewensibus servata definita.* Auctoribus G. Bentham et J. D. Hooker. Londini, 1862–1883. (The Nation, No. 942, July 19, 1883.)

clared, genera of plants, in the sense of scientific botany, were first established. There would probably be more recognition of this dictum if the present work were to be planned anew, and the genera which Linnæus himself admitted as of Tournefort, along with those which modern botanists have restored, could have been attributed to this real founder, without thereby compromising the proper position to hold in respect to herbalistic and ancient names. The second "Genera Plantarum" was that of Linnæus, in 1737, of which the last edition revised by the author himself was that of 1767. The third was that of Jussieu, "secundum ordines naturales disposita," which appeared in the year 1789. That of Endlicher — a monument of literary or bibliographical erudition rather than of botanical research — was brought out in the main between 1836 and 1843, at about the same time with the more unpretending synoptical compilation of Meisner. These were important in their way. But the "Genera Plantarum" of Bentham and Hooker, which began to be issued in the year 1862 and was finished in the spring of the present year, is the lineal successor of the three classical works above mentioned, that of Tournefort representing the botany of the close of the seventeenth century; that of Linnæus the first half, and that of Jussieu the latter part, of the eighteenth century. The present work — increased from the one small octavo of Linnæus to three thick imperial octavo volumes of nearly 1200 pages each — stands in like relation to the nineteenth century, and is based, like them, or even more than they, upon actual investigation, and upon the comparison of a vastly greater number of original types than was formerly possible.

Unlike its predecessors, however, — and in this respect agreeing with the other great botanical work of the century, the "Prodromus" of De Candolle, — the whole of cryptogamic botany is omitted. This vast field must be left to specialists. Fries, of Upsala, who died a few years ago at a ripe old age, was the last phænogamic botanist who was at the same time master of one or two cryptogamic orders; and now even the best of cryptogamists can hardly aspire to more than a general and superficial acquaintance with any other department than the one to which he devotes himself. This inevitable state of things has its disadvantages. The reasons for it do not really apply to the Ferns and their allies, and it was naturally expected, as it is much to be desired, that these should enter into the present work. May we hope that this still may be?

Some idea of the progressive enlargement of the field may be had by a comparison of the number of genera characterized in these successive works. The phænogamous genera of

Linnæus, "Gen. Pl.,"	ed. 1, A. D. 1737,	were	. .	887
" "	ed. 6, A. D. 1764,	"	. . .	1,189
Jussieu, "	. . . . A. D. 1789,	"	. . .	1,707
Endlicher, "	. . . . A. D. 1843,	"	(about)	6,400
Bentham & Hooker, "	. . . . A. D. 1883,	"	. . .	7,585

If the last had been elaborated upon the scale of Endlicher, or with the idea of genera which is still common if not prevalent, the number of genera would have amounted to at least ten thousand. An estimate of the number of known species of each genus and higher group has been made throughout the work — a rough approximation only, mentioning first the number in the books, and the number to which, in the opinion of the authors, these may probably be reduced by botanists who adhere to the Linnæan view of species; from which it appears that upon the very strictest estimate their number, as now known to botanists, is at least 95,620. In round numbers, it may fairly be said that about 100,000 species of phænogamous plants are in the hands of botanists. The five largest orders, as well for genera as for species, are the following, and in this rank: *Compositæ*, *Leguminosæ*, *Orchideæ*, *Rubiaceæ*, *Gramineæ*. The high standing of the Orchid family in the list will be a surprise to many. Linnæus knew only a hundred species; five thousand is now a moderate estimate — about half as many as there are of *Compositæ*, which hold to their proportion of one tenth of the whole. In both families every country and district is largely peculiar in its species and types. The far greater prominence of *Compositæ* over Orchids is owing to the vast number of individuals in the former, their paucity in the latter.

Those who desire to know the respective parts which the two authors have taken in the elaboration of the "Genera Plantarum" may be referred to a short article on the subject in a recent number of the "Journal of the Linnæan Society of London." Great thanks from all botanists are due to them both.

BOTANICAL NOMENCLATURE.[1]

SIXTEEN years have passed since M. De Candolle laid before the International Botanical Congress held at Paris, August 16–26, 1867, a body of Laws of Botanical Nomenclature, which he had drawn up for consideration by that assembly. The code was discussed by a special committee, afterward by the congress in full session, some modifications introduced, and it was then all but unanimously voted, "by about one hundred botanists of all countries:" "That these laws, as adopted by this assembly, shall be recommended as the best guide for nomenclature in the vegetable kingdom." The adopted code, with an extended commentary, was published by De Candolle early in the autumn of the same year; and an English translation, made by the lamented Dr. Weddell, appeared early in 1868. The "Laws," but without the more voluminous explanatory commentary, were reprinted from the English translation in this Journal in July of that year, occupying only twelve pages; and some remarks and suggestions by the present writer were appended. As was then said, the code did not make, but rather declared, the common law of botanists. It announced principles, systematically and perspicuously, and indicated their application in leading cases; but many practical questions, as well as conflicts of

[1] *Nouvelle Remarques sur la Nomenclature Botanique.* Par M. Alphonse De Candolle. Geneva, 1883. (American Journal of Science and Arts, 3 ser., xxvi. 417.)

rules in particular instances, which would inevitably come up, were necessarily left to be settled when they arose. No small discussion upon certain details has indeed ensued, in which our author, naturally appealed to, has taken an active part.

In the first part of the present publication (of 79 pages 8vo), the discussions of the intervening years are summed up and reviewed; and a few changes, which experience has shown the need of, are proposed. In the second part the author takes up certain questions which the Paris Congress left untouched, such as the nomenclature of organs (which he treated in his recent "Phytographie"), the nomenclature of fossils, and the rules according to which names and authorities should be cited when old genera are combined or reconstituted; also some matters of orthography and punctuation are briefly considered. Finally, in the third part, the laws adopted by the Paris Congress are reprinted, with the suggested changes. The alterations and additions are printed in italic type, so that they may be seen at a glance.

We are not sure of an English edition; and in any case it is desirable to make so important a publication as this generally known to our own naturalists. So our abstract and comments may run to some length. They are intended partly to illustrate and reinforce the author's doctrine, in respect to matters upon which there is still diversity of opinion and practice, perhaps occasionally to offer a criticism or suggestion; also as helps and guides to botanists in our own country, who are beginning to take interest in such matters and to feel (or at least to show) the need of giving attention to them.

As respects all the weightier matters of the law, the most experienced phænogamous botanists are in general agreement. Having accepted the code of 1867, they will be ready to accept proposed modifications and interpretations which are in accordance with its principles. Among them all, perhaps no one is so well qualified as De Candolle, by the bent of his mind and course of his studies, his opportunity of leisurely consideration, and his long editorial experience — combining, as it may be said to do, that of two generations, — for declaring what the present consensus of authority is, and for discussing the cases which still need adjustment, some of them of considerable consequence in phytography.

At the head of the publications upon nomenclature emanating from or sanctioned by associations or committees of naturalists which have appeared since the year 1867, De Candolle places the Report to the American Association for the Advancement of Science, at the Nashville meeting, 1877, on Nomenclature in Zoölogy and Botany, prepared by Captain Dall, after much conference with leading American naturalists. The differences between these rules and those of the botanical code relate mainly to questions mooted by the zoölogists, whose systems and views had been comparatively loose and variant. The first alteration in De Candolle's revised draft is adopted from Dall's report, with high commendation, namely: —

"Art. 3. The essential point in all parts of nomenclature is: 1. Fixity in names; 2. Avoidance or rejection of forms or names which may create error or ambiguity or introduce confusion into science." The "fixity in names" is taken from the American code, and is said to supply a real omission. Although merely declaratory, some practical consequences flow from it. The same idea dominates in the report made by Douville, chairman of a committee of the Geological Congress at Bologna in 1881, and which concerned itself with nomenclature in palæontology. This report insists that "The law of priority being fundamental in nomenclature, it appears to be necessary to apply it with all possible generality and to suppress derogations and exceptions to this law. . . . Contradiction between the signification of a name and the characters of a genus or species is no sufficient reason for changing such a name," etc.

Another code referred to is one by a committee of the Zoölogical Society of France, in 1881, M. Chaper, chairman, in which this principle of fixity is said to be less prominent, more exceptions being allowed.

On reviewing the whole field, De Candolle assures us that the tendency during the last sixteen years has been: 1, to an increasing agreement of the zoölogists with the botanists;

and 2, an increasing recognition of the law of priority as the fundamental principle of nomenclature, and as prevailing over considerations of elegance, linguistic purity, and precise meaning. But some recent publications of individual botanists look the other way; notably so two publications by Saint-Lager, "Réforme de la Nomenclature Botanique," and "Nouvelles Remarques sur la Nomenclature Botanique," published at Lyons, in the Annals of the Botanical Society of that city; of which it may be said, that the reflex effect of the multitudinous changes proposed in the view of making old and accepted names better, more classical, or more significant, has greatly strengthened the hands of those who contend for the absolute fixity and unalterableness of published names. Whether and to what extent misspelled or otherwise wrongly formed names may be corrected, and whether in some cases the principle of fixity should not prevail over absolute priority, are matters which may be discussed further on. We take up in the author's order the points which we wish to specify or to comment on.

Article 6 of the code declares that "scientific names should be in Latin. When taken from another language, a Latin termination is given to them except in some cases sanctioned by custom." Here our author asks, "But what is Latin?" He concludes that the Latin of Linnæus should be the model. It is the classical language of botany, and is much more precise than the Latin of antiquity, in which very many words bear two, three, or half a dozen senses, either in the same or in different ages; while in the technical language of botany each word has but one meaning, and each idea or object is expressed by a single term. De Candolle elaborated this point in his "Phytographie," to which he refers for illustrations; and he returns to it in special applications when commenting on article 66, as we shall see. But, in fact, even technical language cannot always avoid ambiguities and the use of words in senses which have to be determined by the context. *Folium*, for example, in botanical description, may mean the blade of a leaf only, or this along with its petiole, or blade with petiole and stipules together, or it may even mean any homologue of the ordinary leaf.

"Art. 15. Each natural group of plants can bear in science but one valid designation, the most ancient, whether adopted or given by Linnæus, or since Linnæus,—provided it be consistent with the essential rules of nomenclature." De Candolle now adds an article 15bis, which is purely explanatory, but has a bearing upon subordinate questions. It is: "The designation of any group, by one or more names, has not for its object the enunciation of its characters or history; it gives merely the name by which we are to call it." He comments upon the tendency which is often shown to mix up the question of name with other considerations. Before Linnæus introduced the binomial system, the names of species were at the same time names and characters. In separating these two things, Linnæus rendered a great service, and we should be careful to preserve this advantage. "A name is a name; characters are characters; the succession of names is synonymy. To mingle such different ideas leads to confusion. In these days there is a disposition to attribute too much importance to the meaning of names, and also to intermix the synonymy—*i. e.*, the bibliographical history of the groups—with the names, at least with the indication of the author, which being commonly annexed comes to be almost a part of the same. Such complications are contrary to the general principle that different ideas should be expressed separately. If this rule is neglected, we may be led into attempts to express in the name, or with the name, the phytogenitic history of a group, that being just now one of the ideas in vogue." Names that have an appropriate meaning are very well, and botanists always endeavor to make such; but experience shows that meaningless names are in some respects better—are generally better than names founded on ideas, which in the progress of knowledge often become false.[1]

De Candolle's main remarks upon article 15 relate to the

[1] Or may be essentially false from the beginning. One of our common Maples has two names, *Acer dasycarpum* and *A. eriocarpum*, both signifying that the fruit is woolly, whereas it is perfectly glabrous; only the ovaries are woolly, yet no botanist has ever proposed to change the received name,—which is remarkable.

point of departure for the law of priority in botany. The names of the two great classes, Monocotyledones and Dicotyledones, are of Ray in 1703, who scientifically distinguished and named them. The counterpart, Acotyledones, is of Jussieu (1789), and to all three names our author would apply the law of priority. That the name Acotyledones fails to express the true character is a small objection. This is only an example of the disadvantage of significant names, which may lose aptness in the advance of knowledge. Far better are such names as Aves, Pisces, Vermes, etc., which time and discovery can never falsify. Cryptogamia as the name of a class (which for meaning is hardly so good as Acotyledones) was introduced by Linnæus in 1735, and has maintained its place.[1] The counterpart, Phanerogamia (or Phænogamia), is not of Linnæus, but much later. Natural Families or Orders date from Jussieu's "Genera Plantarum," in 1789; Cohorts and Tribes, from A. P. De Candolle's "Systema," 1818. Subgenera begin with R. Brown, in 1810, according to our author. There are a few instances (without the name) in the "Prodromus Floræ Novæ Hollandiæ." But it is in the Oudney and Denham-Clapperton paper, in 1826 (p. 16), that their use is discussed, and the mode of designating them in citation by interpolation between generic and specific name in parenthesis, is introduced. Species (as distinguished from varieties) and the actual binomial nomenclature date from 1753, and the first edition of the "Species Plantarum." There is substantial agreement among botanists as to this point of departure; and the fact that the specific phrase of earlier authors is occasionally of a single substantive does not militate against it. *Galega vulgaris*, *Lappa major*, and *Trifolium agrarium* of the old herbalists were in good binomial form; but the adjectives are phrases, not specific names.

Generic names bring in a question of interpretation and usage. In his table De Candolle makes the point of departure for priority, Linnæus, "Genera Plantarum," ed. 1, 1737. All

[1] On p. 58 De Candolle has collected nineteen synonyms of the name Cryptogamia, all of later date, and specified the objections which may be brought against each of them, besides that of want of priority.

agree, or should agree, that no anterior name has right of priority to a Linnæan name or to a name adopted by Linnæus. But as respects generic names adopted by him, are we to follow Linnæus or are we not? He says, "Tournefortius primus characteres genericos ex lege artis condidit." And in the "Genera Plantarum": "Ipsi non immerito inventionis gloriam circa genera concedere debeam," — and so he uniformly accredits to Tournefort the generic names adopted from him; and the same as to "Plumerius, . . . Vaillantius, Dillenius, . . . Michelius et pauci alii," "qui ejus vestigia presserunt." De Candolle remarks that Tournefort had the merit which Linnæus ascribes, but that "he kept a good many adjective names for genera (Acetosa, Bermudiana, etc.)." Since Linnæus did not adopt these, they are out of the present question. Moreover, not to speak here of a score or two of really adjective generic names, Linnæus himself adopted two which Tournefort had discarded, Mirabilis and Impatiens, and deliberately made another, Gloriosa, in place of a proper name, Methonica, of a sort which, though not of the best, is now regarded as next to the best. But it is completely understood that Linnæus is not to be corrected; so Gloriosa, Impatiens, etc., remain.

Are we equally to follow Linnæus in regard to names which he adopted from Tournefort and a few later authors, some of them his own contemporaries? If so, we shall continue to write Salicornia, Tourn., Corispermum, A. Juss., Olea, Tourn., Justicia, Houst., Dianthera, Gronov., Lycopus, Tourn., Linnæa, Gronov. The practice of the leading botanists has been essentially uniform in this respect, from Jussieu down to De Candolle, father and son, even to the latest volume of the "Monographia," published during the current year. It seems perfectly clear therefore — although we believe that the question is not raised in this revision — that such genera are expected still to be cited as of their respective founders. And, as hardly any one doubts that Tournefortian genera suppressed by Linnæus but restored by modern botanists (such as Fagopyrum) are to be cited "Tourn.," it follows that only in a restricted sense do genera begin with Linnæus in the year 1737. This case, indeed, is governed by the principle in citation, so well insisted on by M. De Candolle, that no author is to be represented as saying the thing that he does not say. The alternative course is to write "Linnæa, Linn.," which is certainly what Linnæus has not said. The only authors we know of who have on principle followed this alternative — and a notable exception it is — are Bentham and Hooker in the new "Genera Plantarum," against which a protest was made in this Journal when the first part was issued. They have followed the rule that botanical genera began with Linnæus so strictly as to cite even authors as recent as Gærtner for Tournefortian genera and to ignore botanists like Gronovius, contemporary with Linnæus, and publishing since the year 1737; and it is only by an infraction of their rule that they have avoided writing Linnæa, Linn.

No change of rule 15 seems actually required to bring it into unison with the almost universal practice in citation. We have only to understand that genera adopted by Linnæus from Tournefort, etc., and so accredited, should continue to be thus cited; that the date 1737 (Linnæus, "Genera," ed. 1) is, indeed, the point of departure from which to reckon priority, yet that botanical genera began with Tournefort; so that Tournefortian genera which are accepted date from the year 1700. That is the limit fixed by Linnæus, and it definitely excludes the herbalists and the ancients whose writings may be consulted for historical elucidation, but not as authority for names.

Upon articles 21 and 22, which give rules for the names of orders and other supra-generic groups, our author offers no new remarks. We venture to offer two. It being the general rule that *aceæ* is the proper termination for ordinal names which take their appellation from a typical genus, it is desirable to conform to it as fully as well may be.[1] Since *Saxifragaceæ*, *Myrsineaceæ*, *Styracaceæ*, *Gentianaceæ*, *Nyctaginaceæ*, and even *Lauraceæ* and *Juglandaceæ* were adopted

[1] *Cruciferæ*, *Leguminosæ*, *Umbelliferæ*, *Compositæ*, *Labiatæ*, and the like, are no exception to the rule, rightly stated, as they are not named from typical genera. We shall not have any more of them, but the old ones in use are among the best.

in the "Prodromus," it seems to us retrograde and unadvisable to have gone back in the new "Genera Plantarum" to *Saxifrageæ*, and the like; and this upon no obvious principle, as we have *Samydaceæ*, *Cornaceæ*, etc., brought into harmony with the rule. And if in the "Prodromus" we have *Styracaceæ*, why not also *Salicaceæ?* And if, in the "Genera Plantarum," Styrax could take the *aceæ* termination as *Styraceæ*, why not as *Styracaceæ?* If it be objected that some such terminations have an unclassical aspect, this objection applies all the more to the cases under our second remark: Namely, that too rigid adherence to the rule that names of suborders, tribes, etc., shall end in *-eæ* and the like, gives us nearly unpronounceable words of four or five consecutive vowels, or, when the diphthongs are printed in separate letters, according to a prevalent fashion, one or two more. Of these — the diphthongs written out — *Sauraujeae*, *Spiraeeae*, *Catesbaeeae*, *Jaumeeae*, *Thymleeeae*, and *Moraeeae* are the worst instances, and would justify any infraction of rules.[1] The last, and one of the worst, would have been avoided by writing the ordinal name *Iridaceæ* when that of the tribe would have been *Irideæ*. Names are to be spoken as well as read, and botanists who have to teach think more of these things than those who only write.

At the head of his remarks upon generic names (art. 25 et seq.), our author commends to other naturalists the very clear directions given in the rules for Zoölogical Nomenclature, edited by Dall, for rendering Greek letters into Latin in the construction of generic and specific names. He notes, however, that the rendering of η by *e* is not in full accordance with Latin usage, as witness *bibliotheca*, *dialectica*, *Hecubea*, etc.

[1] Far better to write *Spiræaceæ*, with De Candolle. The use of this termination for tribal names need not be objected to by those who take little pains to use it for orders. And those of us who are careful so to employ it, would prefer its occasional use for tribes and suborders to the concatenation of vowels, which it is not easy to write and almost impossible to pronounce. Some quite unnecessary tribal names in *aceæ*, such as *Vernoniaceæ* and *Eupatoriaceæ*, adopted by De Candolle from Lessing, are kept up, although exceptional.

Article 28 of the code, which is reprinted without alteration or comment, consists of a series of recommendations of points to be attended to in the construction of generic names. As such names are to be "in Latin," it would have been well to recommend that in their formation from the Greek, from which most of them are nowadays drawn, the principles of Latin prosody should not be wholly ignored. Such names as Trichóclădus and Ancistróclădus would have been euphonious as Trichocladia and Ancistrocladia, and very little longer. Acanthópănax, Didymópănax, Dimorphóchlămys, Trigonóchlămys, Aulacócălyx, Pellácălyx, Micróchăris, and the like, are harder to pronounce than the makers probably thought; and so of many others.

One of the actual recommendations is "Eviter les noms adjectifs." This in Weddell's version is translated, "To avoid adjective nouns:" doubtless a wrong translation. Adjective nouns we take to be substantives which are directly formed from adjectives. Not many such are likely to be made for genera; but if such good ones can be constructed as those we already have in Nigella, Amarella, Flaveria, Chlora, Rubia, Leucas, and Hyptis, they will not be objected to. Clearly the recommendation is to avoid adjective names for genera. That may be done for the future, but has not been done in the past. They are contrary to the rule of Linnæus, but not to his practice. Not to refer again to Gloriosa, Mirabilis, and Impatiens, at least two score of obviously adjective names for genera may be counted in the first edition of the "Genera Plantarum," — such as Arenaria, Stellaria, Utricularia, Dentaria, Asperula, Angelica, Trientalis, Pedicularis, Digitalis, and from the Greek such as Polycarpon. Amphicarpum and Mitracarpum are recent names of this kind. To conform the rule to the fact it were better to state that: generic names are either substantives or adjectives which may be used as substantives, the latter mostly feminine in gender. Angelica is understood to be *Planta Angelica*, Sanguinaria, *Planta sanguinaria*, etc.

It is recommended "to avoid making choice of names used in zoölogy." But it has become nearly impossible to follow this advice, nor is it now thought to be important.

Article 33 is suppressed. It was no more than a statement of the custom that personal names for species were to be nouns in the genitive (*e. g.*, *Clusii*) when the person commemorated was a discoverer, describer, or an illustrator of the species, but were in adjective form (*e. g.*, *Clusiana*) when the name was merely complimentary. The rule sometimes worked awkwardly; for many personal names do not take kindly to latinization in the genitive form, which are sufficiently euphonious as adjectives in *-ana* or *-anum;* and it is well to do away with a needless restriction.

Article 34, which recorded the fact that many names of species are substantives, is the subject of a few remarks, called out by the publication of Saint-Lager, in which it is proposed to change all these into adjectives, — a proposition which botanists are not likely ever to adopt. Some of them are among the best of descriptive names, *e. g.*, *Stellaria nemorum*, *Convolvulus sepium*, *Rhus vernix*, *Chamærops hystrix* (and such should be written without a capital initial); those which are proper names, either old names of genera or of the herbalists, are rightly said to be significant either of the absorption of a former genus, or of a transference, or as preserving a native appellation, or as indicating a likeness. "*Digitalis Sceptrum* means a Digitalis which had been called *Sceptrum;*" *Ardisia Pickeringia*, a species of this genus which, mistaken by Nuttall for a new one, had been named Pickeringia; *Rudbeckia Heliopsidis*, a *Rudbeckia facie Heliopsidis*, from its resemblance to a Heliopsis. Linnæus gave us many such names; and no sufficient reason appears either for discarding these, or for forbidding the discreet adoption of new ones. But we cannot commend such a name as *Senecio Bhot* for a species indigenous to Bhotan.

Article 36 consists of a series of recommendations for the formation or adoption of specific names. Its fifth sub-article we may refer to in another connection, namely, along with article 48. The recommendation to "name no species after one who has neither discovered, nor described, nor figured, nor studied it in any way," should be respected; yet there are occasions for departing from it, especially in case of new species in very large genera. Excellent and sometimes needful is the advice to "avoid names designating little known or very limited localities." We are obliged to cite — happily as a synonym — *Helenium Seminariense*, published by a Professor who thought he had discovered a new species of Helenium in the vicinity of the "seminary," in one of our southern States, where he taught botany.

Article 40 suggests that names of varieties originated in cultivation, and still more half-breeds and sports (so important for horticulturists to distinguish), should have only fancy names, generally vernacular, and in some form as different as possible from the Latin specific names of botany, — names which, when needful, may be appended to the botanical name of the species, when that is known, *e. g.*, *Pelargonium zonale*, Mrs. Pollock. This has been seconded by the editor of the "Gardeners' Chronicle" and other judicious experts, and is slowly making its way.

Article 42, treating of the conditions of publicity, is the subject of additional remarks. The rule is, that "Publication consists in the sale or the distribution among the public of printed matter, plates or autographs. It consists, likewise, in the sale or distribution, among the leading public collections, of numbered specimens, accompanied by printed or autograph tickets, bearing the date of the sale or distribution." De Candolle now remarks that distribution among the members of an exchange club, of collections not offered to the public, does not come up to the rule; also that, as Dr. J. Müller states, the distribution of specimens without characters or any indication of the reason for calling it new, is nearly tantamount to announcing a species or genus in a publication, but without characters; which article 46 declares is not publication. But the cases are not quite alike. The possession of the named specimen enables a botanist to ascertain its distinctions. A published description without access to specimens may or may not serve the same purpose, very often does not. Unfortunately an insufficient or even a misleading description — and we have many such to deal with — claims the same right of priority that a good one does.

It is well, therefore, that publication by sufficient distribution of named specimens should be recognized. But the remark is true that, in fact, very few distributed collections fulfill all the requirements of article 42.

Article 47, sect. 2, recommends botanists "to publish no name without clearly indicating whether it is that of an order or of a tribe, of a genus or of a section, of a species or of a variety — in short, without giving an opinion as to the nature of the group to which the name is given." Unless this is attended to, and unless citations are equally precise, — for instance, unless subgeneric names are cited as such and not as generic, and *vice versa*, — much confusion in synonymy and in indexes will ensue.

Article 48, on the citation of the authority for generic and specific names, and matters herewith connected, involves questions which have been more disputed than any other. In the revised article the phrase printed in italic type is interpolated: "For the indication of the name or names of any group to be accurate and complete, and for the ready verification of the date, it is necessary to quote the author who first published the name or combination of names."[1] The statement might

[1] M. De Candolle appears to insist upon this verification (or incipient verification) for higher groups as well as for genera and species, and would deprecate the not unusual custom in compendious Floras, Catalogues, etc., of omitting to cite the authority for orders, sub-orders, tribes, etc. We should agree with him if the omission was held to signify that the names of the groups in question, when thus simply given, were proposed as new. But in fact, authority is omitted, not because the groups are new, but because they are old and entirely familiar. Nobody will ever suppose that *Ranunculaceæ*, *Clematideæ*, etc., nakedly written, are novelties. In this regard, the nature and plan of the publication are to be considered; what is necessary in a Systema or a Genera Plantarum may be superfluous in a local or a compendious work. Indeed it may be nearly impossible to assign the authority for the name of an order correctly, without explanation and extended references. One would wish to write, succinctly, *Cornaceæ*, DC., as De Candolle founded the order, but it was in the form *Corneæ*. A well-known order was instituted as *Onagræ*, Juss., "Genera Plantarum," which the founder altered to *Onagrariæ;* for reasons referred to in this article we may wish to adopt the form *Onagraceæ*. As all this is most familiar matter, yet may not be correctly

be simplified by omitting "or names" and "or combination of names," on the ground that the name of a plant is one, that it has a name not names, its name being the generic followed by the specific appellation. *Ranunculus bulbosus* is one name of two words. Our veteran botanist, Bentham, has insisted upon this; and it has a bearing upon the mooted question of mode of citation of authority.

The governing principle for the citation of authorship, etc., is well declared by De Candolle: "Never make an author say that which he does not say." It is difficult to go wrong when this principle is kept in mind, and when it is also understood that the appended name of an author, or its abbreviation, makes no part of the name of the plant, but is only the initial portion of its bibliography. Those who take a different view seem to have fallen into it by failing to distinguish strictly between name and history, and especially by mixing the history of a preceding with the statement of an actual name. A single example may illustrate this. When we write "*Mathiola tristis*, Brown," we give the name of a certain kind of Stock and the original authority for it; and we may, when needful, complete the citation by adding the name of the book, with the volume and page, where it was first published. If, with some, we write "*Mathiola tristis*, Linn.," we make an untrue statement. Linnæus had a wholly different genus *Mathiola*, and no *M. tristis*. If we add "sp.," and somewhere explain its import to be that the latter half of the name was given by Linnæus, the other half remains unaccounted for. And we have still to seek in the synonymy for the name of the genus under which Linnæus knew the plant, and also for that of the author who transferred it to Mathiola. If, with others, we write "*Mathiola tristis*, Linn. (*Cheiranthus*)," or "*Mathiola tristis*, Linn. (*sub-Cheirantho*)," our longer phrase still wants

stated without somewhat detailed exposition, why not in a local or condensed botanical book write simply *Onagraceæ?* The proper exposition is in place in a Genera Plantarum; and it would have been better if Bentham and Hooker had critically attended to this, instead of referring merely to the preceding work of Endlicher. It would have added somewhat, yet not very much, to their great labor.

the essential part of the citation. If, to secure this, we write "*Mathiola tristis*, Linn. (*Cheiranthus*, Brown)," our name, if it may be so called, now extended to five words and two signs in print, or of seven words when spoken, is still ambiguous and confused. It is a jumble of synonymous names and authorities, which become explicit and clear only when we translate it into "*Mathiola tristis*, Brown (*Cheiranthus tristis*, Linn.)," that is, into name and synonym, with respective authorities. This is clear and literally truthful; the injection of the synonymy into the name is neither. Linnæus reformed nomenclature by freeing the name from the descriptive phrase. The school in question would deform it by rebuilding, in another way (as De Candolle observes), ante-Linnæan phrases, only making them historical instead of descriptive.

The practice of appending the authority to the name whenever the species is mentioned has been so strictly and pedantically adhered to that many take the former to be a part of the name. To obviate this impression, it might be well to treat the names of common plants as we do those of genera; that is, to omit the reference to authorship in cases where there is no particular need of it. Not, however, so as to cause any confusion with the cases referred to in the following paragraph: —

"When a botanist proposes a new name . . . it is impossible for him to cite an author; consequently the absence of such citation suffices to show that the name of the species or other group is new. Linnæus, Lamarck, De Candolle, R. Brown, Martius, etc., followed this course. It is then a useless complication of many modern naturalists to append 'mihi,' 'nobis,' 'sp. nov.,' 'gen. nov.,' etc., to a new name. A large majority of species, genera, and families were published without these wholly personal indications." This is good as a general rule; but the "gen. nov." and an indication of the order or tribe are often needful.

No new comments are made upon article 49, probably because the practice of botanists generally is conformed to it. The article reads: "An alteration of the constituent characters, or of the circumscription of a group, does not warrant the quotation of another author than the one that first published the name. . . . When the alteration is considerable, the words 'mutatis char.,' or 'pro parte,' or 'excl. syn.,' 'excl. sp.,' etc., are added," etc. The translation would have been better worded, "does not warrant the quotation of another author in place of the one that first published the name." For, in fact, the addition of the reforming author's name to the citation is often warranted and helpful, sometimes is almost a necessity, in the case of genera. It appears that R. Brown began, in an oblique way, the practice objected to, and for which there is often a plausible excuse; and the elder De Candolle sometimes followed it. It was only when the practice was systematically carried out by one or two authors that the consequences became apparent; for few genera or species have now their Linnæan limits or signification, and the new rule was practically proved to be a necessity.

Among the recommendations contained in article 36 was the following: "Readily adopt unpublished names found in travellers' notes or in herbaria, unless they be more or less defective." Guided by the practice of the elder De Candolle and his contemporaries, it used to be thought a duty, or at least a part of common courtesy, to do this, in all cases in which the author's approval could fairly be supposed. But certain inconveniences and misunderstandings have resulted, especially as to mode of citation, which have suggested its withdrawal or modification. In this Revision, M. De Candolle only adds the restriction, "or unless the author has not in advance approved the publication." This does not alter the case except for living authors: their approval ought to be obtained or counted on; and in respect to authors no longer living a botanist takes up only such names as in his opinion ought to be published, and which he supposes the posthumous author would have approved. On the whole it were probably better not to take up names left unpublished by a deceased botanist; and De Candolle assigns good reasons for letting them alone. If he had modified the article decidedly in this sense he would have more fully expressed his own view, and probably have been sustained by prevalent opinion. Moreover, he might have distinguished this practice from an essentially different one, namely: —

The case of plants sent under manuscript names by a discoverer or an investigator to some botanist engaged in publication, and with a view to their publication; of which the sending by Nuttall of new *Umbelliferæ* to the elder De Candolle when elaborating that order for the "Prodromus," is a marked and not unusual instance. For this is a practice that need not be discouraged. Any small inconvenience that may arise as to mode of citation is counterbalanced by the greater concentration of publication, new genera and species thus appearing in monographs, floras, or in the papers of leading botanists, which otherwise would have dimly seen the light in obscure or local periodicals. And they are more likely to have proper characters assigned to them, instead of vague descriptions, by incompetent or unpractised hands, such as often try a botanist's patience.

Article 50 treats of the mode of dealing with such names as the above mentioned after they have been published, *i. e.*, "names published from a private document, an herbarium, a non-distributed collection, etc." It declares that such names "are individualized (Fr. *precisés*) by the addition of the name of the author who publishes them, notwithstanding the contrary indication that he may have given." This is found to mean that, although the elder De Candolle gives us "Eulophus, Nutt.," as the name of a genus communicated by Nuttall, with a specimen, for the purpose of its being so published in the fourth volume of the "Prodromus," yet subsequent writers, looking only to the work it was published in, are to cite it as Eulophus, DC. And that the genus which Linnæus published as "Linnæa, authore Clariss. Dr. Gronovio," we are to cite as Linnæa, Linn. This is not only quite contrary to the practice of botanists from Linnæus down to De Candolle and later, but is also contrary to the golden rule of citation, already referred to, never to make an author say something different from or opposed to that which he does say.

Appreciating this, the author of the code has now recast

article 50, so as to read, "When an inedited name has been published (by another botanist), in attributing it to its author, those who afterwards mention it ought to add the name of the person who published it; for example, Leptocaulis, Nutt. in DC.; *Oxalis lineata*, Gillies in Hook."

This is reasonable, and in the first instance such names will almost of necessity be so cited, must always be so cited when work, volume, etc., are specified. But De Candolle remarks that the addition will soon vanish; for instance, "*Cynoglossum ciliatum*, Douglas, Mss.," published by Lehmann in "Pugillus," etc., and in Hooker's "Flora Boreali Americana," will soon come to be quoted simply as "*Cynoglossum ciliatum*, Dougl.," that is, just as other names are quoted. And why not? Because, it is said, the name dating only from the publication, it is necessary to know when and where this vicarious publication was effected. For this "Nutt. in DC." may fairly serve, nearly all names published by De Candolle being contained in the "Prodromus." Not so, however, with "Gillies in Hook." Sir William Hooker published very widely, in periodicals, in the "Botanical Magazine," and in numerous independent works. In such cases the double citation gives little help. The experienced botanist may know where to look; the inexperienced must turn to indexes at once; for both these must be the final and the usual resort; and in them the double has little if any advantage over the single citation. Moreover, if this principle is fully applied, the number of double-cited names may be inconveniently numerous. The first volume of Torrey and Gray's "Flora of North America" abounds in species and genera published by them for Nuttall. If these have all to be permanently quoted "Nutt. in Torr. & Gray," why not also the many species published, say by Bentham in De Candolle's "Prodromus," in the "Flora Brasiliensis," etc., and even the species published by Brown in the second edition of the "Hortus Kewensis," and elsewhere? On the whole it seems probable that these double citations will be used only in first or in early quotations, or in special instances; that it will not be deemed necessary to retain them when the names become settled in Floras or general works, except in the bibliography or full reference; when of course the "*Leptocaulis inermis*, Nutt. in DC. Coll. Mem. v. 39, x. 10, et Prodr. v. 107," will fully appear. But so long as the abbreviated citation of the author and publisher together is requisite, the mode or citation recommended by De Candolle is the one to be employed.

A quite different case is that of citing, as authority for a genus or species, the name of a botanist which is not upon the record. There is reason to believe that L. C. Richard edited the "Flora Boreali-Americana" of Michaux, and drew up the excellent generic and specific characters of the new plants in it. There is equal reason to believe that he purposely withheld his name. Upon no just principle of citation, therefore, can the name of Richard be quoted, as the younger Richard and Kunth essayed to do. The same holds for the work of Solander in the first edition of Aiton's "Hortus Kewensis." And if it does not hold for the contributions of Brown to the second edition, it is because he claimed them in his lifetime, rather than because they have been collected and republished under his name since his death. Only confusion will come from the admission of hypothetical constructive authorship. The old rule, that what does not appear is no better than non-existent, must apply to all such cases.

In the comments upon article 52, the duty of abbreviating authors' names in the normal way is insisted on, and the bad practice of doing so by leaving out the vowels is deprecated. "Michx." for Michaux, which is partially shortened in this way, was a necessity on account of the ancient botanist Micheli. But "Crn." for Crouan is intolerable. Such a name need not be abbreviated at all. Monosyllabic names should rarely if ever be curtailed. "R. Br." has so long been used for Robert Brown that it may continue to be used, although "Brown" is better. In the other form, it may be counted among the few cases in which initial letters are used instead of the first syllable and first consonant of the second, — cases which should probably be restricted to the "L." for Linnæus, "DC." for De Candolle, "H. B. K."

for Humboldt, Bonpland, and Kunth. We are not sure that De Candolle would favor the latter.

A series of remarks is made upon articles 29–66, taken together, — relating to names which are to be rejected or modified, and those which are to be maintained notwithstanding certain faults. As already mentioned, the tendency among working naturalists is to preserve names in spite of faults; while a few linguistic reformers, such especially as M. Saint-Leger, propose changes which would affect 733 recognized names of species in Europe alone, and ten or twelve thousand in the vegetable kingdom at large, and "this after all the endeavors of botanists for the past half century to establish the law of priority and to have more stability of names. *Sagittaria sagittifolia* and *Psamma arenaria* must be changed, forsooth, because they are pleonasms; all substantive specific names, because a great majority of specific names are adjectives, and many others because they are not sufficiently classical." For instance, Dianthus he would change to Diosanthus, Mentha to Minthe, *Hydrocotyle brevipes* to *H. brevipedata*, Cactus to Cactos, Arum to Aron, and so on. De Candolle adds, that Cicero was not so particular in Latinizing Greek words, as witness "barbarus," "machina," "emporium." As to Pirus, our author insists that even if Pyrus is not Latin, it is the botanical name of the genus as adopted by Linnæus (from Tournefort and from all the herbalists, and it is old enough to be entered as an alternative form in the dictionaries), and so is to be preserved under the law of priority. There is little danger that the reform of Saint-Leger will prevail. There is some danger that the reaction will so stiffen the rule of priority as to forbid the correction of obvious mistakes. See, for instance, the form in which article 60 is now recast by De Candolle: "A generic name should subsist just as it was made, although a purely typographical error may be corrected. The termination of a Latin specific name may be changed to bring it into accordance with its generic name." From this it would seem that a slip of the pen and a mistaken orthography of a man's name may not be corrected. We trust that, when the change would not sensibly affect the place of a name in an index, such obvious corrections as of Wisteria to Wistaria may prevail. We may assume that the error was typographical; for Dr. Wistar was at the time too well known in Philadelphia for Nuttall to have been ignorant of the orthography of the name. The correction of Balduina into Baldwinia brings it into accordance with the rule that personal names used for genera should be written as near as may be with the original orthography of the person's name. "*Astragalus aboriginorum*" is neither a typographical nor a clerical error. It is a hard rule that forbids us to write "aboriginum," still retaining Richardson's name as authority.

Botanists may take more kindly to the rule when applied to such names as Eleocharis and Aplopappus, in the formation of which the Greek aspirate was neglected. We cannot well suppose this to have been a typographical or clerical oversight on the part of Robert Brown or of Cassini. Perhaps a majority of botanical authors have preserved the original orthography, on the ground that the right of priority, like that of a certain king, is *super grammaticum*, — while the remainder have written Heleocharis and Haplopappus; whence some confusion in the indexes. The requirement to preserve the original form of the generic and specific names and to abide by the Latin of Linnæus and his contemporaries, notwithstanding classical faults, enables us to retain such familiar names as *Ranunculus acris*, *Lathyrus palustris* and *sylvestris* (instead of *R. acer*, *L. paluster* and *silvestris*), and to keep up "laevis" for smooth, — probably to the disgust of classical scholars.

M. De Candolle has a note on Diclytra of Borckhausen, changed into Dielytra to make it conformable to a conjectured meaning, and then into Dicentra that it might agree with the etymology given by Borckhausen himself: he gives it as a case in which an excess of erudition has loaded the genus with three names in place of one; and he concludes, as do we, that it were better to have kept the original orthography, and have treated it as a name which had, through some mistake, failed of meaning. But the name having been changed into

Dicentra on the ground that the right word κέντρον, and not the impossible word κλύτρον, must have been intended by Borckhausen, we think it should now be maintained, although it might have been left in the original form. Moreover, the doctrine that names must not be mended and that sense is unimportant, however good and needful, is so recent that it must not be too rigidly applied to long-standing cases.

This consideration should not be wholly overlooked in the case of old and long-established genera, especially those of numerous species for which some obscure older name has come to light. Since it is impossible to make rules for the infraction of a rule, such cases must be left to sound discretion. In our opinion such discretion would forbid the transference of the name Stylidium from Swartz's genus to Marlea, and the revival of Labillardière's transient first Candollea for Swartz's Stylidium.

The fourth section of article 60, which enjoined the rejection "of names formed by the combination of two languages," is now suppressed. Nothing is put in its place; but let us hope that we shall not be driven to the acceptance of the specific name "acuticarpum" which one of our fellow-botanists has recently perpetrated. Although hybrid names are to be avoided, yet, as De Candolle remarks, they cannot consistently be outlawed by people who accept "centimetre," "decimetre," "bureaucracy," "terminology," and the like, nor by botanists who raise no objection to "ranunculoides," "scirpoides," "linnæoides," "bauhinioides," etc.

Names of identical meaning but of different orthography, as our author insists, may well enough co-exist. In a vast genus it might be neither inconvenient nor harmful to maintain species named respectively "fluviorum," "fluvialis," and "fluviatilis," at least if they belonged to different parts of the world.

We pass to some brief annotations upon the second part of the publication before us, which deals with questions not taken up by the congress of 1867.

The first topic is that of the nomenclature of organs, which was treated with some fullness in the "Phytographie." The remark is here repeated that the greater part of the so-called names of organs are only terms, that is, names indicative of the condition of organs or parts of the plant. For some of these substantive names are necessary or highly convenient, yet most were better provided with adjective terms only, which belong to terminology, not to nomenclature. Doubtless principles of fixity and the rule of priority should apply to these, both to names and to terms. But it seems unlikely that the phytotomists will at present heed the counsels of the phytographers in this matter. Yet the latter may insist that established names used in descriptive botany shall not be displaced on the pretense of getting more appropriate ones. For instance, the long recognized name "testa" for the outer seed-coat is to be discarded, because, forsooth, this covering is not always or even not generally a shell, or of the texture of earthenware. As well ask the French to discard the word "tête" (or "teste"), because the human head, or the skull which gave the name, does not really resemble a brick or earthen pot.

The second is upon the nomenclature of fossils. And the rule is that they are named according to laws which apply to living plants. The Bologna congress of palæontologists ordained that, to secure priority for specific names of fossils, they should not only be described but figured. De Candolle, after consultation with Heer (whose recent death we have to deplore), concludes that this rule is too absolute. It seems to us that so long as a large part of the names of fossil plants are merely tentative and provisional, we should be content with a general approximation to the received rules of botany.

The nomenclature of groups inferior to species (varieties, sub-varieties, variations and sub-variations) is considered; but no new rules are proposed; nor is the question of sub-species discussed.

Although it is not exactly a matter of nomenclature, we should have liked that our author had considered the two modes of disposing of varieties, and had expressed an opinion as to whether the character of the species should or should not completely cover the variety or varieties assigned to it. In the former case there is a variety α, followed by β, etc. In the latter, the species is defined upon its type, without any special regard to the appended variety or varieties, which are then characterized as to the points in which they differ from the type. We prefer the latter method, as being on the whole clearer, and as a saving in names; avoiding the awkwardness or the superfluity of a varietal name for the type of the species.

Some noteworthy observations are introduced in respect to the plight which systematic botany is threatened with by what De Candolle would call micromorphic botanists, like Jordan and Gandoger, who abandon the Linnæan idea of species altogether, and give this name and rank to what ordinary botanists take for sub-varieties. For example, we are informed that M. Gandoger divides the Roses of Europe and North America into 4600 species, or groups provided with names similar to those of species, under numerous subgenera, which in effect take the place of genera. Mentha has already undergone a similar micro-metamorphosis. If this goes on, and the names should be written every one, I suppose that even the world itself could not contain the books (or indexes) that should be written. The obvious and only remedy is to relegate this kind of botany to a world of its own, with which the legitimate science need have nothing to do.

Questions having been raised as to the proper use of capital initials in certain specific names, M. De Candolle has devoted two or three pages to this and related topics. Linnæus used capital initials only for substantive names; Lamarck employed them for personal and some geographical names, seemingly without system. A. P. De Candolle used the initial capital systematically for all three, and even for "Alpina" when used to designate a plant of the Alps. His example has generally been followed until recently; and this is in accordance with the custom of the English language. To the objection that it is contrary to the customs of the Latin language, our author replies at some length, substantially as follows. He finds that in the matter of orthography, etc., classical writers distinguish nine phases or periods of the Latin language, of which the most classical is the seventh period, that of Augustus; and there is no foundation in classical Latin for either punctuation (the points distinguishing words, not phrases) or accentuation by signs, and that the distinction between capitals and small letters was made since the dark ages by scholars whom a purist of our day might tax with ignorance of the proper way of writing Latin; that the object and result of all these and other innovations was greater clearness and precision; that the question is not at all one of ancient latinity, but of modern usages, both of the philologists and the naturalists; and these have happily modified classical Latin into a medium of greater precision and clearness and better adapted to the needs of science.

Finally, we have a brief discussion of the question: "When an author has comprehended one genus in another without naming a species, can he be cited for the names of the species which implicitly result from this union?" The answer he decisively gives is: "This would be neither right nor possible, nor practically convenient. To be correct, one should attribute to an author exactly what he has published. When it is said that the genus B should be united with the genus A, this is not saying that the species of the genus B should be called by such and such names in the genus A. To name them correctly it is necessary to examine them one by one. A glance at the Genera of Bentham and Hooker, or at the works of Baillon, will show how impossible it is to attribute the designation of the species to the authors who have changed the names of the genera, without an explanation under each species." This is illustrated by the supposed case of three genera combined into one, each of which has a species "lanceolata"; by the case of a species "minor" transferred to a genus of which it may be the largest species; and by reference to the state of all large and many small genera, full of obscure, misunderstood, or debatable species, the arrangement and naming of which can be effected only by patient and prolonged study. When this work has not actually been done by the reformers of genera, it should be left to monographers and the editors of Floras. If, by article 45, "a species is not looked upon as named unless it has a generic name as well as a specific one," neither is it named unless a specific as well as a generic name is assigned to it. Besides the instances in which the old specific name is impossible under another genus, there are very many in which it would be improper or questionable, and in respect to which particular consideration is required. Between these cases and the plain ones in which implied naming could not go wrong, who is to draw the line? Perhaps it might be drawn at monotypic new genera with old specific names. But how to do even this upon recognized principles is a problem.

A fatal objection to the principle of names by implication is that all such names, if they are existent, must be indexed in the new "Nomenclator Botanicus" now in preparation. To transcribe under Senecio the specific names pertaining to all the genera which Bentham has referred to that already vast genus is no small matter, and a part of the work will prove superfluous if — as we suppose to be the case — some of these genera, such as Cacalia, ought to be maintained. But that is only the beginning. A more recent author, Baillon, has reduced the genera of *Compositæ* nearly one half. For example, to Helenium he has referred Gaillardia, Actinella, Cephalophora, etc.; to Tagetes he has referred Dysodia, Nicolettia, Hymenatherum, and others; to Helianthus, a greater number of genera, most of them prolific in species. In all probability, most of these reductions will not be approved. Yet, if the principle of constructive naming is adopted, the "Nomenclator" must burden its columns with these hosts of inchoate specific names of Baillon, either as received names or as synonyms. It is plain that the principle referred to, besides its incongruity with the leading ideas of received nomenclature, breaks down with its own weight. There are, nevertheless, taking arguments in its favor, which need not here be recapitulated; and the common system has its disadvantages and liability to abuse; yet it appears to be the only workable system. As already intimated, the right assignment of specific names in reconstructed genera requires particular knowledge and careful investigation. And the botanist who reconstructs genera should himself adjust and state the specific names as far as he can.

BALL'S FLORA OF THE PERUVIAN ANDES.

THE personal observations and the collections upon which this essay[1] was founded were made in April, 1882, in an excursion by railway from Lima up to Chicla, which although only seventy-five miles in distance, is at the elevation of 12,220 English feet. Much to his surprise, Mr. Ball found that at this elevation he had not yet reached the alpine region, which really begins about 2000 feet higher. This is three or four thousand feet higher than Grisebach had placed it, on the authority of Tschudi and Humboldt; yet is only what we should expect, since the proper alpine vegetation of the Rocky Mountains in lat. 40° N. hardly descends below 10,000 feet, and the oscillations of temperature in the Peruvian Andes are small.

Equally mistaken, Mr. Ball suspects, must be the common view that the flora of the tropical Andes is scanty in species as compared with high-mountain floras in general. He makes some comparisons from which he infers that the paucity is apparent rather than real, and may be attributed mainly to the paucity of collections in the Andes, since these vast regions have been visited at very few points and far between.

About a quarter of the Andean phænogams is of *Compositæ*, which is double their ratio in North America, which again is greater than that of any other continent. The characteristic and the most abundant Andean *Compositæ* are the *Mutisiaceæ*. Mr. Ball, referring to Bentham's indication of the complex affinities of this group, ventures "to believe that under *Mutisiaceæ* are included very many different lines of descent, but that among them there are some minor groups distinguished by very high relative antiquity." And in another connection he opines "that the arguments that have led some distinguished botanists to consider the great family

[1] *Contributions to the Flora of the Peruvian Andes, with Remarks on the History and Origin of the Andean Flora.* By John Ball; Journal Linnæan Society, xxii. London, 1885. (American Journal of Science and Arts, 3 ser., xxxi. 231.)

of *Compositæ* as of comparatively recent origin to appear to me altogether inconclusive. When I consider the vast variety of forms which it includes, the degree in which some large groups are localized in different regions of the earth, while others, such as Senecio, have representatives in every zone, I shrink from the conclusion that their origin can be, even in geological language, at all recent. It is, of course, not inconceivable that plants which we class together under the name *Compositæ* may have come into existence by different lines of descent through gradual modification from different ancestral types. But when we consider the general agreement in the structure and arrangement of the essential organs, I think that the balance of probability inclines decidedly toward the belief in a community of origin of all the various existing forms. Be that as it may, we are, I think, justified in looking to the mountain region of South America as the original home of many large groups, such as the genus Baccharis, most of the *Mutisiaceæ*, and many genera of other tribes."

As to these two suggestions, although it is practically convenient, and perhaps necessary, to bring all the Labiatiflorous *Compositæ* under one tribe, as Bentham has done, it seems to us altogether probable that the existing forms are descended from different lines of ancestry. Indeed, by such a conception we can more naturally understand their diverse affinities. But as to the great order they belong to, if there is any large group in which the structure suggests community of origin, it is the *Compositæ*. And we suppose that systematic botanists of large experience would entirely agree with Mr. Ball, that the wide differentiation and distribution of this vast order indicated its high antiquity. Our author has assigned some strong reasons for this opinion. The only argument to the contrary that we know of is an ideal one, based upon two suppositions: one, that Dicotyledons culminate in the *Compositæ* and in such-like orders; the other, that the highest ideal type of plants must be of the latest evolution. But, indeed, the vegetable commonwealth shows no tendency to culminate in any one group or set of groups;

and it is a questionable morphology which would promote the capitulum of a Thistle or a Dandelion to the head of the class.

We remember an interesting lecture, in which, recognizing the dominant part which the northern hemisphere and its boreal lands, with their favorable configuration, have undoubtedly played in floral distribution, it was inferred that the rôle of the southern had always been comparatively insignificant. But a great deal may have happened in the austral regions before this boreal supremacy was established. Mr. Ball several years ago brought forward his doctrine of the very high antiquity of our actual temperate and alpine floras, of their coexistence in highly elevated regions of low latitudes even in early times. So now, applying his former conclusions to the southern hemisphere, and to "a period remote even in geological language," he notes that "the special generic types of the antarctic flora" "belong without exception to the great groups or natural orders which are now almost universally diffused throughout the world; and the ancestral types from which they originated were probably carried to that region at a remote period, when the physical conditions of the earth's surface were widely different from those now prevailing." "Various considerations tend to the conclusion that the dispersal of the chief cosmopolitan genera of plants may have coincided with the period of the older secondary rocks; and at that period physical agencies far transcending those of our experience prevailed throughout the earth. If the ancestors of the antarctic types of vegetation were then established in a south polar continental area, and were developed from them by gradual modification, I see no difficulty in believing that they may have maintained themselves through successive gradual changes of physical conditions within the same region, and even that some may still survive within the Antarctic Circle."

Whether or not one accepts the idea of such high geological antiquity which Mr. Ball claims for what he calls Cosmopolitan types, we must wholly agree with him in his use of this name for them, in preference to that of the Scandinavian. The latter term was used by Hooker before the relation of the present flora of our temperate zone to a former high-northern vegetation was made clear, and before the types in question could "with more reason be referred to North America than to Scandinavia." Mr. Ball's remark that, as to many of them, "the balance of evidence points to an original home in the high mountains of lower latitudes" chimes in with his favorite and original doctrine. And this indeed seems likely to gain ground the more it is considered and applied, as he is applying it, to the explanation of actual distribution.

The interesting problem is to discriminate, as well as may be, the two commingled elements of the northern temperate floras, one of arctic, the other of more endemic mountain origin. An interesting presentation, as concerns central Europe, is made in Heer's "Nival Flora of Switzerland," a posthumous work published by the Société Helvetique des Sciences Naturelles, of which a summary is given in "Nature" for December 31, 1885.

The following idea is extremely suggestive. "In a zoölogical as well as a botanical sense Brazil is one of the most distinct and separate regions of the earth. It is in large part a granitic region, from which vast masses of superincumbent strata have been denuded, and where the granite itself has undergone a great amount of decay and ablation. We there see the ruins of one of the greatest mountain masses of the earth, where a very ancient flora and fauna were developed, of which portions were able to migrate to a distance, while others have been modified to adapt themselves to the gradual changes of the environment. Many vegetable groups, which are but slightly represented in the higher region of the Andes, such, for instance, as the *Melastomaceæ*, probably had their origin in the mountains of ancient Brazil."

We are now only beginning to reach some conception of the rôle which the Andes and their prolongation through Mexico have taken in determining the character of no small part of the North American flora. Following up some ideas which were touched upon in this Journal (vol. xxvii., Nov., 1884, p. 340) and elsewhere, Mr. Ball writes: —

"When we consider that, although subsidence has probably at various times separated the two portions of the continent, the highlands of Mexico and Central America have, in all probability, served during long periods as a bridge over which some portions of the mountain vegetation may have been transferred from north to south, and *vice versa*, we are led to feel surprise rather at the separations now existing than at the presence of many genera and of a few identical species in the flora of the Andes and that of the Rocky Mountains. It is true that I have reckoned as Andean genera and species many that extend northward as far as Mexico; and it may well be that that region, so rich in varied forms of vegetation, is the original home of some that now appear to be more fully developed in the mountain ranges of western North America. Among the widespread American types we must note two natural orders whose original home may with some confidence be placed in the northwestern part of the continent. The *Polemoniaceæ*, of which about 140 species belong to that region, are represented in the Andes by five species of Gilia, one of Collonia, and by the endemic genus Cantua. They have sent to the Old World two or three species of Phlox in northern Asia [we believe only one, and that not far over the border], and a single emigrant which has reached Britain, — the Jacob's Ladder of old-fashioned gardens, — which maintains a struggling existence in several isolated spots in Europe. The other specially American family is that of the *Hydrophyllaceæ*, of which 12 genera are known in North America, but which is represented in the Andean chain by only four species of Phacelia." The *Loasaceæ* illustrate the opposite course of migration.

A list of the plants which Mr. Ball collected in the upper valley of the Rimac in the Peruvian Andes, with various annotations and the characters of some new species, concludes the present interesting contribution to Andean Botany. We believe that a second paper upon the subject may be expected. Two or three comments upon individual plants of the list will bring our review to a close.

Erodium cicutarium. — Although Mr. Ball notes the wide diffusion of this Old World species in South America, and that it attends the distribution of cattle, he seems at a loss to account for its presence in the Peruvian Andes at the height of 12,500 feet. He supposes that it has not shown the same readiness to establish itself in North America. This is true of the Atlantic but not of the Pacific side. In California and the adjacent districts the Alfilaria, as it is popularly called, has taken such full possession that we can hardly convince even the botanists that it is an introduced plant. The authors of the "Botany of California" speak of it as "more decidedly and widely at home throughout the interior than any other introduced plant, and, according to much testimony, it was as common throughout California early in the present century as now. . . . It is a valuable and nutritious forage plant, reputed to impart an excellent flavor to milk and butter." At Santa Barbara and other parts of southern California it is used for lawns around dwellings, and it seems to be the only resort. It makes a passable substitute for grass so long as the rainy season lasts or irrigation is kept up. It must have been brought in with the earliest cattle, and have found on the Pacific coast a perfectly suitable climate.

Caldasia of Lagasca, Mr. Ball shows us, must be restored as the name of the genus named Oreomyrrhis by Endlicher.

Relbunium, upon a general survey of the species, will in our opinion be found quite untenable as a genus.

Phacelia circinata, which extends almost from one end to the other of the American continent, is said to be singularly constant, exhibiting no marked varieties. But we have in North America a remarkable diversity of forms, the extremes of which, by themselves, no botanist would refer to one species, although intermediate forms inextricably combine them.

INDEX.